Environmental Chemistry and Toxicology of Aluminum

Timothy E. Lewis

Library of Congress Cataloging-in-Publication Data

Environmental chemistry and toxicology of aluminum.

Proceedings of a symposium held during the 194th Annual Meeting of the American Chemical Society in New Orleans, La. on August 30 through September 4, 1987.

Includes bibliographies and index.
1. Aluminum--Toxicology--Congresses. 2. Aluminum--Environmental aspects--Congresses. I. Lewis, Timothy E. II. American Chemical Society. Symposium (194th : 1987 : New Orleans, La.)
RA1231.A5E58 1989 615.9'25673 89-8117
ISBN 0-87371-194-7

LEWIS PUBLISHERS, INC.
121 South Main Street, Chelsea, Michigan 48118

PRINTED IN THE UNITED STATES OF AMERICA

This volume is dedicated to
Henry and Elizabeth Lewis,
my beloved parents.

PREFACE

Lift up your eyes to the sky,
Then look to the earth beneath;
For the sky will vanish like smoke,
And the earth will wear out like a garment,
And its inhabitants will die in like manner

Isaiah 51:6

The chapters comprising this work offer the reader a broad overview of the sources, transport, and fate of aluminum in the environment. Each chapter represents a detailed summary of platform discussions presented at a three-day symposium entitled "The Environmental Chemistry of Aluminum," held during the 194th Annual Meeting of the American Chemical Society in New Orleans, Louisiana, August 30 through September 4, 1987. The symposium, co-sponsored by the U.S. Environmental Protection Agency and the Environmental Chemistry Division of ACS, brought together international scientists investigating a wide diversity of topics in aluminum chemistry and toxicology. It was a unique opportunity for formal participants and for scientists and laymen in the audience to interact. Considerable knowledge was exchanged during those three days, both in the formal sessions and while enjoying the many attractions New Orleans has to offer.

To further disseminate the scientific findings presented at the symposium the speakers prepared more in-depth manuscripts, each of which represents a chapter in the book. The scope of the book is broad and touches on perhaps every scientific discipline. Every scientist who has dealt with aluminum in its many forms will undoubtedly find additional information, generated from disciplines he or she had previously thought were not engaged in such research. For those not actively involved in either aluminum research or in other scientific endeavors, the book will serve to enlighten them as to the importance of this once-believed-innocuous element in the environment.

Each book chapter was peer reviewed by at least two scientists knowledgeable in the particular subject area. Although they must remain anonymous, I would like to thank each one for critically reviewing the manuscripts, and especially for the quick turnaround most reviewers provided. Not all papers were suitable for publication after the peer review process was complete. Nevertheless, I greatly appreciate the time and effort the authors and reviewers expended. Each author was given great freedom with respect to manuscript format, style, and content. Some chapters present results of original research, while others offer a review of the scientific literature. Some speculative and philosophical commentary was permitted in hopes of providing a stimulus for future research. All in all, the blend of various formats makes for very interesting and thought-provoking reading.

There are many people who generously contributed their time and effort, from when the approval to organize the symposium was obtained to when the last chapter was submitted to the publisher. I would like to extend sincere thanks to Edward M. Heithmar (U.S. EPA), co-chairman of the symposium. He helped organize the sessions and generally kept the squeaky wheels of bureaucracy well oiled so that the symposium could move ahead as smoothly as possible. Many thanks go to Steve Simon (Lockheed-ESC) who offered constant encouragement and support whenever the genesis of a symposium looked bleak. Without the guidance and support of all those in the Purchasing Department (LESC), particularly Norma Dorn and Nadine Gaston, the symposium would not have been possible. I am very grateful to Peggy Oakes (LESC) for her assistance in putting out brush fires at every bend in the road. I would like to gratefully acknowledge the diligent efforts of the CSC Word Processing Department and the Pipkin Graphic Arts Department for their assistance in manuscript preparation. I would like to thank Dolores Macchiaverna for assistance in the book cover design. Also, I am very grateful to Jerry Stephens and his staff at Lazer Graphix for their friendly service and cooperation.

Sincere thanks are extended to all the session chairmen--Johan Bjorksten, Peter Campbell, D.R. Crapper McLachlan, and Mark Peden--for keeping the proceedings on track and on time.

It has been an enjoyable and enlightening experience organizing the symposium and editing the book, the fruition of the efforts of all involved. Final thanks go to Ed and Jon Lewis of Lewis Publishers, Inc. for their guidance and support during the entire book preparation process. I appreciate their assistance in disseminating these very important ideas and insights concerning aluminum in the environment.

Although some of the work described in this book has been funded wholly, or in part, by various government agencies, including the U.S. Environmental Protection Agency, it has not been subjected to review by these various federal agencies and therefore does not necessarily reflect the views of these agencies and no official endorsement should be inferred. Mention of trade names or commercial products does not constitute endorsement or recommendation for use

CONTENTS

Environmental Chemistry and Toxicology of Aluminum

INTRODUCTION

Aluminum is the most abundant metal and the third most abundant element, and composes approximately 8% of the earth's crust. Its ubiquitous nature is understandable and if one reviews the literature on aluminum sources, transport, and fate in the environment, this becomes even more apparent. Aluminum occurs only in the oxidized form, Al(III). In aqueous solution the ion hydrolyzes (protolyzes) surrounding water molecules, forming hydroxo complexes and decreasing the pH of the solution. Aluminum forms many inorganic and organic complexes in soil and surface waters.

In the early part of this century the phytotoxicity of aluminum in agroecosystems was well known. Elaborate measures were taken to ameliorate aluminum toxicity by lime application and by selection of aluminum-tolerant species or cultivars. While some plant species are highly sensitive, others are capable of thriving in high-aluminum soils. The tea plant, *Symplocos spicata*, is able to grow in very acidic soils, where aluminum is readily available for uptake by the roots. The tea leaves serve as a sink, accumulating over 10,000 ppm aluminum in some instances.

The historical data, for the most part, have been neglected. The recent (last two decades) flurry in aluminum research has focused on the effects of increased mobilization of aluminum by acidic deposition. Increased aluminum concentrations and corresponding reductions in fish populations have been attributed to the mobilization of aluminum from the edaphic to the aquatic environment. Most of the research in aquatic ecosystems is contemporary, although much information on the terrestrial effects of aluminum dates back to the early 1900s. Many aquatic chemists and toxicologists have overlooked this important source of data. This is a prime example of the general lack of interdisciplinary communication prevalent in the chemical and toxicological research of aluminum.

Another factor hindering progress in understanding the effects of aluminum in the environment is the lack of analytical methods for determining the multitude of forms in which this element occurs. It is now well understood that the toxicity of aluminum in aquatic and terrestrial systems is not well correlated with total aluminum concentrations, but is a function of the concentration of the biologically active fraction of the total. In terms of acute toxicity, the inorganic monomeric form of aluminum is believed to be the most toxic. However, organically bound species may be capable of crossing biological membranes and contribute to chronic bioaccumulation. In order to ascertain which species of aluminum are acutely or chronically toxic, chemists must supply toxicologists with the necessary analytical tools. Existing methods are usually operationally defined (e.g., filterable, acid-extractable, nonexchangeable, labile, and nonlabile) and their biological relevance is not well known.

The implication of aluminum in several human disorders such as Alzheimer's disease and senile dementia has been of public concern and scientific controversy in recent years. Dietary sources are suspected. Aluminum is added to drinking water as a clarifying agent. Aluminum is also present in small quantities in foods and is added to insoluble and soluble compounds for thickening and anti-caking. Aluminum has been shown to be released from aluminum cookware. Dietary intake may be a critical issue in the future.

Effects of aluminum in the environment are highly dependent upon the form in which the element enters the system. The forms in which aluminum occurs are as numerous as the scientific disciplines studying this element. In many instances, however, there is a general lack of interdisciplinary communication. The necessity for interaction between scientists working in all facets of aluminum chemistry and toxicology is essential for setting wise future research goals and making sound regulatory decisions.

CHAPTER 1

COORDINATION CHEMISTRY AND SPECIATION OF Al(III) IN AQUEOUS SOLUTION

Arthur E. Martell and Ramunas J. Motekaitis

INTRODUCTION

The aquo Al(III) ion is a highly hydrated "hard" metal ion with considerable affinity for "hard" bases such as RO^- (R = alkyl or H) > $C_6H_5O^-$ >> $-COO^-$. Effective bidentate combinations of these donor groups are present in several well-known ligands which are recognized as effective chelating agents for Al(III), and other trivalent (hard) metal ions.[1] Such ligands, in approximate order of decreasing hardness and decreasing affinity for Al(III) are glycols, catechols, aliphatic hydroxy acids, and aromatic hydroxy acids. More complex chelating agents which are effective for trivalent metal ions such as Al^{3+}, Fe^{3+}, Ga^{3+}, In^{3+}, etc. contain various combinations of these functional groups. Common examples are citric acid, gluconic acid, and tartaric acid. The well-known aminopolycarboxylic sequestering agents such as nitrilotriacetic acid, NTA, ethylenediaminetetraacetic acid, EDTA, diethylenetriaminepentaacetic acid, DTPA, etc., contain amino nitrogen and carboxylate functional groups, and would be expected to have somewhat lower affinity for Al(III) than do the hydroxy acids and catechols. In the latter compounds, however, the lower "hardness" of the amino and carboxylate groups may be compensated for, or even superseded by, large chelate effects resulting from their multidentate character.

An important consideration in the design of effective ligands for hard metal ions such as that of Al(III) is competition with hydrogen ions in aqueous solution. The hard donor groups of the ligands described above have very high basicities and hence large protonation constants. The values for all aliphatic hydroxyl groups as well as for many phenols and catechols are too high to be measured by potentiometric methods (which are restricted to a working pH range of approximately 2-12). If the ionic strength is to be maintained at 0.100 M, it is not possible to measure protonation constants by any experimental method (e.g., spectrophotometrically) if the protonation constant is higher than 10^{13}. Thus, the most effective ligands for hard metal ions such as Al^{3+} frequently contain hydroxy or phenolic donor groups, which do not deprotonate unless coordinated to the metal ion. In view of the fact that the highest protonation constants of the ligand cannot be determined, it is necessary to express metal-ligand equilibria as proton displacement reactions rather than as the standard formation constant.

Many ligands with very basic aliphatic alkoxy or aromatic phenoxy donor groups do not bind metal ions at low or even weakly alkaline pH because of competition with hydrogen ions. Thus, ligands may be too basic to be effective in aqueous solution, while ligands with more weakly basic donor groups bind metal ions quite effectively. This factor is described with examples in more detail below.

The strong tendency of the aquo Al(III) ion to form hydrolytic species[2] in weakly acid solutions, and the insolubility of $Al(OH)_3$, are interferences which discouraged many of the early experimentalists from measuring the stability constants of Al(III) complexes. Many of the stability constants reported over a decade ago were measured at low pH (ca. 3 or lower) where hydrolysis of the metal ion is minimal. At that time it was not possible to handle the large number of equilibria involving simultaneous formation of metal complexes and hydrolytic species that may be present in the pH range 5-8. Because of the simplified methods of processing equilibrium data employed at that time, hydrolysis and olation reactions frequently were not considered in the calculations and the stability constants reported were therefore inaccurate. In recent work on the stabilities of Ga(III) chelates[3,4,5] the authors found that previous investigators had a tendency to assume the formation of species such as $ML(OH)_2$ to account for the potentiometric data, whereas in fact the metal dissociates from the ligand to form gallate ($Ga(OH)_4^-$) and free ligand. Part of the confusion resulted from the fact that, with fairly effective ligands, the Ga(III) ion is still complexed in the pH range where $Ga(OH)_3$ would otherwise precipitate, whereas at higher pH the still more stable tetrahydroxogallate would be formed but not detected because of its solubility. Similar reactions may also occur with Al(III) complexes, which may disproportionate to form the soluble aluminate ($Al(OH)_4^-$) species at high pH even though the ligand is effective enough to prevent the precipitation of $Al(OH)_3$ in the intermediate pH range.

In this chapter the reliable quantitative data currently available on the Al(III) complexes formed in solution are examined, and the nature of the complex species probably formed are discussed. Discrepancies in the stability constants reported previously are pointed out, and possible misinterpretations of potentiometric data are suggested. The scope of the chapter is limited to the catechols and hydroxy carboxylates, which are the most effective ligands, and the aminocarboxylates, which are expected to be somewhat less effective. The extent of treatment is limited somewhat by the small amount of data on Al(III) complexes.

EXPERIMENTAL

Most of the equilibrium constants presented in this chapter were measured potentiometrically, except for a few dissocation constants of the catechols, which were measured spectrophotometrically. Detailed procedures may be found in the publications cited. All protonation constants are expressed in terms of concentrations of the species involved. In cases where mixed (Bronsted) constants were reported, they have been corrected by converting hydrogen ion activities to concentrations, as has been the practice for the Critical Stability Constants compilation.[6] All equilibrium constants employed were measured at 0.100 M ionic strength and 25.0°C, except in a few cases where such data are not available.

When constants available in the literature involved conditions which did not correspond to the desired conditions, they were adjusted accordingly. Constants were adjusted to the standard temperature when ΔH° was known. Corrections for ionic strength were made empirically from variations of equilibrium constants for the same ionic types recorded in Critical Stability Constants.[6] The criteria for selection of these constants from the values given in the published literature are described in detail in the introduction of each volume.

The constants presented for citrate complexes were calculated from p[H] (-log [H^+]) profiles by the use of the program PKAS[7] for protonation reactions and with program BEST[8] for metal ion-ligand reactions. Species distributions were obtained with program SPE[9], which employs the algorithm taken from BEST, adapted to our VAX 11/780 computer system for graphical display of species distribution curves (degree of species formation as a function of p[H]). This program has now been modified for heterogeneous systems containing one or more precipitates, provided the corresponding solubility products are known. This modification consists of subtraction of the precipitated species (to meet K_{sp} requirements) from the solution component concentrations, followed by recalculation of concentrations of species in solution. The resulting distribution curves indicate the quantities of precipitates formed when equilibrium is reached.

RESULTS AND DISCUSSION

The Al(III) aquo ion is the "hardest" of the trivalent ions commonly found in the environment and in biological systems. Its (crystal) ionic radius of 0.51 Å is considerably smaller than other commonly encountered trivalent metal ions such as Fe^{3+} (0.64 Å), as well as other metal ions of Group III, Sc^{3+} (0.73 Å), Ga^{3+} (0.62 Å) and In^{3+} (0.81 Å). Trivalent boron, with an ionic radius of 0.23 Å, does not occur as an electropositive ion in aqueous solution.

Because of its small ionic radius, Al(III) polarizes solvated water molecules very strongly, giving rise to several hydrolytic species, which are formed in accordance with the equilibrium constants listed in Table 1. In addition to the simple monohydroxo aquo ion, $Al(OH)^{2+}$, formed in the first step, several other hydrolytic species have been detected: $Al_2(OH)_2^{4+}$, $Al(OH)_2^+$, $Al(OH)_3$ (insoluble), $Al(OH)_4^-$, and the unique polynuclear complex $Al_{13}(OH)_{32}^{7+}$, indicated in Table 1 as the partially dehydrated species $Al_{13}O_4(OH)_{24}^{7+}$. The constants listed in Table 1 are reasonably reliable, having been checked by many investigators with various techniques. A complete analysis of Al(III) hydrolysis as been published by Baes and Mesmer.[2] The distribution curves for

Al(III) based on the equilibrium constants in Table 1 are presented in Figures 1 and 2 for 10^{-3} and 10^{-6} molar concentrations, respectively.

It should be emphasized that the distribution curves in Figures 1 and 2 represent equilibrium conditions that include the solid phase $Al(OH)_3$. Thus, some of the species listed in Table 1 that do not appear in Figures 1 and 2, are formed in solutions which are supersaturated with respect to $Al(OH)_3$. Although such complexes have been detected potentiometrically, and their equilibrium constants have been determined, they would not exist in the presence of solid $Al(OH)_3$ and in that sense may be considered metastable complexes.

Table 1. Equilibrium constants for hydroxo and fluoro Al(III) complexes

Ligand	Equilibrium Quotients and Log K Values				Ref.
OH^-	$\frac{[AlOH^{2+}]}{[Al^{3+}][OH^-]}$	8.21[a]	$\frac{[Al(OH)_2^+]}{[Al^{3+}][OH^-]^2}$	19.0[a]	11, 12
	$\frac{[Al(OH)_3]}{[Al^{3+}][OH^-]^3}$	27.0[b]	$\frac{[Al(OH)_4^-]}{[Al^{3+}][OH^-]^4}$	31.4[b]	2
	$\frac{[Al_3(OH)_4^{5+}]}{[Al^{3+}]^3[OH^-]^4}$	41.35[a]	$\frac{[Al_{13}O_4(OH)_{24}]^{7+}}{[Al^{3+}]^{13}[OH^-]^{32}}$	33.02[a]	11
	$[Al^{3+}][OH^-]^3$	-33.5[c]			
F^-	$\frac{[AlF^{2+}]}{[Al^{3+}][F^-]}$	6.43[b]	$\frac{[AlF_2^+]}{[AlF^{2+}][F\text{-}]}$	5.20[b]	13, 14, 15
	$\frac{[AlF_3]}{[AlF_2][F^-]}$	3.9[b]	$\frac{[AlF_4^-]}{[AlF_3][F^-]}$	2.8[b]	13, 14, 15

[a] t = 25°C, 0.6 M; [b] t = 25°C, 0.1 M; [c] K_{sp} s,@ 25.0°C, 0 M.

It is seen that at 10^{-3} M concentrations, the predominant species at equilibrium between p[H] 4 and 11 is precipitated $Al(OH)_3$. Of the soluble hydrolytic species listed in Table 1, only a small amount of $Al(OH)^{2+}$ appears in the p[H] range, 3-4.3. At p[H] 10 the $Al(OH)_4^-$ complex begins to form and predominates above p[H] 12. For the 10^{-6} M system (Figure 2) the distribution diagrams are considerably different. The partially-hydrolyzed mononuclear complexes ($Al(OH)^{2+}$ and $Al(OH)_2^+$) form in appreciable concentrations between p[H] 3 and 6, while solid $Al(OH)_3$ begins to appear when the p[H] reaches 5, and is the predominant species in the p[H] range 5.2-8.8. The soluble aluminate $Al(OH)_4^-$ begins to form at p[H] 7, predominates above p[H] 9, and is the only species present above p[H] 10. At these concentrations, the presence of $Al(OH)_3$ as a soluble complex (5 x 10^{-8} M) is apparent.

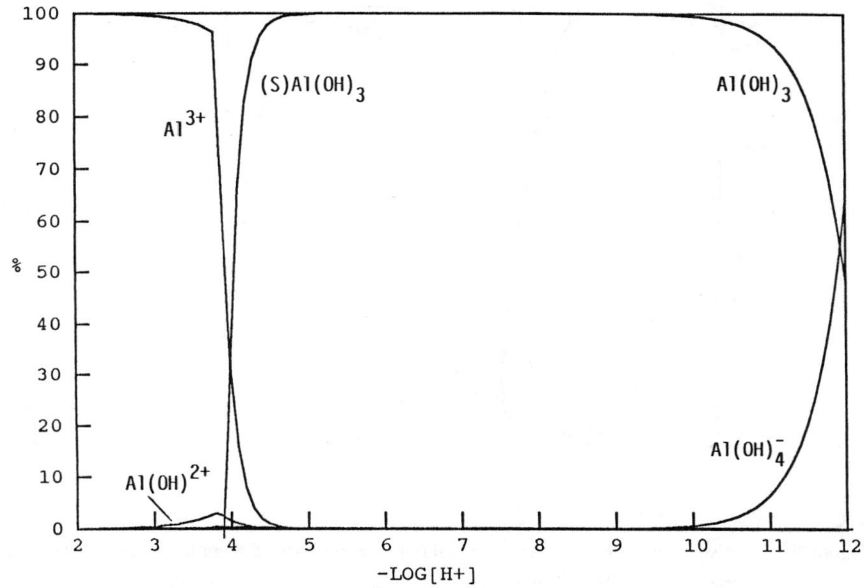

Figure 1. Relative molar concentrations of hydrolytic species formed from aquo Al(III) ion as a function of -log [H^+]. Total analytical concentration of Al(III) species is 1.00 x 10^{-3} M. (S)$Al(OH)_3$ represents the relative quantity of precipitated $Al(OH)_3$, t = 25.0°C, μ = 0.100 M.

Stability constants of Al(III) complexes of polyphenols, hydroxycarboxylic acids, and aminopolycarboxylic acids are listed in Tables 2, 3, and 4, respectively. While original references are cited for the equilibrium constants listed, some have been adjusted in the selection process employed for the development of the Critical Stability Constants compilation.[6] The references cited for the years 1980-1985 appear in Volume 6 of the Critical Stability Constant series, recently published.

The highest stability constants for Al(III) complexes are those of catechol and its analogs, but the effectiveness of these ligands in complexing Al^{3+} is mitigated by the lower chelate effect inherent in the formulation of the 3:1 complexes, and by the very high ligand pK's, which are characteristic of ligands with hard oxygen donor groups. Therefore, the formation of catechol complexes involves very strong hydrogen ion competition, so that the ligands in Table 1 are most effective at very high p[H] and tend to be ineffective at low p[H]. It is noted that the equilibrium constants expressed as hydrogen ion displacement reactions are more accurate than those expressed as formation constants, whereby the metal ion combines directly with the completely deprotonted ligand. Calculation of the latter type of constant requires the protonation constants of the ligands, at least one of which is too high to be measured accurately, as explained above.

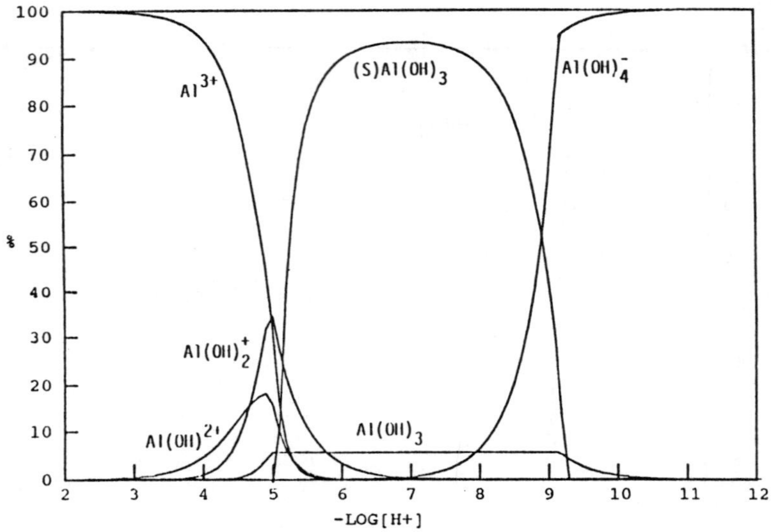

Figure 2. Relative molar concentrations of hydrolytic species formed from the aquo Al(III) ion as a function of -log $[H^+]$. Total analytical concentration of Al(III) species is 1.00 x 10^{-6} M. (S)$Al(OH)_3$ represents the relative quantity of precipitated $Al(OH)_3$, t = 25.0°C, μ= 0.100 M.

The aminopolycarboxylate ligands, on the other hand, contain less strongly coordinating carboxylate groups, held together by coordinating tertiary amino nitrogens which have lower affinity for hard metal ions. On the other hand the strong chelate effects of these multidentate ligands increase their metal ion affinities relative to bidentate ligands such as the catechols, especially in dilute solution. A comparison of the effectiveness of EDTA and catechol is provided by the distribution curves in Figure 3 for a 1:3:1 molar ratio of EDTA:catechol:Al(III) as a function of p[H]. Here it is seen that Al^{3+} is very strongly complexed by EDTA at low p[H]. Over two thirds of the Al^{3+} present is complexed at p[H] 2, and the concentration of free aquo Al^{3+} drops to near zero at p[H] 3. As the p[H] is increased toward the alkaline region, EDTA complexes predominate, aided considerably by stabilization of the monohydroxo complex $Al(OH)L^{2-}$. Catecholate complexes begin to appear at p[H] 7 as the p[H] is increased further, but do not predominate over EDTA complexes until the p[H] is 8.5 and higher. The catecholate complexes reach their highest concentration at about p[H] 10.

Table 2. Al(III) complexes of catechol and its derivatives

Ligand	Equilibrium Quotients and Log K Values				Ref.
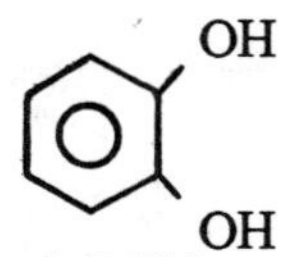 Catechol, H_2L	$\frac{[HL^-]}{[L^{2-}][H^+]}$	13.7[a]	$\frac{[H_2L]}{[HL^-][H^+]}$	9.37[a]	16
	$\frac{[AlL^+]}{[Al^{3+}][L^{2-}]}$	16.9[a]	$\frac{[AlL_2^-]}{[AlL^+][L^{2-}]}$	13.6[a]	16
	$\frac{[AlL_3^{3-}]}{[AlL_2^-][L^{2-}]}$	8.9[a]			16
	$\frac{[AlL^+][H^+]^2}{[Al^{3+}][H_2L]}$	-6.08[b]	$\frac{[AlL_2^-][H^+]^2}{[AlL^+][H_2L]}$	-9.18[b]A	10
	$\frac{[AlL_3^{3-}][H^+]^2}{[AlL_2^-][H_2L]}$	13.52[b]	$\frac{[AlHL_2]}{[AlL_2^-][H^+]}$	6.05[b]	10
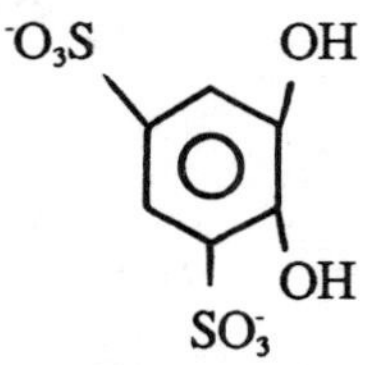 3,5-disulfocatechol (tiron), H_2L^{2-}	$\frac{[HL^{3-}]}{[L^{4-}][H^+]}$	12.5[b]	$\frac{[H_2L^{2-}]}{[HL^{3-}][H^+]}$	7.65[b]	17
	$\frac{[AlL^-]}{[Al^{3+}][L^{4-}]}$	16.6[b]	$\frac{[AlL_2^{5-}]}{[AlL^-][L^{4-}]}$	13.4[b]	18
	$\frac{[AlL^-][H^+]^2}{[Al^{3+}][H_2L^{2-}]}$	-3.26[b]	$\frac{[AlL_2^{5-}][H^+]^2}{[AlL^-][H_2L^{2-}]}$	-6.43[b]	18, 19
	$\frac{[AlL_3^{9-}][H^+]^2}{[AlL_2^{5-}][H_2L^{2-}]}$	-10.75[b]			18, 19
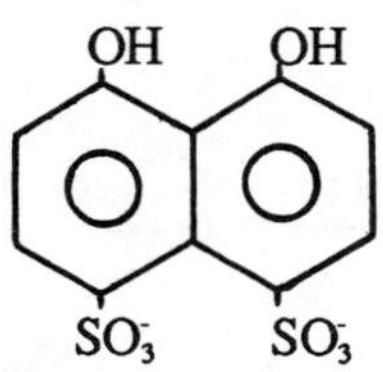 1,8-dihydroxy-4,5-disulfonapthalene (chromotropic salt), H_2L^{2-}	$\frac{[HL^{3-}]}{[L^{4-}][H^+]}$	15.6[a]	$\frac{[H_2L^{2-}]}{[HL^{3-}][H^+]}$	5.33[a]	16
	$\frac{[AlL^-]}{[Al^{3+}][L^{4-}]}$	17.1[a]	$\frac{[AlL_2^{5-}]}{[AlL^-][L^{4-}]}$	12.7[a]	16, 20, 21
	$\frac{[AlL^-][H^+]^2}{[Al^{3+}][H_2L^{2-}]}$	-3.82[a]	$\frac{[AlL_2^{5-}][H^+]^2}{[AlL^-][H_2L^{2-}]}$	-8.25[a]	16, 20, 21

[a] t = 20°C, 0.10 M; [b] t = 25°C, 0.10 M.

Table 3. Stabilities of Al(III) complexes of hydroxy carboxylic acids

Ligand	Equilibrium Quotients and Log K Values				Ref.
OH HOOCCH$_2$CHCOOH Malic acid, H_2L	$\frac{[HL^-]}{[H^+][L^{2-}]}$	4.48[a]	$\frac{[H_2L]}{[H^+][HL^-]}$	3.16[a]	17
	$\frac{[AlL^+]}{[Al^{3+}][L_{2-}]}$	4.32[b]			22
OH HOC$_2$HCHCHCHCHCOOH HO OH	$\frac{[HL]}{[H^+][L^-]}$	3.44[a]	$\frac{[AlL^{2+}]}{[Al^{3+}][L^-]}$	1.98[a]	10
Gluconic acid, HL	$\frac{[AlH_{-1}L^+][H^+]}{[AlL^{2+}]}$	-2.87[a]	$\frac{[AlH_{-3}L^-][H^+]^2}{[AlH_{-1}L^+]}$	-9.29[a]	10
OH HOOCCH$_2$ CCOOH CH$_2$COOH	$\frac{[HL^{2-}]}{[H^+][L^{3-}]}$	5.71[a]	$\frac{[H_2L^-]}{[H^+][HL^{2-}]}$	4.39[a]	10
Citric acid, H_3L	$\frac{[H_3L]}{[H^+][H_2L^-]}$	3.03[a]	$\frac{[AlL]}{[Al^{3+}][L^{3-}]}$	7.98[a]	10
	$\frac{[AlHL^+]}{[AlL][H^+]}$	2.94[a]	$\frac{[AlH_{-1}L^-][H^+]}{[AlL]}$	-3.31[a]	10
HO OH HOOCCHCHCOOH	$\frac{[HL^-]}{[H^+][L^{2-}]}$	3.93[a]	$\frac{[H_2L]}{[H^+][HL^-]}$	2.79[a]	10
D,L-tartaric acid, H_2L	$\frac{[AlH_{-1}L][H^+]}{[Al^{3+}][L^{2-}]}$	1.18[a]	$\frac{[AlH_{-2}L^-][H^+]}{[AlH_{-1}L]}$	-5.15[a]	10
	$\frac{[AlL_2^-]}{[Al^{3+}][L^{2-}]^2}$	7.65[a]	$\frac{[AlH_{-1}L_2^{2-}][H^+]}{[AlL_2^-]}$	-3.72[a]	10

Table 3. (Continued...)

Ligand	Equilibrium Quotients and Log K Values				Ref.
	$\frac{[AlH_{-3}L_2^{4-}][H^+]^2}{[AlH_{-1}L_2^{2-}]}$	-12.67[a]	$\frac{[AlH_{-4}L_3^{7-}][H^+]^4}{[Al^{3+}][L^{2-}]^3}$	15.92[a]	10
	$\frac{[AlH_{-5}L_3^{8-}][H^+]}{[AlH_{-4}L_3^{7-}]}$	-10.89[a]	$\frac{[AlH_{-6}L_3^{9-}][H^+]}{[AlH_{-5}L_3^{8-}]}$	12.7[a]	10
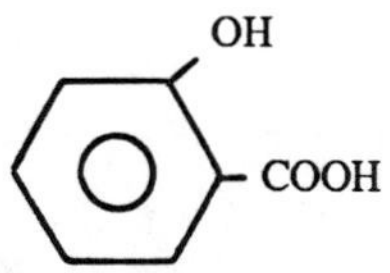	$\frac{[HL^-]}{[L^{2-}][H^+]}$	13.1[c]	$\frac{[H_2L]}{[HL^-][H^+]}$	2.92[c]	16
Salicylic acid, H_2L	$\frac{[AlL^+]}{[Al^{3+}][L^{2-}]}$	12.9[c]	$\frac{[AlL_2^-]}{[AlL^+][L^{2-}]}$	10.3[c]	16, 23
	$\frac{[AlL_3^{3-}]}{[AlL_2^-][L^{2-}]}$	6.6[c]			16, 23
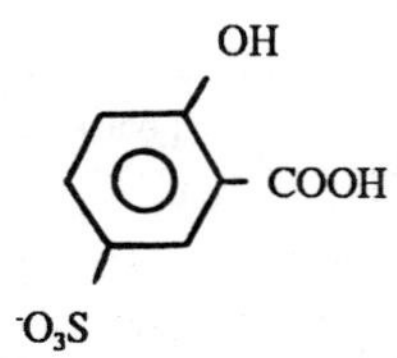	$\frac{[HL^{2-}]}{[L^{3-}][H^+]}$	12.0[c]	$\frac{[H_2L^-]}{[HL^{2-}][H^+]}$	2.60[c]	16
5-Sulfosalicyclic acid, H_2L^-	$\frac{[AlL]}{[Al^{3+}][L^{3-}]}$	12.3[c]	$\frac{[AlL_2^{3-}]}{[AlL][L^{3-}]}$	9.7[c]	16, 21, 24
	$\frac{[AlL_3^{6-}]}{[AlL_2^{3-}][L^{3-}]}$	5.8[c]			16, 21, 24, 25

[a] t = 25°C, 0.10M; [b] t = 29°C, 0.10 M; [c] t = 20°C, 0.10M.

Table 4. Stabilities of Al(III) complexes of aminopolycarboxylic acids

Ligand	Equilibrium Quotients and Log K Values				Ref.
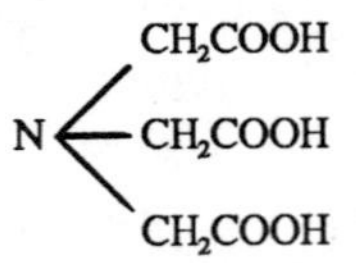 Nitrilotriacetic acid, NTA, H_3L	$\frac{[HL^{2-}]}{[L^{3-}][H^+]}$	9.67[a]	$\frac{[H_2L^-]}{[HL^{2-}][H^+]}$	2.52[a]	17
	$\frac{[H_3L]}{[H_2L^-][H^+]}$	1.9[a]			17
	$\frac{[AlL]}{[Al^{3+}][L^{3-}]}$	11.37[b]	$\frac{[AlHL^+]}{[AlL][H^+]}$	1.90[b]	26
	$\frac{[AlL]}{[Al(OH)L^-][H^+]}$	5.09[b]	$\frac{[Al(OH)L^-]}{[Al(OH)_2L^{2-}][H^+]}$	8.28[b]	26
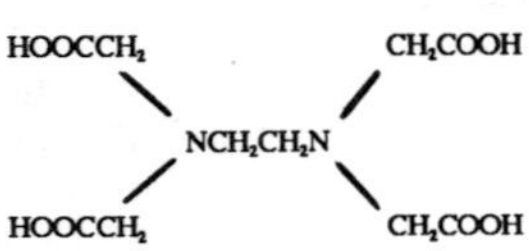 Ethylenediaminetetraacetic acid, EDTA, H_4L	$\frac{[HL^{3-}]}{[L^{4-}][H^+]}$	10.19[a]	$\frac{[H_2L^{2-}]}{[HL^{3-}][H^+]}$	6.13[a]	17
	$\frac{[H_3L^-]}{[H_2L^{2-}][H^+]}$	2.69[a]	$\frac{[H_4L]}{[H_3L^-][H^+]}$	1.99[a]	17
	$\frac{[AlL^-]}{[Al^{3+}][L^{4-}]}$	16.3[a]	$\frac{[AlHL]}{[AlL^-][H^+]}$	2.6[a]	26 28 29
	$\frac{[AlL^-]}{[Al(OH)L^{2-}][H^+]}$	5.87[a]	$\frac{[Al(OH)L^{2-}]}{[Al(OH)_2L^{3-}][H^+]}$	10.3[a]	26 27 28 29
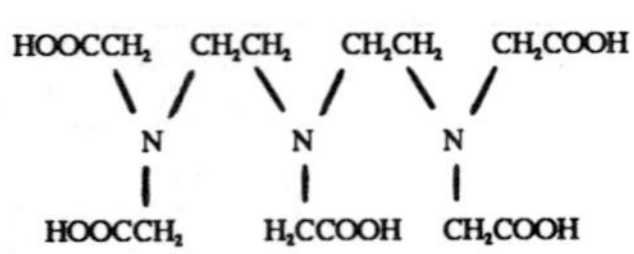 Diethylenetriaminepentaacetic acid, DTPA, H_5L	$\frac{[HL^{4-}]}{[L^{5-}][H^+]}$	10.48[a]	$\frac{[H_2L^{3-}]}{[HL^{4-}][H^+]}$	8.60[a]	17
	$\frac{[H_3L^{2-}]}{[H_2L^{3-}][H^+]}$	4.28[a]	$\frac{[H_4L^-]}{[H_3L^{2-}][H^+]}$	2.6[a]	17
	$\frac{[H_5L]}{[H_4L^-][H^+]}$	2.0[a]	$\frac{[AlL^{2-}]}{[Al^{3+}][L^{5-}]}$	18.6[a]	28 29
	$\frac{[AlHL^-]}{[AlHL^-][H^+]}$	4.4[a]	$\frac{[AlL^{2-}]}{[Al(OH)L^{3-}][H^+]}$	7.2[a]	28 29

[a]t = 25°C, 0.10M; [b]t = 25°C, 0.2M.

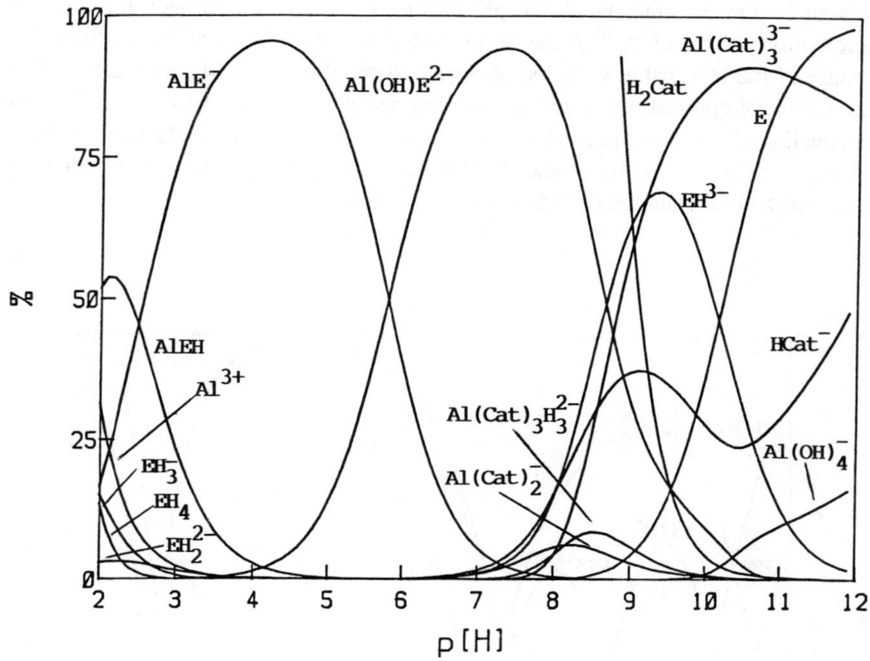

Figure 3. Relative molar concentrations of complex species formed as a function of -log [H^+] in a solution containing the total analytical concentrations, Al(III), 1.00 x 10^{-3} M; pyrocatechol, 3.00 x 10^{-3} M; EDTA, 1.00 x 10^{-3} M; t = 25.0°C; μ = 0.100 M; Cat = dinegative catechol anion and E is tetranegative EDTA anion.

Below that p[H] the EDTA complexes compete strongly for Al^{3+}. Above p[H] 10 the aluminate ion begins to form, reaching at p[H] 12 a relative concentration of over 10% of the Al^{3+} present. At extremely high p[H], 13-14, it is expected that the aluminate ion would predominate. It appears that there is no tendency for $Al(OH)_3$ to precipitate at any p[H] in this system, so that complete sequestration of Al^{3+} is achieved by these two chelating agents, one of which is effective at low and neutral p[H] while the other is effective at high p[H].

Motekaitis and Martell[10] have shown that many α-hydroxycarboxylic acids are very effective ligands for Al^{3+}, and that strong coordination of Al^{3+} is due, to a large extent, to dissociation of hydrogen ion from the α-hydroxyl groups and chelate formation. Thus, many ligands found in nature, such as citrate, tartrate, and malate, form stable complexes with Al^{3+} . Many workers in the past have been reluctant to assume dissociation of the alpha hydroxyl group of hydroxy acids in complex formation, and had assigned complexes of ligands such as citrate, L^{3-}, formulas such as $Al(OH)L^{2-}$, the hydroxide being formed by dissociation of a hydrogen ion from a coordinated water molecule. It was demonstrated,[10] however, that this was not the case by a simple comparison of deprotonation constants. It is seen from Figure 2 that dissociation of hydrogen ion from a water molecule in the coordination sphere of Al^{3+} occurs with a pK of about 5. On the other hand, the pK for dissociation of a hydrogen ion from the coordination sphere of Al^{3+} in the aluminum citrate complex was found to be 3.3. Because of the fact that any degree of coordination of Al^{3+} would increase the pK of dissociation of coordinated water, it was reasoned that the hydrogen ion is displaced from the ligand. Further support for this conclusion resulted from the fact that many α-hydroxy acids behave in this manner, but that such facilitated hydrogen ion dissociation was not observed when the α-hydroxyl group is absent.

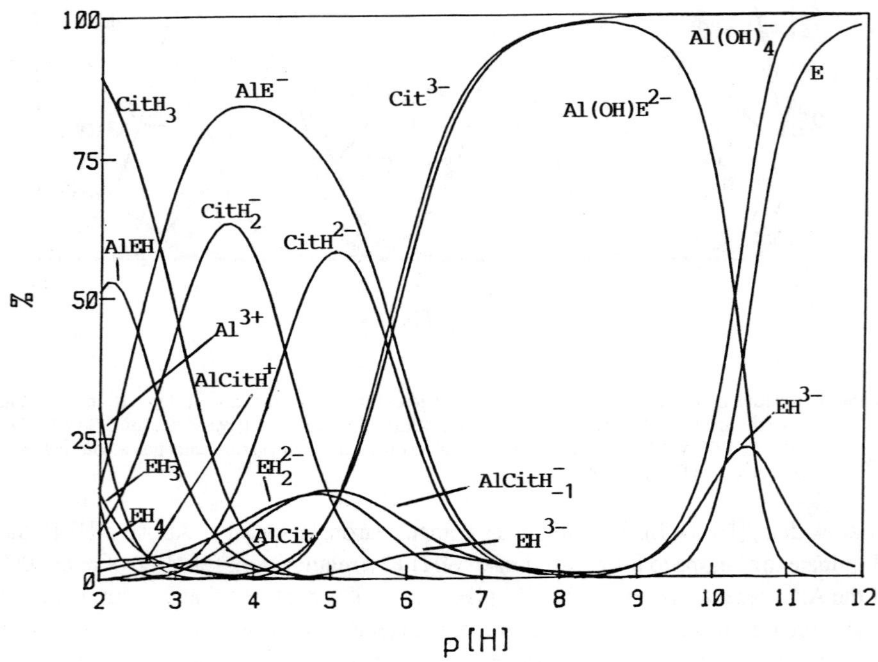

Figure 4. Relative molar concentrations of complex species formed as a function of -log $[H^+]$ in a solution containing the total analytical concentrations, Al(III), 1.00 x 10^{-3} M; citrate, 1.00 x 10^{-3} M; and EDTA, 1.00 x 10^{-3} M; t = 25.00°C; μ = 0.100 M; Cit = trinegative citrate anion; E = tetranegative EDTA anion.

An inspection of the data in Table 3 reveals the inadequacy of the constant listed for malic acid. By analogy with other hydroxy acids such as citric, gluconic, and tartaric acids, it is clear that the malate ion forms one or more stable complexes involving dissociation of the α-hydroxyl group, a fact which has been completely overlooked in literature publications.

Figure 4 illustrates the competition between citrate ion and EDTA for Al^{3+} . Here it is seen that citrate is not effective in competing with EDTA, with the maximum concentration of its principal complex, $AlH^{-1}L^{-}$, never reaching 20% of the Al(III) present. It is also seen that conversion of Al(III) to $Al(OH)_4^-$ occurs at a much lower p[H], with only EDTA as the effective ligand, compared to Figure 3, where catechol provided more effective complexation at high p[H].

From the data in Figures 3 and 4, it is readily predicted that in a system containing catechol and citrate, the Al(III)-citrate complexes would predominate at low p[H], and would be converted to catechol complexes at higher p[H], but would do so at p[H] values considerably below those in Figure 3, which involves competition between catechol and a much stronger ligand.

Since aluminum is such an important element, and since it has only one oxidation state, its activity, reactivity, and transport are related mainly to its coordination chemistry. In this chapter we have critically selected most of the currently available stability constants as a starting point for quantitative measures of coordination reactions of Al(III) in aqueous solution, for the several classes of ligands found both in nature, *in vivo*, and in important analytical applications. The ligand-ligand competition distribution diagrams provide examples of the use of accurate stability constant data to describe the nature of the complexes formed as a function of p[H] and other solution conditions.

REFERENCES

1. Pitt, C. G. and A. E. Martell. "The design of chelating agents for the treatment of iron overload," *ACS Symp. Ser.* 140:279-312 (1980).

2. Baes, C. F. and R. E. Mesmer. *The Hydrolysis of Cations* (New York: John Wiley and Sons, Inc., 1976)

3. Harris, W. R. and A. E. Martell. "Aqueous complexes of gallium(III)," *Inorg. Chem.* 15:713-720 (1976).

4. Motekaitis, R. J. and A. E. Martell. "Gallium complexes of multidentate ligands in aqueous solution," *Inorg. Chem.* 19:1646-1651 (1980).

5. Letkeman, P., A. E. Martell, and R. J. Motekaitis. "Complexes of gallium(III) and hydroxyaromatic ligands," *J. Coord. Chem.* 10:47-53 (1980).

6. Smith, R. M. and A. E. Martell. *Critical Stability Constants*, Vol. 1, 2, 3, 4, 5 (New York: Plenum Publishing Corp., 1974, 1975, 1977, 1976, 1982).

7. Motekaitis, R. J. and A. E. Martell. "Program PKAS: A novel algorithm for the computation of successive protonation constants," *Can. J. Chem.* 60:168-173 (1982).

8. Motekaitis, R. J. and A. E. Martell. "BEST - A new program for rigorous calculation of equilibrium parameters of complex multicomponent systems," *Can. J. Chem.* 60:2403-2408 (1980).

9. Motekaitis, R. J. and A. E. Martell. "SPE - A New Species Distribution Program," to be published.

10. Motekaitis, R. J. and A. E. Martell. "Complexes of aluminum(III) with hydroxy carboxylic acids," *Inorg. Chem.* 23:18-23 (1984).

11. Öhman, L. O. and W. Forsling. "Equilibrium and structural studies of silicon(IV) and aluminum(III) in aqueous solution. 3. A potentiometric study of aluminum(III) hydrolysis and aluminum(III) hydroxo carbonates in 0.6 M Na(Cl)," *Acta Chem. Scand. A.* 35:795-802 (1981).

12. Öhman, L. O., S. Sjöberg, and N. Ingri. "Equilibrium and structural studies of silicon (IV) and aluminum(III) in aqueous solution. 7. Redox, hydrolysis and complexation equilibrium in the system Al^{3+}-1,2-naphthoquinone-4-sulfonate/1,2,-dihydroxy-naphthalene-4-sulfonate-OH^- : A potentiometric study in 0.6 M Na(Cl)," *Acta. Chem. Scand. A* 35:561-568 (1983).

13. Agarwal, R. P. and E. C. Moreno. "Stability constants of aluminum fluoride complexes," *Talanta* 18:873-880 (1971).

14. Baumann, E. W. "Determination of stability constants of hydrogen and aluminum fluorides with a fluoride-selective electrode," *J. Inorg. Nucl. Chem.* 31:3155-3162 (1969).

15. King, E. L. and P. K. Gallagher. "The thermodynamics of aluminum(III) fluoride complex ion reactions: The graphical evaluation of equilibrium quotients from $n([X])^1$," *J. Phys. Chem.* 63:1073-1076 (1959).

16. Havelkova, L. and M. Bartusek. "Potentiometric investigation of the complexes of aluminum with *o*-diphenols, chromotropic acid and *o*-phenol carboxylic acids," *Coll. Czech. Chem. Comm.* 34:3722-3731 (1969).

17. Smith, R. M. and A. E. Martell. *Critical Stability Constants, Vol.6* (New York: Plenum Publishing Corp., 1989).

18. Nasanen, R. and J. Veivo. "The stability of aluminum and iron(III) chelates of catecholdisulphonic acid in aqueous solution," *Suomen Kem.* B29:213-217 (1956).

19. Nasanen, R. "Studies on the stability of aluminum chelates of catecholdisulphonic acid in aqueous solution," *Acta. Chem. Scand.* 11:1308-1312 (1957).

20. Jantti, O. "Potentiometrische Untersuchung der Metallchelate der Chromotropsaure," *Suomen Kem.* B30:136-137 (1957).

21. Athavale, V. T., N. Mahadevan, P. K. Mathur, and R. M. Sathe. "Potentiometric study of the complexes of malonic, 5-sulphosalicyclic and chromotropic acids with some metal ions," *J. Inorg. Nucl. Chem.* 29:1947-1951 (1967).

22. Misra, M. K. and R. K. Nanda. "Spectrophotometric study of malate complexes of copper and aluminium," *J. Indian. Chem. Soc.* 42:267-271 (1965).

23. Das, B. and S. Adiyta. "Spectrophotometric study of aluminium-salicylic acid complex," *J. Indian. Chem. Soc.* 36:473-478 (1959).

24. Nanda, R. K. and S. Aditya. "Spectrophotometric study of metal chelates: Aluminum sulphosalicylate," *J. Indian. Chem. Soc.* 34:577-580 (1957).

25. Nanda, R. K. and S. Aditya. "Spectrophotometric studies on metal chelates. aluminum sulphosalicylate," *J. Indian Chem. Soc.* 40:660-666 (1963).

26. Bhat, T. R., R. R. Das, and I. Shankar. "Complexes of aluminium with some polyaminopolycarboxylic acids," *Indian. J. Chem.* 5:324-327 (1967).

27. Schwarzenbach, G., R. Gut, and G. Anderegg. "Komplexone XXV. Die Polarographische Untersuchung von Austauschgleichgewichten: Neue Daten der Bildungskonstanten von Metallkomplexen der Athylendiamintetraessigsaure und der 1,2-Diaminocyclohexan-Tetraessigsaure," *Helv. Chim. Acta* 37:937-957 (1954).

28. Moeller, T. and S. K. Chu. "The stabilities and thermodynamic functions for the formation of aluminium and mercury(II) chelates of certain polyaminepolyacetic acids," *J. Inorg. Nucl. Chem.* 28:153-159 (1966).

29. Aikens, D. A. and F. J. Bahbah. "Potentiometric characterization of aluminum aminopolycarboxylate chelonates," *Anal. Chem.* 39:646-649 (1967).

CHAPTER 2

A SPECIATION METHOD FOR PARTITIONING MONONUCLEAR AND POLYNUCLEAR ALUMINUM USING FERRON

P. M. Jardine and L. W. Zelazny

INTRODUCTION

The chemical partitioning of aqueous Al into its inorganic and organic, mononuclear and polynuclear forms has received considerable attention in recent years.[1,2,3,4,5] Determination of the relative phytotoxicity of the various Al forms to plants and aquatic life requires accurate speciation methods.[6,7,8,9] Water treatment processes are particularly concerned with the speciation of polynuclear Al for the analytical control of toxic F^- levels in water.[10,11] Numerous techniques have been used in an effort to partition mononuclear and polynuclear Al, several of which consider the kinetics of Al complexation with a colorimetric reagent.[12,13,14] It is assumed that the colorimetric reagent reacts most quickly with mononuclear Al and more slowly with polynuclear Al. Speciation of these Al forms is then based on arbitrary time separations.

Investigating the kinetic reaction of Al with 8-hydroxyquinoline, Bloom et al.[12] and David and Driscoll[15] assumed a 15-min reaction duration between the latter reagent and total mononuclear inorganic and organic Al. Bloom et al.,[1] James et al.,[2] and Hodges[5] have since suggested a 15-s reaction duration for estimating total inorganic mononuclear Al. The more rapid reaction time was chosen in an effort to minimize release of organic-Al complexes to the reagent. James et al.[2] have found, however, that organic-Al complexes are partially reactive with 8-hydroxyquinoline within a 15-s reaction duration. Driscoll[3] has assumed that this short reaction time is sufficient for 8-hydroxyquinoline to degrade all mononuclear organic and inorganic Al complexes.

The kinetic reaction of Al with ferron (8-hydroxy-7-iodo-5-quinoline sulphonic acid) has also been utilized to partition mononuclear from polynuclear Al.[16,17,10] Smith[18] assumed that mononuclear Al reacts instantaneously with ferron, intermediate polymers react by pseudo first-order kinetics, and large polymers or initial solid phases react with ferron extremely slow or are totally inert. Bersillion et al.[19] suggested that at least four categories of soluble Al species exist in partially neutralized solutions with mononuclear Al distinguished from the others by assuming a 30-s reaction with ferron. Distinction of monomoric from polymeric and organic Al assuming a 30-s reaction duration between the former Al type and ferron has also been suggested by others.[20,21,22] Hodges[5] has suggested, however, that ferron may degrade organic-Al complexes within this 30-s period.

The following study will introduce a method that eliminates the arbitrary separation of mononuclear and polynuclear Al by modeling the differential kinetic reactions of ferron with partially neutralized inorganic and organic Al solutions.

MATERIALS AND METHODS

Experimental

Stock solutions 0.008 mol L^{-1} with respect to Al were prepared from the Cl^-, NO_3^- and ClO_4^- salts. Stock solutions of 0.004 mol L^{-1} $Al_2(SO_4)_3$ were also prepared. The solutions were preacidified with acids of the corresponding anion so that their pH levels were <3, which ensured that $Al(H_20)_6^{3+}$ was the predominant species. Solutions $\simeq 8 \times 10^{-5}$ mol L^{-1} with respect to Al were prepared from the stocks. Diluted $AlCl_3$ solutions, containing $\simeq 8 \times 10^{-5}$ mol L^{-1} Al, were also mixed with various quantities of NaF (6×10^{-5} to 1.5×10^{-3} mol L^{-1}), NaH_2PO_4 (6×10^{-5} to 1.5×10^{-3} mol L^{-1}), and known quantities of various organic acids (organic ligand/Al molar ratios ranging from 0.08 to 7.5). The concentration of Al, organic acids, and the lower concentrations of P and F used in this study are typical of natural soil solutions.[15,23,24,25] Ten-milliliter aliquots of the Al anion, solutions were then partially neutralized under a N_2 atm with $\simeq 4.0 \times 10^{-3}$ mol L^{-1} NaOH at a rate of 0.031 mL min^{-1} using a Radiometer automatic titration system. The slow rate

Table 1. Quantity of precipitated Al in freshly prepared partially neutralized Al solutions of varying inorganic anion composition

	n_o			
	0.5	1.0	1.5	2.0
	Fraction precipitated Al x 100			
$Al(ClO_4)_3$	0.0	0.0	0.0	0.0
$AlCl_3$	0.0	0.0	0.0	0.0
$Al(NO_3)_3$	0.0	0.0	0.0	0.0
$Al_2(SO_4)_3$	0.0	0.0	8.0	29.9
Al-P (6×10^{-5} mol L^{-1} P)	0.0	0.0	5.8	19.1

of base addition results in metastable mixtures of mononuclear and polynuclear Al which have no precipitated Al or small quantities of precipitated Al (Table 1). It is assumed that precipitated Al is undetectable by ferron after 2 h.

The interaction of ferron with the unneutralized and partially neutralized Al inorganic and organic solutions involved the addition of a 0.200 ml sample to a quartz cuvette to which 1.000 mL of ferron was rapidly injected. Absorbance changes were monitored at 370 nm and 298 ± 0.5 K as a function of time using a Beckman DU-7 UV-VIS spectrophotometer. The unneutralized Al solutions were equilibrated at least 24 h prior to analysis with ferron, and the partially neutralized Al solutions were monitored immediately following their preparation. The ferron solution was prepared by methods described by Jardine and Zelazny[26] and was used 5-7 days after its final preparation.

The interaction of Al with fulvic acid was also investigated. The water soluble fulvic acid fraction was obtained from the Oe horizon of a Black Creek soil from New York (Typic Haplorthod), which had 3.8 x 10^{-4} mmol (+) mL^{-1} negative charge at pH 6.0. Various concentrations of the fulvic acid were added to ≃8 x 10^{-5} mol L^{-1} $AlCl_3$ and their kinetic reaction with ferron investigated. The background fulvic acid color and the reaction of ferron with residual ions on the fulvic acid were corrected for as follows. Fulvic acid solutions containing no added Al were equilibrated in a quartz cuvette to which 1.000 mL of ferron was added. The spectrophotometer was immediately zeroed which eliminated the background fulvic color. A fresh sample of the fulvic acid solution was again mixed with ferron as above, and a small but significant absorbance increase was monitored with time. This absorbance increase was the result of ferron reactions with residual ions on the fulvic acid. Subtracting this interaction from the reaction of ferron with fulvic acid containing added Al eliminated the residual ion effect.

To estimate the fraction of Al complexed by the various organic and inorganic anions, solutions were speciated using a modified version of GEOCHEM.[27,28,29] The equilibrium constants of the data base were checked with literature values and refined as needed. Constants for Al complexation with malonic, tartaric, and formic acids were obtained from Sillen and Martell,[30] and constants for malic and oxalic acids were obtained from Perrin.[31] Equilibrium constants for Al complexation with salicylic acid were derived from Perrin[31] and Perlmutter-Hayman and Tapuhi,[32] and values for citric acid were obtained from Lopez-Quintela et al.[33] The Al-SO_4 and -F equilibrium constants were obtained from Lindsay,[34] and the Al-P equilibrium constants were obtained from Hogfeldt[35] and Stumm and Morgan.[36]

Theoretical

Partially neutralized Al solutions have been shown to exist as a binary mixture of mononuclear and polynuclear Al.[16] The interaction of these species with the common reagent ferron results in an irreversible bimolecular reaction expressed as:

$$nA + R \xrightarrow{k_a} C \tag{1a}$$

$$mB + R \xrightarrow{k_b} C \qquad (1b)$$

where A and B are concentrations in milligrams per liter of mononuclear and polynuclear Al, n and m are the moles of A and B or the order of the reaction with respect to A and B, R is the concentration of the reagent, k_a and k_b are the forward rate constants for monomers and polymers, respectively, and C is a common product. If conditions are adjusted so that a 50-fold or greater excess of reagent to reactant is imparted, the rate equations for the two competing irreversible reactions can be described as:[37]

$$\frac{-dA}{dt} = k_a A^n \qquad (2a)$$

$$\frac{-dB}{dt} = k_b B^m \qquad (2b)$$

where t is time. Integration of Eqns. (2a) and (2b) and separating variables yields, for n and m $\neq$ 1

$$A = \left(\frac{1}{k_a t(n-1) + 1/A_o^{n-1}} \right)^{\frac{1}{n-1}} \qquad (3a)$$

$$B = \left(\frac{1}{k_b t(m - 1) + 1/B_o^{m-1}} \right)^{\frac{1}{m-1}} \qquad (3b)$$

where A_o and B_o are the initial mononuclear and polynuclear Al concentrations in solution, respectively. Since A and B react with ferron to form a common product, then

$$C_t = C_\infty - A - B \qquad (4)$$

where $C_\infty = A_o + B_o$ is the concentration of product at equilibrium and C_t is the concentration at any time *t*. Substituting Eqns. (3a) and (3b) into Eqn. (4) yields:

$$C_\infty - C_t = \left(\frac{1}{k_a t(n-1) + 1/A_o^{n-1}} \right)^{\frac{1}{n-1}} + \left(\frac{1}{k_b t(m-1) + 1/B_o^{m-1}} \right)^{\frac{1}{m-1}} \qquad (5)$$

which expresses two simultaneous general order reactions that have a common product. By regressing $C_\infty - C_t$ (unreacted Al) as a function of *t* using nonlinear least squares inversion, we can obtain parameters A_o, B_o, k_a, k_b, n, and m. In this study k_a, n, and m are usually fixed, with the remaining parameters fit using Eqn. 5. In some cases k_a is unknown, and this parameter is fit as well. Equation 5 holds for all values of n and m, integer or fraction, with the exception of n or m = 1. For situations where first-order

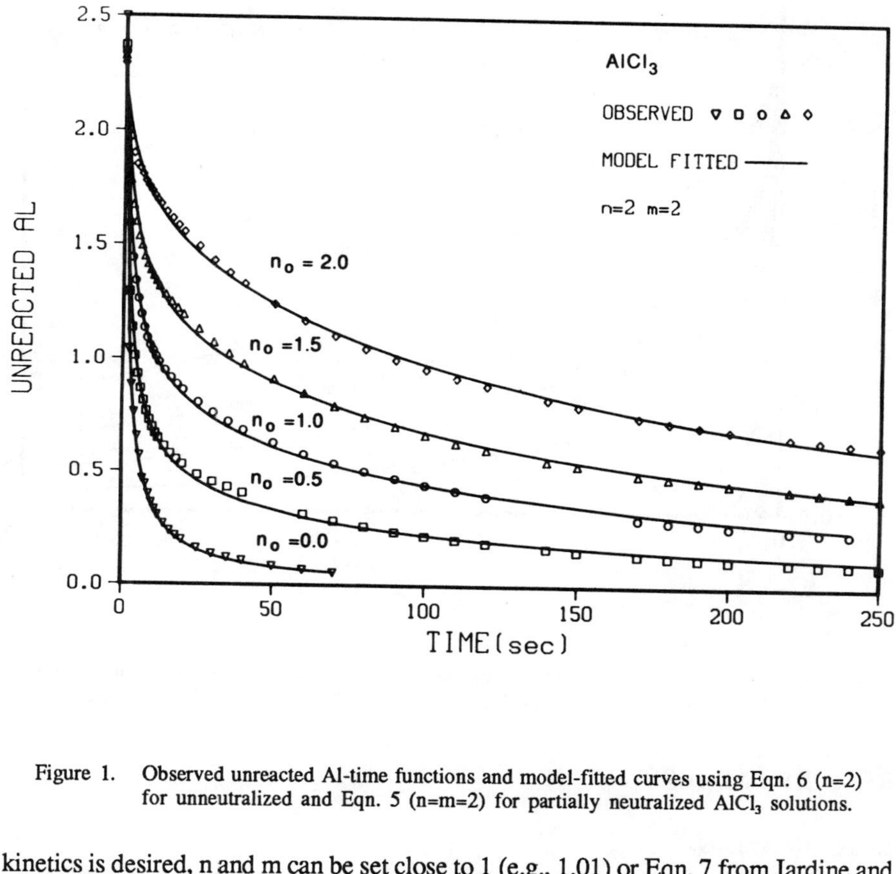

Figure 1. Observed unreacted Al-time functions and model-fitted curves using Eqn. 6 (n=2) for unneutralized and Eqn. 5 (n=m=2) for partially neutralized $AlCl_3$ solutions.

kinetics is desired, n and m can be set close to 1 (e.g., 1.01) or Eqn. 7 from Jardine and Zelazny[26] may be used. Consistent with previous work,[26] only the first 250 s of observed $C_\infty - C_t$ vs. t data was fit to the binary species function [(e.g., Eqn. 5)] with the tertiary reaction of large Al polymers, with ferron being ignored for simplicity. We have assumed that the tertiary reaction is described by a single first-order equation as shown by Jardine and Zelazny.[26] The reaction of ferron with mononuclear Al solutions was described using a degenerate form of Eqn. 5 that considers only one species:

$$C_\infty - C_t = \left(\frac{1}{k_a t(n-1) + 1/A_o^{n-1}} \right)^{\frac{1}{n-1}} \tag{6}$$

where $C_\infty = A_o$ is again the concentration of product at equilibrium.

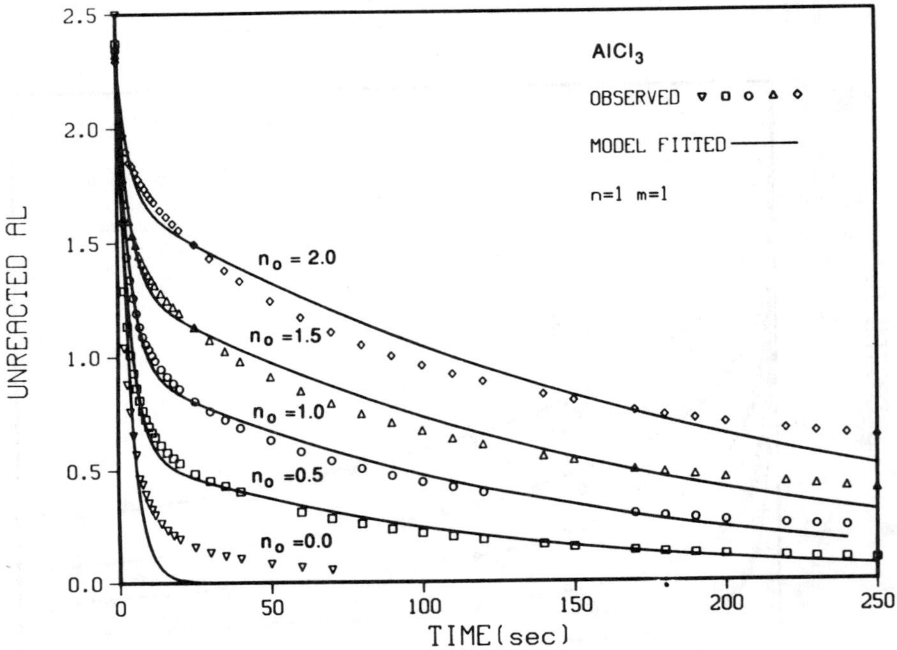

Figure 2. Observed unreacted Al-time functions and model-fitted curves using Eqn. 6 (n=1) for unneutralized and Eqn. 5 (n=m=1) for partially neutralized $AlCl_3$ solutions.

RESULTS AND DISCUSSION

Inorganic Anions

6 The interaction of ferron with unneutralized mononuclear Al solutions (n_o = OH/Al = 0.0) containing various inorganic anions was best modeled by a single second-order reaction (Figure 1). No other fractional or integer order described the observed data as well as a second-order function. For example, mononuclear Al solutions were poorly described by first-order kinetics (Figure 2), which has been used previously to model these functions.[26] Second-order conformity may result from the predominance of ferron^{2-} reacting simultaneously with two Al monomers.[38] Mononuclear Al solutions containing ClO_4^-, Cl^-, NO_3^-, SO_4^{2-}, $H_2PO_4^-$, and F^- with anion to Al molar ratios of 13, 21, 15, 5, 0.75, and 0.08, respectively, expressed similar reactions with ferron which provided model-fitted k_a values (rate of monomer interactions with ferron) of 11 to 15 $\pm$0.7 L mg^{-1} min^{-1} (Figs. 1, 4, 6). Variations in k_a resulted from experimental variability and were not related to the type of anion at these concentrations. Predicted speciation of the mononuclear Al solutions using a modified version of GEOCHEM,[27,28,29]

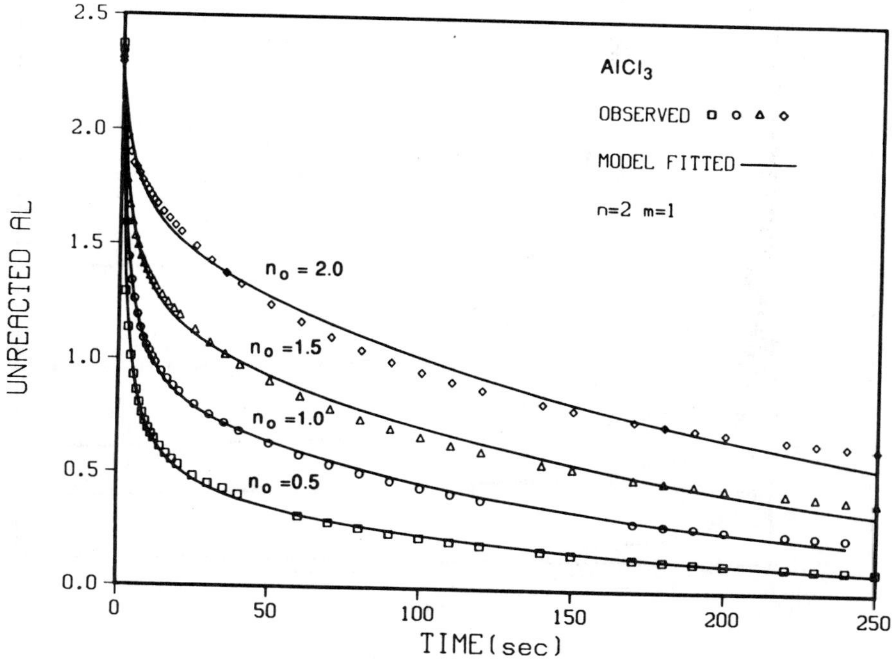

Figure 3. Observed unreacted Al-time functions and model-fitted curves using Eqn. 5 (n=2, m=1) for partially neutralized $AlCl_3$ solutions.

suggested 12% Al complexed with SO_4^{2-}, 7% Al-F complexes, and 36% Al-P complexes. It appears that ferron easily degrades these complexes, which react identically with uncomplexed mononuclear Al solutions. Chloride/Al molar ratios as high as 13,000 have been found to not affect the rate of Al reaction with ferron.[26] This scenario most likely occurs as well for high concentrations of NO_3^- and ClO_4^-.

Ferron interactions with partially neutralized Al solutions of varying basicity (n_o = OH/Al) were best described by two simultaneous second-order reactions (Figs. 1 and 4). Although not shown, unneutralized and partially neutralized $Al(NO_3)_3$ and $Al(ClO_4)_3$ solutions produced observed and model-fitted kinetic curves similar to those for $AlCl_3$ (Figure 1), and Al-P solutions (P/Al molar ratio $\simeq$ 1) produced curves similar to those for $Al_2(SO_4)_3$ (Figure 4). Consecutive first-order functions did not describe the observed data well (Figure 2) nor did assuming mononuclear and polynuclear Al to react with ferron via second- and first-order kinetics, respectively (Figure 3).

Since k_a values for mononuclear Al reactions with ferron have been well established in this chapter, this value was fixed at 15.0 L mg^{-1} min^{-1} in Eqn. (5). As an added check, values of k_a were frequently fit in Eqn. (5) and generally converged very close to the fixed value. The simultaneous second-order reaction function modeled the observed

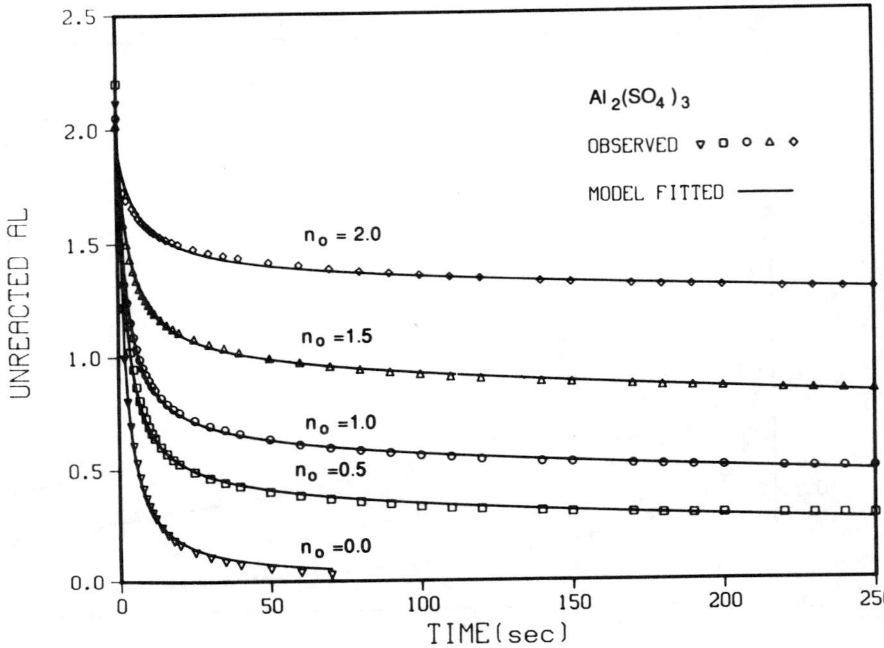

Figure 4. Observed unreacted Al-time functions and model fitted curves using Eqn. 6 (n=2) for unneutralized and Eqn. 5 (n=m=2) for partially neutralized $Al_2(SO_4)_3$ solutions.

data extremely well and provided values for the rate of polynuclear Al reactions with ferron, k_b, and the fraction of mononuclear Al in solution, $f_a = A_o/(A_o + B_o)$ (Figs. 1, 4, Table 2). As n_o increased, model-fitted k_b and f_a decreased for all Al-anion solutions studied (Table 2). The partially neutralized $Al(ClO_4)_3$ system had the most rapid kinetic reaction with ferron, suggesting possible complex formation between Al-polymers and the anions Cl^-, NO_3^-, SO_4^{-2}, and $H_2PO_4^-$. These various anions may in turn influence the size and stability of the polymers formed. The inclusion of Cl^- and NO_3^- into the structure of Al-polymers has been shown by Ross and Turner.[39] These authors further suggest that ClO_4^- had little or no effect on the crystallization of $Al(OH)_3$. The bond stability of Cl^- and NO_3^- complexes is not exceptionally large, however, judging from their rapid rate of degradation with ferron (Table 2). The SO_4^{-2} and $H_2PO_4^-$ systems expressed the slowest kinetics of polymer degradation by ferron due to very strong anion interactions with the polymers. Overall, the magnitude of k_b decreased with increasing n_o, suggesting polymers of a larger molecular size or greater stability at higher n_o values (Table 2).

Table 2. Model-fitted k_b and f_a values for the reaction of ferron with partially neutralized Al solutions of varying inorganic anion composition

	n_o			
	0.5	1.0	1.5	2.0
	$^a k_b$ (L mg^{-1} min^{-1})			
$Al(ClO_4)_3$	4.48 ± 0.41[b]	1.31 ± 0.08	0.730 ± 0.048	0.198 ± 0.015
$AlCl_3$	1.75 ± 0.26	0.719 ± 0.046	0.419 ± 0.027	0.251 ± 0.012
$Al(NO_3)_3$	1.04 ± 0.05	0.740 ± 0.050	0.382 ± 0.025	0.262 ± 0.013
$Al_2(SO_4)_3$	0.286 ± 0.063	0.100 ± 0.019	0.0415 ± 0.0042	0.0108 ± 0.0039
Al-P (6 x 10^{-5} mol L^{-1} P)	0.252 ± 0.038	0.0501 ± 0.0050	0.0116 ± 0.0028	0.0082 ± 0.0004
	$^a f_a$			
$Al(ClO_4)_3$	0.77 ± 0.01[b]	0.61 ± 0.02	0.45 ± 0.03	0.32 ± 0.03
$AlCl_3$	0.82 ± 0.03	0.65 ± 0.02	0.48 ± 0.02	0.29 ± 0.03
$Al(NO_3)_3$	0.76 ± 0.01	0.59 ± 0.02	0.40 ± 0.03	0.27 ± 0.03
$Al_2(SO_4)_3$	0.83 ± 0.02	0.70 ± 0.02	0.52 ± 0.01	0.29 ± 0.02
Al-P (6 x 10^{-5} mol L^{-1} P)	0.68 ± 0.03	0.47 ± 0.02	0.25 ± 0.02	0.097 ± 0.004

[a] Fitted with two simultaneous second-order reactions, with k_a = 15.0 L mg^{-1} min^{-1}.
[b] Confidence interval = 0.95

Model-fitted f_a's for the ClO_4^-, Cl^-, and NO_3^- systems were similar over the range of n_o investigated (Table 2). Unexpectedly high f_a values were predicted for partially neutralized $Al_2(SO_4)_3$ solutions, possibly because SO_4^{-2} may compete well with OH^- and limit the amount of polymer formed (Table 2). Such anion shielding has been suggested by others.[39,40] Polynuclear Al species that do form in the SO_4^{-2} system, however, are more difficult to denature with ferron relative to the ClO_4^-, Cl^-, and NO_3^- systems. Model-fitted f_a's for partially neutralized Al-P solutions were much lower than for the other Al-anion systems (Table 2). The strong interaction of P with partially neutralized Al may serve as a template to enhance the amount of polymer formed.[41,42]

The kinetic model was also tested on partially neutralized Al solutions that were monitored with age using ferron. The observed data were well described by two simultaneous second-order reactions (Figure 5) with the curves suggesting increasing polymerization or enhanced stability of preexisting Al polymers as the solution aged. This scenario is supported by decreasing OH/Al ratios and pH values, as well as increasing quantities of precipitated Al with solution age (Table 3). Model-fitted f_a values, however, remained essentially constant as the partially neutralized Al solution aged. Constant mononuclear Al concentrations in aging partially neutralized solutions have also been suggested by others.[43,44]

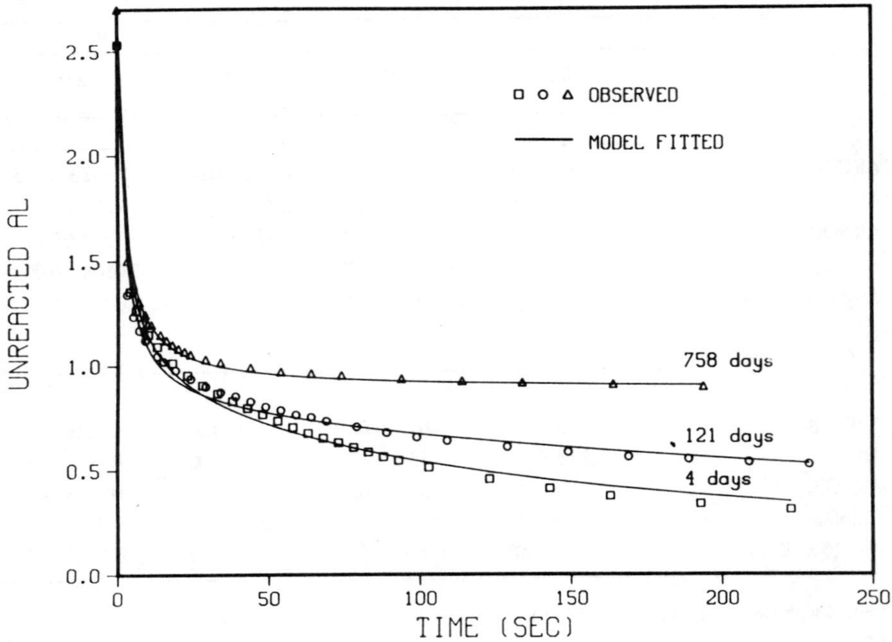

Figure 5. Observed unreacted Al-time functions and model-fitted curves using Eqn. 5 (n=m=2) for an Al solution with initial OH/Al = 0.84 and subsequently aged.

Table 3. Chemical characteristics and model-fitted f_a using Eqn. 5 for the reaction of ferron with aged partially neutralized Al

Age of partially neutralized Al (days)	OH/Al	pH	precipitated x 100	Fraction of Al f_a
4	0.84	4.74	0.0	[a]0.65 ± 0.04
121	0.59	4.70	7.8	0.69 ± 0.04
758	0.38	4.56	21.0	0.67 ± 0.03

[a] Confidence interval = 0.95.

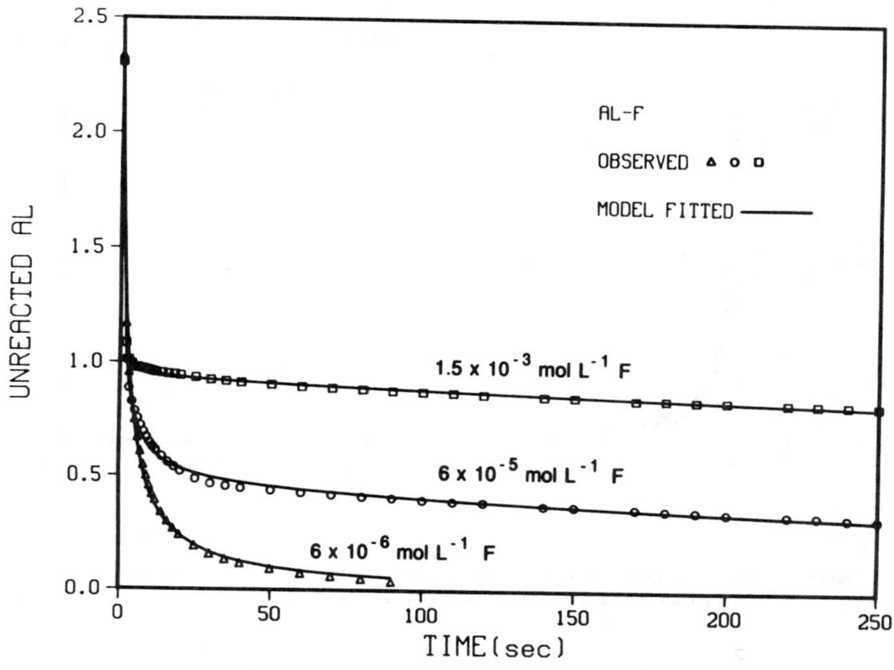

Figure 6. Observed unreacted Al-time functions and model-fitted curves using Eqns. 5 and 6 for solutions $\simeq$ 8 x 10^{-5} mol L^{-1} $AlCl_3$ with varying concentrations of F^-.

Model-fitted f_a's (Table 1) are believed to be more accurate than those proposed by Jardine and Zelazny[26] and more closely agree with values established by Bersillion et al.[19] and Bertsch et al.[45] for partially neutralized $AlCl_3$ solutions where $n_o < 1.5$. These latter authors distinguished between mononuclear and polynuclear Al assuming a 30-s reaction duration between the former Al type and ferron. Observed data from this study also indicate agreement between model-fitted f_a's and those determined assuming a 30-s reaction duration with ferron for solutions where $n_o < 1.5$. An arbitrary termination time of 30 s for the detection of mononuclear Al in partially neutralized solutions of low n_o appears to serve as a good approximation for distinction of monomeric and polymeric Al. However, caution must be used when employing constant separation times for partially neutralized Al solutions with $n_o > 1.5$ since the duration of monomer interactions with ferron depends on n_o.[26]

Partially neutralized Al-F solutions were not prepared since the complexes of F^- with Al are strongly favored over OH^- until high pH values are attained. As noted earlier, $AlCl_3$ solutions with F^-/Al molar ratios of 0.08 conformed to second-order kinetics with fitted k_a values very near those of the other Al anion solutions (Figure 6). A tenfold increase in the concentration of F^-, resulting in 66% complexed Al (as

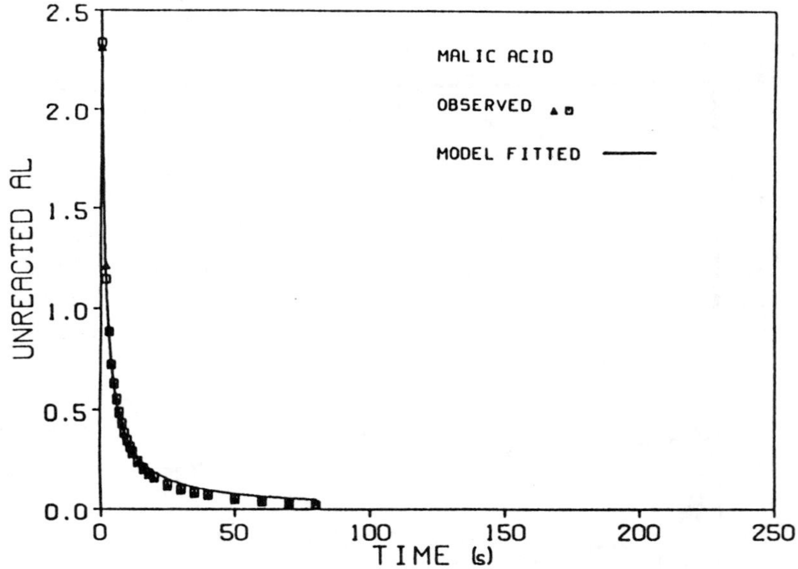

Figure 7. Observed unreacted Al-time functions and model-fitted curves using Eqn. 6 (n=2) for malate/Al molar ratios of 3.8 (Δ) and 0.38 ( ).

estimated by GEOCHEM), produced kinetic reactions with ferron that were described by two simultaneous second-order reactions (Figure 6). Increasing the F^- concentration to 1.5 x 10^{-3} mol L^{-1} resulted in 100% complexed Al which again produced kinetic reactions with ferron that were described by two simultaneous second-order reactions (Figure 6). It appears that when the F^-/Al molar ratio of 0.08 is exceeded, Al-F interactions with ferron can no longer be quantitatively distinguished as rapidly reacting mononuclear Al. In mixed systems a F^-/Al molar ratio >0.08 will make it difficult, if not impossible, to separate mononuclear Al from polynuclear Al. Increasing the $H_2PO_4^-$/Al molar ratio above ≃ 1.0 also results in complex kinetic curves for unneutralized and partially neutralized Al with ferron (not shown) which again will cause difficulty with the speciation of Al.

Organic Anions

The interaction of ferron with mononuclear $AlCl_3$ solutions containing various concentrations of selected organic anions generally conformed to second-order kinetics (Figs. 7-9, Table 4). The concentrations of Al and organic anions used in this study cover the range typically found in soil solutions.[46,47,4] Mononuclear Al solutions containing malate, glutarate, succinate, salicylate, formate, and malonate with ligand/Al molar ratios ranging from 0.4-7.5 reacted rapidly with ferron and were modeled using a single second-order equation. Similarly, mononuclear Al solutions containing citrate, tartarate, and oxalate with ligand/Al molar ratios ≤ 0.13, 0.08, and 0.38, respectively, also reacted rapidly with ferron and were modeled using a single second-order expression. Model-fitted k_a values for these organic-Al solutions were similar to k_a values determined for uncomplexed $AlCl_3$ solutions without added organic ligands (Table 4). Predicted speciation of these organic-Al solutions with GEOCHEM and potentiometric titration analysis indicated that a large fraction of solution Al may be complexed by malic, salicylic, glutaric, and succinic acids, while these ligands were found to have no effect on the kinetics of ferron interaction with Al (Table 4, Figure 7). This implies that predictions of complex stability are not necessarily indicative of the kinetic lability of Al in the complex.[48,2] Such observations conflict with the assumptions of several researchers who have suggested that Al-organic complexes react slowly with colorimetric reagents.[45] A large fraction of the formic acid exists as a free anion and remained uncomplexed with Al, explaining the similarity of model-fitted k_a values with those of $AlCl_3$ without added organic anions (Table 4). Increasing concentrations of citrate, tartarate, and oxalate resulted in a larger fraction of complexed Al which reacted

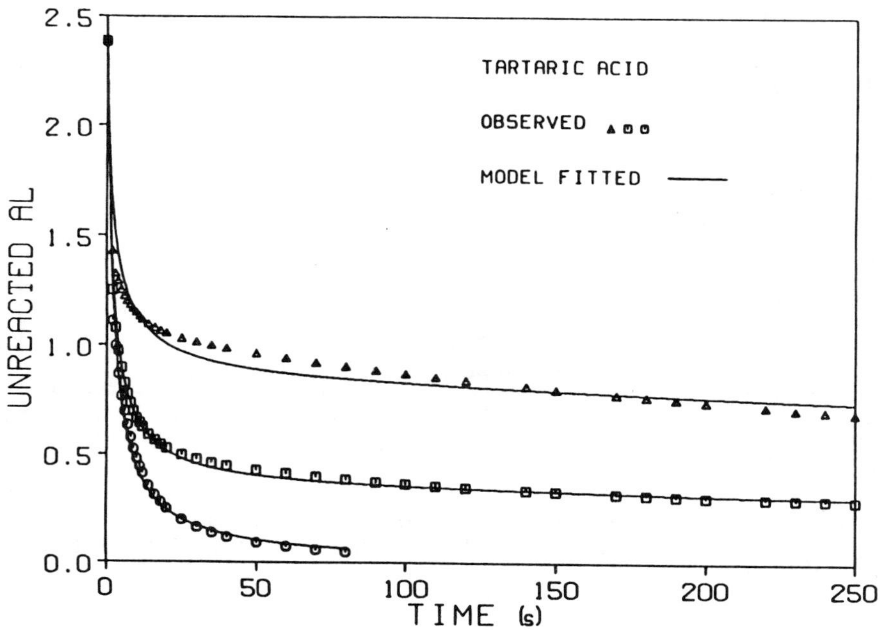

Figure 8. Observed unreacted Al-time functions and model-fitted curves using Eqn. 6 (n=2) for the tartarate/Al molar ratio of 0.08 (o) or Eqn. 5 (n=m=2) for molar ratios of 3.8 (Δ) and 0.38 ( ).

more slowly with ferron (Table 4). Interactions of ferron with Al containing increasing concentrations of tartarate were modeled using two simultaneous second-order reactions (Figure 8, Table 4). These data are very similar to ferron interactions with Al-F complexes (Figure 6); however, no F^- was detected in the tartaric acid stock solution. The slow kinetic reactions observed are the result of stable complexation products between Al and tartaric acid. Ferron interactions with Al containing increasing concentrations of citrate and oxalate also resulted in slow kinetics that could no longer be described by second-order reactions (Figure 9, Table 4). The observed data were well described by first-order kinetic equations which suggests a change in the

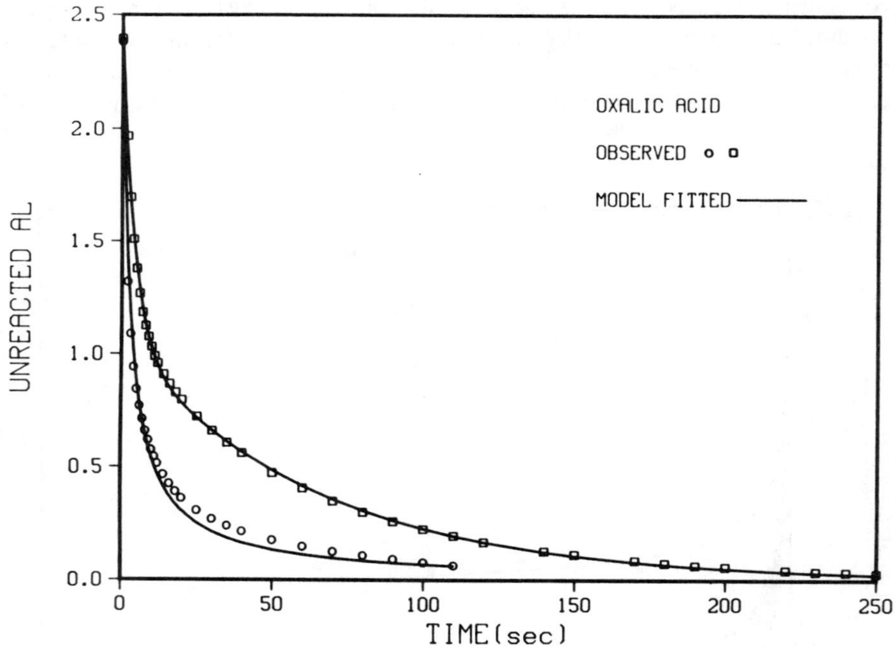

Figure 9. Observed unreacted Al-time functions and model-fitted curves using Eqn. 6 (n=2) for oxalate/Al molar ratio of 0.38 (o) or Eqn. 5 (n=m=1) for the molar ratio of 3.8 ( ).

mechanism by which ferron degrades these complexes at the higher organic anion concentration. Thus, when the citrate/Al, tartarate/Al, and oxalate/Al molar ratios exceed 0.13, 0.08, and 0.38, respectively, their interactions with ferron can no longer be quantitatively distinguished as rapidly reacting mononuclear Al. In mixed systems these high molar ratios will limit the credibility of accurate distinction between mononuclear and polynuclear Al by ferron.

Numerous mononuclear Al solutions containing a mixed array of organic and inorganic anions were prepared using concentrations identical to those in this manuscript. The interactions of ferron with these solutions supported the observations for ferron interactions with the single inorganic organic ligand systems. That is, when the $H_2PO_4^-$, F^-, citrate, tartarate, and oxalate to Al molar ratios exceed the critical limits of ≃1.0, 0.08, 0.13, 0.08, and 0.38, respectively, the separation of mononuclear and polynuclear Al becomes difficult, if not impossible. There was further indication that

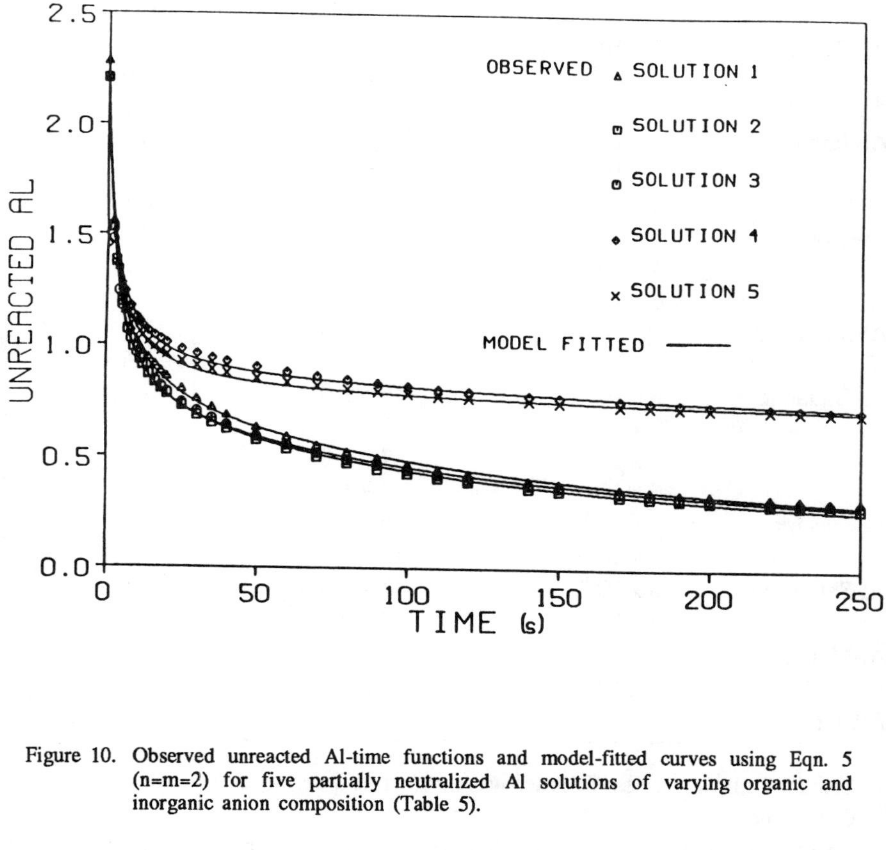

Figure 10. Observed unreacted Al-time functions and model-fitted curves using Eqn. 5 (n=m=2) for five partially neutralized Al solutions of varying organic and inorganic anion composition (Table 5).

no critical molar ratio limits exist for the remaining organic ligands under the conditions of this study (Table 4).

Several of the above Al solutions with ligand/Al mole ratios ≤ the critical values mentioned were partially neutralized with NaOH until a basicity (n_o) of 1.0 was attained (Figure 10, Table 5). The particular set of experiments described here show whether significant interactions exist between polynuclear Al and added citrate or formate. The kinetic reaction of ferron with these solutions was modeled using two simultaneous second-order reactions, suggesting the presence of mononuclear and polynuclear Al.

Table 4. Chemical characteristics and rate constants for the reaction of ferron with $\approx 8 \times 10^{-5}$ mol L^{-1} $AlCl_3$ solutions of varying organic anion composition

Organic anion	Organic anion concentration	k_a	Fraction of Al complexed by organic anion x 100
	mol L^{-1}	L mg^{-1} min^{-1}	
Al-Citrate	1×10^{-4}	[1]2.6 ± 0.1[2]	94.9
	2×10^{-5}	6.6 ± 0.3	22.6
	1×10^{-5}	8.3 ± 0.6	11.9
Al-Tartarate	3×10^{-4}	[3]49.8 ± 3.6	94.8
	3×10^{-5}	[3]18.6 ± 0.3	31.2
	6×10^{-6}	10.8 ± 0.4	6.5
Al-Oxalate	3×10^{-4}	[4]14.0 ± 0.9	99.8
	3×10^{-5}	8.5 ± 0.4	17.1
	6×10^{-6}	10.2 ± 0.6	3.5
Al-Malonate	3×10^{-4}	9.9 ± 0.8	11.6
	3×10^{-5}	13.7 ± 0.6	2.6
Al-Glutarate	3×10^{-4}	16.3 ± 0.4	–[5]
	3×10^{-5}	15.1 ± 0.4	–[5]
Al-Formate	6×10^{-4}	17.0 ± 0.6	2.1
	6×10^{-5}	14.9 ± 0.3	0.2
Al-Succinate	6×10^{-4}	16.9 ± 0.5	–[5]
	6×10^{-5}	14.0 ± 0.4	–[5]
Al-Salicylate	6×10^{-4}	15.7 ± 0.6	63.8
	6×10^{-5}	14.3 ± 0.4	15.3
Al-Malate	3×10^{-4}	14.3 ± 0.8	29.2
	3×10^{-5}	14.3 ± 0.5	7.9
$AlCl_3$ (no organic)		13.7 ± 0.5	0.0

[1] Units in min^{-1} since best fit order was nearly one

[2] Confidence interval = 0.95

[3] Reaction fit by two simultaneous 2nd-order reactions where $k_b = 0.121 \pm 0.004$ and 0.285 ± 0.028 and $f_a = 0.56 \pm 0.01$ and 0.83 ± 0.01 for 3×10^{-4} and 3×10^{-5} mol L^{-1} tartarate, respectively.

[4] Units are in min^{-1} since reaction fit by two simultaneous first-order reactions where $k_b = 0.922 \pm 0.046$ and $f_a = 0.58 \pm 0.02$.

[5] Unreliable or unavailable equilibria data. However, significant complexation was shown using potentiometric titration analysis.

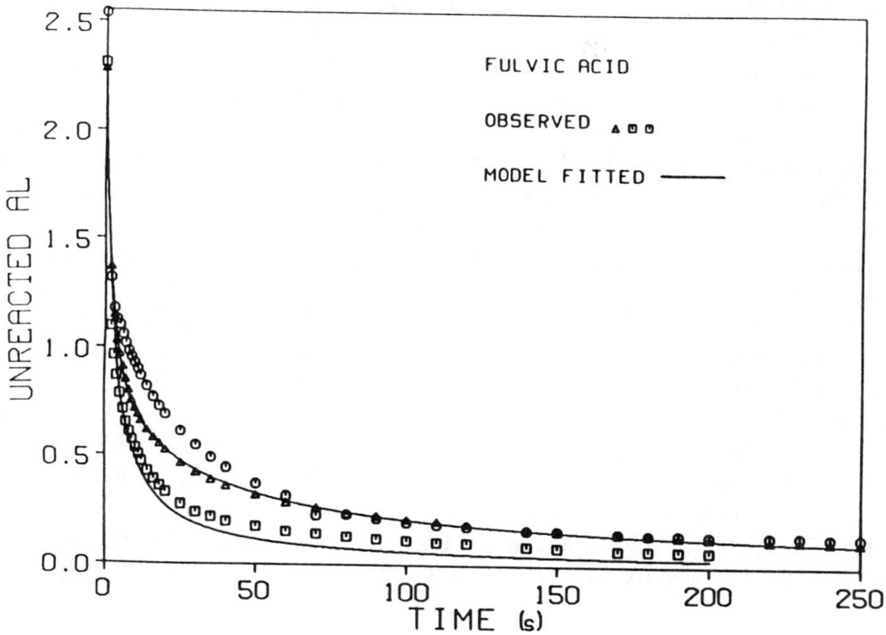

Figure 11. Observed unreacted Al-time functions and model-fitted curves using Eqn. 6 (n=2) for the fulvate/Al molar ratio of 0.48 ( ) or Eqn. 5 (n=m=2) for the molar ratio of 4.8 (Δ). An observed unreacted Al-time function for the $H_2PO_4^-$/Al molar ratio of 18.8 is also included (o).

The addition of base and corresponding rise in pH for Al solutions with added citrate and formate (Solns. 2 and 3, respectively) did not significantly affect the interaction of ferron with the partially neutralized organic-Al solutions (Figure 10). Model-fitted k_b and f_a values were similar to those established for partially neutralized $AlCl_3$ without added organics (Table 5, Solns. 1-3). If organic-Al polymers do form, their interaction with ferron is not significantly different from Al-polymers in the absence of organic ligands.

The interaction of ferron with partially neutralized Al solutions containing a mixture of organic and inorganic anions resulted in kinetic curves that were significantly different from partially neutralized Al solutions containing no added anions (Figure 10. Table 5, Solns. 1 and 4). Observed kinetic data for solution 4 were very similar to partially neutralized Al solutions containing only $H_2PO_4^-$ (Figure 10, Table 5, Soln. 5). It seems that added $H_2PO_4^-$ is responsible for the lower k_b and f_a values for solutions 4 and 5. The presence of organics in these partially neutralized Al solutions did not adversely affect the formation of Al-polymers. Polymer growth is enhanced by the strong interaction of P with partially neutralized Al[42]; however, there is no evidence to suggest the formation of Al-organic phosphate complexes as proposed by others.[49,50]

Soil solutions contain many organic anions other than the low molecular weight compounds discussed here. Many complex polymeric organic constituents exist in the soil solution, of which fulvic acid appears most important in the transport and mobility of ions in soil.[51] The interaction of ferron with $AlCl_3$ solutions containing various concentrations of natural water soluble fulvic acid was in turn investigated. Ferron reactions with solutions containing fulvate/Al molar ratios of 0.48 were modeled by a single second-order kinetic expression, while solutions containing fulvate/Al molar ratios of 4.8 were modeled using two simultaneous second-order reactions (Figure 11). Total chemical analysis of the fulvic acid using Inductively Coupled Plasma analysis indicated that the extract contained 3.1×10^{-4} mol L^{-1} P. With this in mind, Al solutions containing 1.5×10^{-3} mol L^{-1} P (enough to cause 100% complexed Al) were analyzed kinetically with ferron and compared with kinetic curves where the fulvate/Al molar ratio was 4.8 (Figure 11). The observed data were somewhat similar, suggesting that Al-P complexes were responsible for the slow kinetics and not Al-fulvate complexes. This is also suggested for solutions with the low fulvate/Al molar ratio, whose interaction with ferron was similar to that of uncomplexed mononuclear Al solutions (k_a=9.7 ± 0.7 L mg^{-1} min^{-1}). Young and Bache[52] have also shown that the chemical affinity of fulvic acid for Al is no greater than a weak complexant, such as adipic or acetic acid, which ferron can easily degrade.

Table 5. Model-fitted k_b and f_a values for the reaction of ferron with partially neutralized Al solutions of varying organic and inorganic anion composition

Solution Code	Solution with n_o = 1.0	$^a k_b$	$^a f_a$
	mol L^{-1}	L mg^{-1} min^{-1}	
1	8×10^{-5} $AlCl_3$	0.586 ± 0.031[b]	0.65 ± 0.02
2	8×10^{-5} $AlCl_3$ 1×10^{-5} Citrate	0.634 ± 0.017	0.68 ± 0.01
3	8×10^{-5} $AlCl_3$ 6×10^{-5} Formate	0.559 ± 0.031	0.68 ± 0.01
4	8×10^{-5} $AlCl_3$ 1×10^{-5} NaH_2PO_4 6×10^{-6} Citrate 3×10^{-5} Formate 6×10^{-6} NaF 1×10^{-5} Malonate	0.0474 ± 0.0131	0.59 ± 0.02
5	8×10^{-5} $AlCl_3$ 1×10^{-5} NaH_2PO_4	0.0331 ± 0.0150	0.61 ± 0.01

[a] Fitted with two simultaneous 2nd-order reactions, with k_a = 15.0 L mg^{-1} min^{-1}.
[b] Confidence interval = 0.95.

CONCLUSIONS

Distinction of mononuclear and polynuclear Al in solutions containing various inorganic and organic anions is possible using ferron if the ligand composition of the solutions is known. Arbitrary separation of mononuclear and polynuclear Al was avoided using a binary species rate function that accurately separates the two forms of Al based on differential kinetic reactions with ferron. In general, mononuclear Al solutions were modeled best with a single second-order function, while mixed mononuclear and polynuclear Al solutions were described with two simultaneous second-order reactions. Numerous inorganic and organic Al complexes are rapidly degraded by ferron regardless of the ligand concentration (e.g., SO_4^{2-}, malate, salicylate, glutarate, succinate) and express similar kinetics to that of uncomplexed mononuclear Al. Such organic- and inorganic-Al complexes have been shown to be nontoxic to plants.[4,9,53] Similarly, mononuclear Al solutions containing $H_2PO_4^-$, F^-, citrate, tartarate, and oxalate with critical ligand/Al molar ratios $\leq$ 0.75, 0.08, 0.13, 0.08, 0.38, respectively, also react rapidly with ferron and express kinetic reactions similar to that of uncomplexed mononuclear Al. Such complexes that are rapidly decomposed by ferron must be accounted for, possibly using geochemical models, if an estimate of phytotoxic mononuclear Al is desired.

REFERENCES

1. Bloom, P. R., M. B. McBride, and R. M. Weaver. "Aluminum organic matter in acid soils: Buffering and solution aluminum activity, "*Soil Sci. Soc. Am. J.* 43:488-493 (1979).

2. James, B. R., C. J. Clark, and S. J. Riha. "An 8-hydroxyquinoline method for labile and total aluminum in soil extracts," *Soil Sci. Soc. Am. J.* 47:893-897 (1983).

3. Driscoll, C. T. "A procedure for fractionation of aqueous aluminum in dilute acidic waters," *Intern. J. Environ. Anal. Chem.* 16:267-283 (1984).

4. Hue, N. V., G. R. Craddock, and F. Adams. "Effect of organic acids on aluminum toxicity in subsoils," *Soil Sci. Soc. Am. J.* 50:28-34 (1986).

5. Hodges, S. C. "Aluminum speciation: A comparison of five methods. *Soil Sci. Soc. Am. J.* 51:57-64 (1987).

6. Adams, F., and Z. F. Lund. "Effect of chemical activity of soil solution aluminum on cotton root penetration of acid subsoils," *Soil Sci.* 101:193-198 (1966).

7. Pavan, M. A., F. T. Bingham, and P. F. Pratt. "Toxicity of aluminum to coffee in ultisols and oxisols amended with $CaCO_3$, $MgCO_3$, and $CaSO_4$ x $2H_2O$," *Soil Sci. Soc. Am. J.* 46:1201-1207 (1982).

8. Wagatsuma, T., and Y. Ezoe. "Effect of pH on ionic species of aluminum in medium and on aluminum toxicity under solution culture," *Soil Sci. Plant Nutr.* 31:547-561 (1985).

9. Parker, D. R., L. W. Zelazny, and T. B. Kinraide. "Chemical speciation and plant toxicitv of aqueous aluminum," abstract presented at the 194th National Meeting, Division of Environmental Chemistry American Chemical Society, New Orleans, LA, (1987).

10. Buffle, J., N. Parthasarathy, and W. Haerdi. "Importance of speciation methods in analytical control of water treatment processes with application to fluoride removal from waste waters. *Water Res.* 19:7-23 (1985).

11. Parthasarathy, N., and J. Buffle. "Study of polymeric aluminum (III) hydroxide solutions for application in waste water treatment. Properties of the polymer and optimal conditions of preparation," *Water Res.* 19:25-36 (1985).

12. Bloom, P. R., R. M. Weaver, and M. B. McBride. "The spectrophotometric and fluorometric determination of aluminum with 8-hydroxyquinoline and butyl acetate extraction. *Soil Sci. Soc. Am. J.* 42:713-716 (1978).

13. Blamey, F.P.C., D. G. Edwards, and C. J. Asher. "Effects of aluminum, OH:Al and P:Al molar ratios, and ionic strength on soybean root elongation in solution culture," *Soil Sci.* 136:197-207 (1983).

14. Seip, H. M., L. Muller, and A. Naas. "Aluminum speciation: Comparison of two spectrophotometric analytical methods and observed concentrations in some acidic aquatic systems in southern Norway," *Water Air Soil Pollut.* 23:81-95 (1984).

15. David, M. B., and C. T. Driscoll. "Aluminum speciation and equilibria in soil solutions of a Haplorthod in the Adirondack Mountains (New York). *Geoderma* 33:297-318 (1984).

16. Smith, R. W., and J. D. Hem. "Effect of aging on aluminum hydroxide complexes in dilute aqueous solutions," U.S. Geol. Surv., Water Supply Paper 1827-D, p. 51 (1972).

17. Schwan, von M., H. Gorz, and S. Schonherr. "Ein beitrag zur auswertuang der kinetik der aluminium-ferron-reaktion for strukturunter-suchungen an basischen aluminium-salzlosugen," *Z. Phys. Chem.* 262:953-957 (1981).

18. Smith. R. W. "Relations among equilibrium and nonequilibrium aqueous species of aluminum hydroxy complexes," *Am. Chem. Soc. Adv. Chem.* 106:250-279 (1971).

19. Bersillion, J. L., P. H. Hsu, and F. Fiessinger. "Characterization of hydroxy aluminum solutions," *Soil Sci. Soc. Am. J.* 44:630-634 (1980).

20. Hodges, S. C., and L. W. Zelazny. "Interaction of dilute, hydrolyzed aluminum solutions with clays, peat, and resin," *Soil Sci. Soc. Am. J.* 47:206-212 (1983).

21. Snodgrass, W. J., M. M. Clark, and C. R. O'Melia. "Particle formation and growth in dilute aluminum (III) solutions," *Water Res.* 18:479-488 (1984).

22. Jardine, P. M., L. W. Zelazny, and J. C. Parker. "Mechanisms of aluminum adsorption on clay minerals and peat," *Soil Sci. Soc. Am. J.* 49:862-867 (1985).

23. Adams. F. Soil Solution. In E. W. Carson (ed.). *The Plant Root and its Environment.* (Charlottesville, VA: University Press of VA, 1974).

24. Adams, F. and P. J. Hatchcock. "Aluminum toxcity and calcium deficiency in acid subsoil horizons of two coastal plain soil series," *Soil Sci. Soc. Am. J.* 48:1305-1309 (1984).

25. Ares, J. "Identification of aluminum species in acid forest soil solution on the basis of Al:F reaction kinetics: 2. An example at the solling area," *Soil Sci.* 142:13-19 (1986).

26. Jardine, P. M.. and L. W. Zelazny. "Mononuclear and polynuclear aluminum speciation through differential kinetic reactions with ferron. *Soil Sci. Soc. Am. J.* 50:895-900 (1986).

27. Sposito, G., and S. V. Mattigod. "GEOCHEM: A computer program for the calculation of chemical equilibria in soil solutions and other natural water systems," The Kearney Foundation of Soil Science, Univ. of CA, 1980.

28. Sposito, G. "Corrections to the program, GEOCHEM." *Soil Sci. Soc. Am. J.* 50:270 (1986).

29. Parker, D. R., L. W. Zelazny, and T. B. Kinraide. "Improvements to the program, GEOCHEM." *Soil Sci. Soc. Am. J.* 51:488-491 (1987b).

30. Sillen, L. G., and A. E. Martell. "Stability complexes of metal-ion complexes," spec. Pub. No. 25, 865 paper presented at the Chemical Society, Burlington House, London, 1971.

31. Perrin, D. D. 1979. *Stability Constants of Metal-Ion Complexes, Part B, Organic ligands*, IUPAC Chemical Data Series No. 22, 1263 (New York: Pergamon Press, 1979).

32. Perlmutter-Hayman, B., and E. Tapuhi. "Complex formation between aluminum ion and 5-nitrosalicylic acid. Mechanism and equilibrium constant," *Inorg. Chem.* 18:875-877 (1979).

33. Lopez-Quintela, M. A., W. Knoche, and J. Veith. "Kinetics and thermodynamics of complex formation between aluminum (III) and citric acid in aqueous solution." *J. Chem. Soc. Faraday Trans.* 80:2313-2321 (1984).

34. Lindsay, W. L. *Chemical Equilibrium in Soils* (New York: John Wiley and Sons, 1979), p. 449.

35. Hogfeldt, E. 1982. *Stability Constants of Metal-Ion Complexes, Part A: Inorganic Ligands*, IUPAC Chemical Data Series No. 21, 310 (New York: Pergamon Press, 1982).

36. Stumm, W., and J. J. Morgan. *Aquatic Chemistry* (New York: John Wiley and Sons, 1981), p. 780.

37. Frost, A. A., and R. G. Pearson. *Kinetics and Mechanisms*, (New York: John Wiley and Sons, 1961) p.162-165.

38. Goto. K., H. Tamura, M. Onodera, and M. Nagayama. "Spectrophotometric determination of aluminum with ferron and a quaternary ammonium salt", *Talanta* 21:183-189 (1974).

39. Ross, G. J., and R. C. Turner. "Effect of different anions on the crystallization of aluminum hydroxide in partially neutralized aqueous aluminum salt systems," *Soil Sci. Soc. Am. Proc.* 35:389-392 (1971).

40. Serna, C. J., J. L. White, and S. L. Hem. "Anion-aluminum hydroxide gel interactions," *Soil Sci. Soc. Am. J.* 41:1009-1013 (1977).

41. Bartlett, R. J., and D. C. Riego. "Toxicity of hydroxy aluminum in relation to pH and phosphorus," *Soil Sci.* 114:194-200 (1972).

42. White, R. E., L. 0. Tiffin, and A. W. Taylor. "The existence of polymeric complexes in dilute solutions of aluminum and orthophosphate," *Plant Soil*. 45:521-529 (1976).

43. Hem, J. D., and C. E. Roberson. "Form and stability of aluminum hydroxide complexes in dilute solution," U.S. Geol. Survey, Water-Supply Paper 1827-A, U.S. Government Printing Office (1967).

44. Tsai, P. P., and P. H. Hsu. "Studies of aged OH-Al solutions using kinetics of Al ferron reactions and sulfate precipitation," *Soil Sci. Soc. Am. J.* 49:1060-1065 (1984).

45. Bertsch, P. M., W. J. Layton, and R. I. Barnhisel. "Speciation of hydroxy-alumlnum solutions by wet chemical and aluminum-27 NMR methods," *Soil Sci. Soc. Am. J.* 50:1449-1454 (1986).

46. Brenes, E., and R. W. Pearson. "Root responses of three Gramineae species to soil acidity in an oxisol and an ultisol," *Soil Sci.* 116:295-302 (1973).

47. Adams, F., and B. L. Moore. "Chemical factors affecting root growth in subsoil horizons of coastal plain soils," *Soil Sci. Soc. Am. J.* 47:99-102 (1983).

48. Cotton, F. A., and G. Wilkinson. *Advanced Inorganic Chemistry* (1980). Wiley-Interscience, New York. p. 1184.

49. Ng Koe Kwong, K. F., and P. M. Huang. "Surface reactivity of aluminum hydroxide precipitated in the presence of low molecular weight organic acids," *Soil Sci. Soc. Am. J.* 43:1107-1113 (1979).

50. Traina, S. J., G. Sposito, D. Hesterberg, and V. Kafkafi. "Effects of pH and organic acids on organophosphate solubility in an acid, montmorillonitic soil," *Soil Sci. Soc. Am. J.* 50:45-52 (1986).

51. Geering, H. R., and J. F. Hodgson. "Micronutrient cation complexes in soil solution: III. Characterization of soil solution ligands and their complexes with Zn^{2+} and Cu^{2+}, *Soil Sci. Soc. Am. Proc.* 33:54-77 (1969).

52. Young, S. D., and B. W. Bache. "Aluminum-organic complexation: Formation constants and a speciation model for the soil solution," *J. Soil Sci.* 36:261-269 (1985).

53. Kinraide, T. B., and D. R. Parker. "The nonphytotoxicity of the aluminum sulfate ion, $AlSO_4^+$. *Physiol," Plant* (In Press) (1987).

CHAPTER 3

THE STABILITY OF ALUMINUM SPECIES IN A NATURAL AUDIT SAMPLE: POSSIBLE APPLICATION AS A QUALITY CONTROL SOLUTION

L. J. Arent and T. E. Lewis

INTRODUCTION

Mobilization of toxic elements from soils and sediments has been of growing concern as a result of acidic deposition.[1,2] Aluminum, a ubiquitous element comprising nearly 8% (w/w) of the earth's crust, is subject to increased mobility at lower pH levels in soil and surface waters. Aluminum has been known to be highly phytotoxic since the turn of the century.[3] Recently, a correlation has been established between increased Al concentration in lakes and streams and decreases in fish populations.[4,5,6,7] Toxicity is not directly related to total Al concentration, but is correlated with inorganic monomeric Al.[8,9] In an attempt to quantify the biologically active fraction of Al, a multitude of methods have been devised. Given the complexity of Al chemistry and the instability of many Al species in aqueous systems, obtaining reliable precision and accuracy estimates for a given Al analytical scheme is difficult. Synthetic standard solutions can readily be prepared for the precise and accurate determination of total Al. However, when fractionation and speciation are essential, preparation of standard solutions with known Al species in known amounts is difficult, if not impossible.

A survey was conducted by the U.S. EPA in 1986 to characterize the chemical status of surface waters in the northeastern U.S. A rigorous quality assurance and quality control (QC) program was instituted for this synoptic study.[10] An important component of the QC regimen was the use of natural waters as audit samples. These samples were collected from various representative lakes, filtered, homogenized, split into aliquots, and stored at 4°C. Twenty-five water quality parameters were measured, including the fractionation and speciation of monomeric Al. The natural audit sample results were used to evaluate the precision among sample batches analyzed under separate instrument calibrations. The objective of this study was to evaluate the use of one of the natural audit samples as a QC sample for the determination of monomeric Al species. The stability of two Al fractions, total monomeric Al and nonexchangeable monomeric Al, were examined in a single natural audit sample over the 38-day synoptic lake study. The remaining sample from the homogenized batch was stored in polyethylene containers at 4°C for 11 months. A short-term (16-day) experiment was run using the natural audit sample after the 11-month holding time. Additionally, the stability of these Al species in two different storage containers was assessed during the short-term study.

MATERIALS AND METHODS

Audit Sample Collection

Water was collected from the epilimnion of Big Moose Lake (Adirondack Mountains, New York) in August 1986. The sample was placed in 20-liter polyethylene containers and shipped to a laboratory where it was filtered (0.45-µm pore size membrane filters) and homogenized in 55-gallon polyethylene drums. The bulk sample was split into aliquots and stored in 2-liter amber polyethylene bottles which were held at 4°C until analysis. The Big Moose audit sample was chosen for use in this study because it contains both exchangeable and nonexchangeable monomeric Al species.

Analytical Method

A colorimetric method involving the complexation of monomeric Al by pyrocatechol violet (PCV) was used on a two-channel flow injection system. Channel 1 measured total monomeric Al (TRx), which reacted with PCV. Fractionation was accomplished by a cation exchange column (CEC) which removed the exchangeable monomeric Al species.[11] The CEC was incorporated into the Channel 2 flow stream such that nonexchangeable monomeric Al (NEx), which reacted with PCV, was measured directly. The nonexchangeable fraction is believed to consist primarily of organic-Al complexes. The exchangeable fraction is calculated as the difference between TRx and NEx and is operationally defined as inorganic monomeric Al.

Sample aliquots were introduced into the flow injection system by 50-mL polypropylene syringes. The nonacidified samples were analyzed after passage through acid-washed polycarbonate filters (0.45-µm pore size, Nucleopore). Calibration standards were prepared from a commercially available certified Al standard (1000 mg/L). A calibration curve was generated daily for each channel. A 75 µg Al/L solution was prepared daily from a separate Al standard (1000 mg/L) and employed as a synthetic QC standard, to check the accuracy of the calibration procedure. The QC standard was analyzed once without the CEC to compare the results from channels 1 and 2, and once with the CEC to determine if the CEC removed the inorganic monomeric Al present in the standard. The photosensitive PCV solution was prepared daily. The flow injection-PCV method described by Royset[12] was modified for use during the synoptic lake study.[13]

EXPERIMENTAL DESIGN

Synoptic Lake Study

The natural audit sample was used as a daily QC sample for the determination of monomeric Al species. Two 2-liter bottles were used in succession during the 38-day survey. Samples were analyzed on 26 days of the survey period (October 9-November 15, 1987). There were from two to five sample results per day. During the synoptic survey, the calibration standards were prepared daily. Other reagents were prepared as needed.

Short-Term Study

This experiment was designed to examine the stability of TRx and NEx in the natural audit sample. In addition, the effects of two sample storage containers, a 2-liter amber polyethylene bottle and a 50-mL polypropylene syringe, were evaluated. Three source bottles of natural audit sample were used for this study. The short-term experiment spanned 16 days (June 17-July 2, 1987), and samples were analyzed on seven nonconsecutive days. Figure 1 illustrates the design of the short-term experiment. To evaluate the use of the 2-liter bottle as a sample storage container, two sample aliquots were withdrawn from each of the three source bottles and analyzed immediately on each analysis day. The sample aliquots from each syringe were analyzed in duplicate for a total of 12 bottle-held results per day. To evaluate the use of the 50-mL polypropylene syringe as a sample storage container, 14 aliquots of natural audit sample were withdrawn from each of the three source bottles on Day 1 of the experiment and stored at 4°C. Two syringes from each bottle were randomly selected and a sample aliquot from each syringe was analyzed in duplicate for a total of 12 syringe-held results per day. All calibration standards and reagents were prepared in large batches on Day 1 of the experiment.

RESULTS

Sample Characteristics

General water quality characteristics for the natural audit sample are shown in Table 1. The homogeneity of the sample among the three source bottles selected for the short-term study was examined. The 95% confidence interval for each source bottle was calculated from the Day 1 results of the short-term experiment (Figure 2). The confidence intervals overlapped significantly for both TRx (channel 1) and NEx (channel 2), indicating that the three samples were homogeneous with respect to the Al species measured.

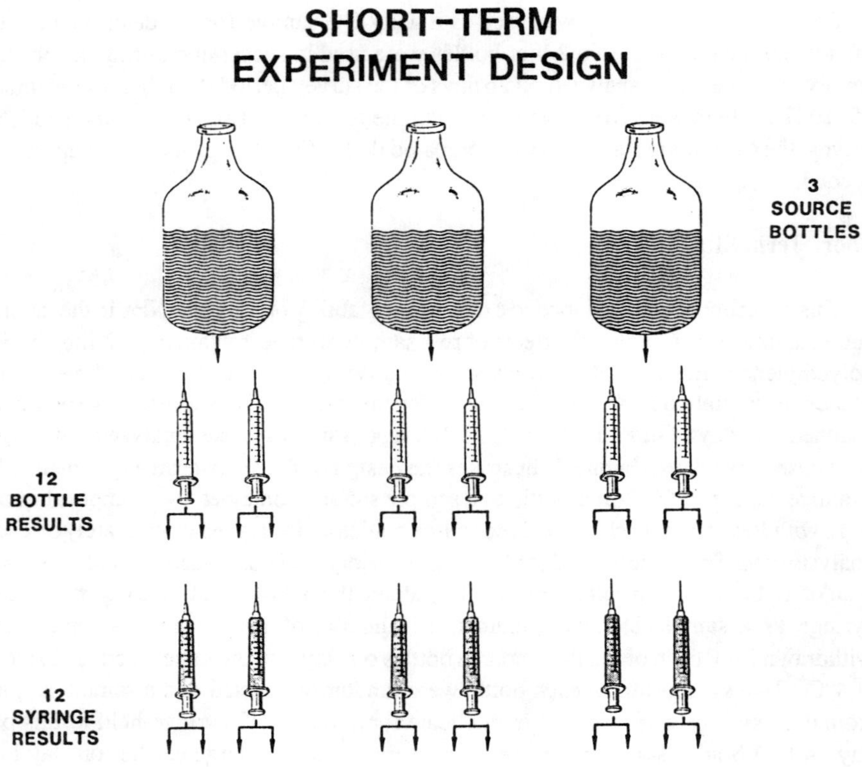

Figure 1. Design of the short-term experiment. Each arrow from each syringe represents a sample analysis result. For each day, there were 12 bottle results and 12 syringe results.

Table 1. Sample characteristics for the natural audit sample collected in August 1986

Parameters	Mean	Standard Deviation	Method
Metals, Dissolved (mg/L)			
Al, extractable	0.12	0.02	[14]
Al, total	0.24	0.05	EPA 202.2 [15]
Ca	1.84	0.03	EPA 215.1 [15]
Fe	0.03	0.004	EPA 236.1 [15]
K	0.27	0.04	EPA 258.1 [15]
Mg	0.30	0.01	EPA 242.1 [15]
Mn	0.07	0.004	EPA 243.1 [15]
Na	0.56	0.14	EPA 273.1 [15]
Major Anions, Dissolved (mg/L)			
Cl^-	0.34	0.01	[16]
F^-, total	0.08	0.01	EPA 340.2 (mod) [15]
NO_3^-	0.95	0.02	[16]
$SO_4^=$	5.81	0.14	[16]
Other Parameters			
Acidity (μeq/L)	61.07	10.76	[17]
Alkalinity (μeq/L)	-2.23	1.42	[17,18]
Color (Pt-Co units)	12.5	6.89	EPA 110.2 [15]
Conductivity (μS/cm)	23.4	1.36	EPA 120.1 [15]
Inorganic carbon, dissolved (mg/L)	0.41	0.04	[17]
Organic carbon, dissolved (mg/L)	3.25	0.13	EPA 415.2 [15]
NH_4^+, dissolved (mg/L)	0.04	0.02	EPA 350.1 [15]
P, total-dissolved (mg/L)	0.002	0.001	USGS I-4600-78 [19]
pH, closed system (pH units)	5.18	0.04	[17]
Silica, dissolved (mg/L)	3.79	0.06	USGS I-2700-78 [19]
Turbidity (NTU)	0.17	0.05	EPA 180.1 [15]

Note: Results from the Synoptic Lake Study (N = 6).

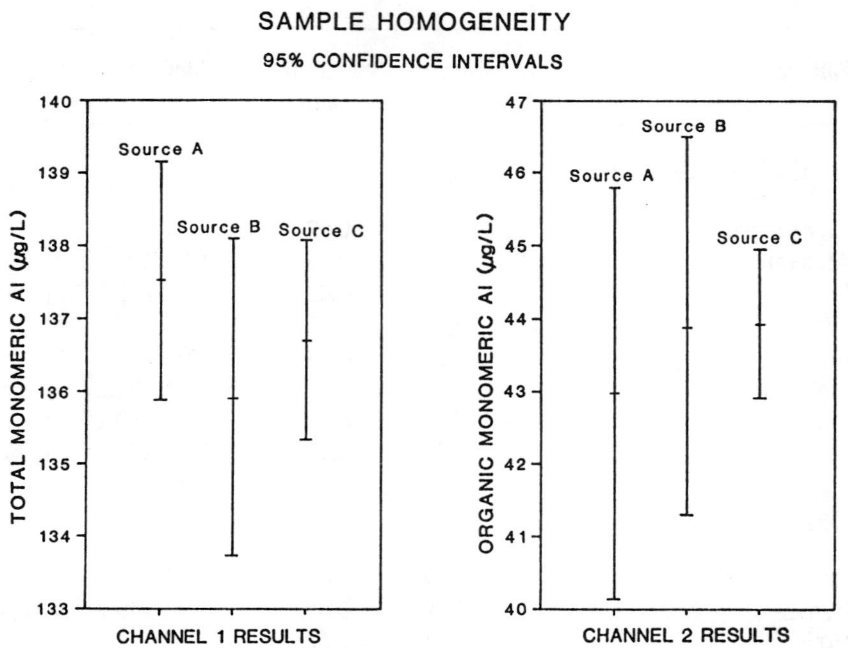

Figure 2. Comparison of the 95% confidence intervals for three source bottles of natural audit sample used for the short-term study for TRx and NEx (N = 8 for each error bar).

Method Characteristics

A control chart for the synthetic QC standard (75 µg Al/L) is shown in Figure 3. The mean ± one standard deviation of the two daily QC standard results was calculated and plotted for each channel in order to demonstrate the accuracy and precision attainable using this method. The synthetic QC standard results are shown as an indication of the method variability among days. The CEC was not connected to the sample flow stream, thus channel 1 and channel 2 both measured TRx. Control lines are shown at ± 10% of the nominal concentration of the QC standard, the target window selected for this method.

The detection limit, defined as three times the standard deviation of 14 nonconsecutive deionized water samples, was 7.2 µg Al/L for the short-term experimental period.

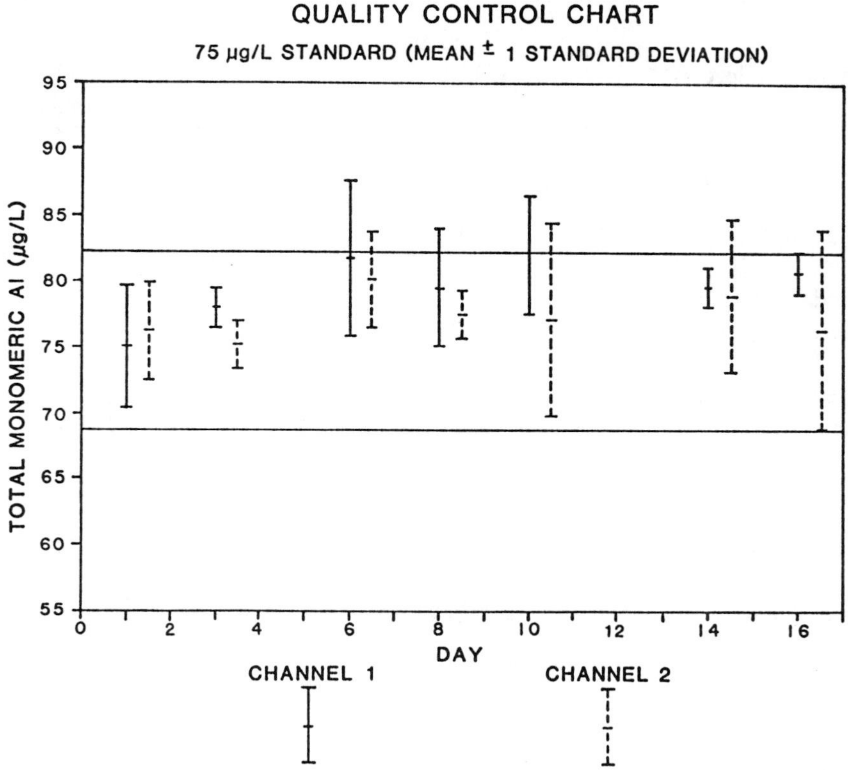

Figure 3. Quality control chart for the synthetic QC standard (TRx versus time). The mean ± one standard deviation is plotted for each channel (N = 2 for each error bar). Control lines are shown at 75 ± 7.5 µg Al/L.

Synoptic Lake Study Results

The results of the analysis of the natural audit sample, which was used as a daily QC sample for the synoptic survey, are shown in Table 2. The average TRx result was 150 ± 9.9 µg/L and the average NEx result was 44.2 ± 6.5 µg/L.

Short-Term Results

For each analysis day, the mean of the syringe-held results from each of the three source bottles of natural audit sample was calculated for TRx. The grand mean and standard error of the means were calculated foı each analysis day. The same calcula-

Table 2. Monomeric aluminum results for the natural audit sample

Aluminum Species	Synoptic Lake Study	Short-term Study (Syringe)	Short-term Study (Bottle)
Total Monomeric Al			
Mean (μg/L)	150	140	141
Standard Deviation (μg/L)	9.9	3.4	2.8
95% Confidence Interval (μg/L)	147-151	139-141	140-142
N	88	84	84
Calibrations	26	7	7
Observations per Day	2-5	12	12
Organic Monomeric Al			
Mean (μg/L)	44.2	43.9	44.2
Standard Deviation (μg/L)	6.5	2.8	3.4
95% Confidence Interval (μg/L)	42.8-45.6	43.3-44.5	43.5-44.9
N	88	84	84
Calibrations	26	7	7
Observations per Day	2-5	12	12

tions were done for the bottle-held samples. Figure 4 is a plot of the TRx data with the syringe and bottle data juxtaposed for comparison. The NEx results are shown in Figure 5. The graphs for TRx and NEx show the observed changes in Al concentration among days. No trend is apparent in either case. The syringe results were pooled and the bottle results were pooled for each species of Al. The statistical results are presented in Table 2.

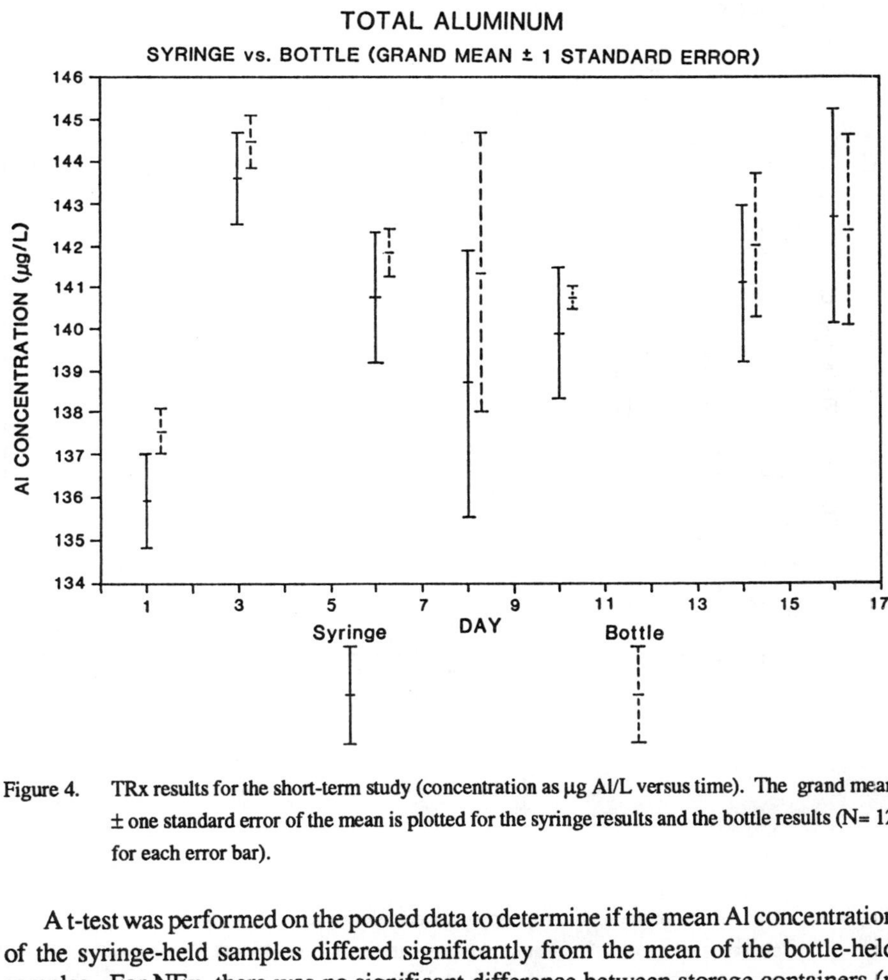

Figure 4. TRx results for the short-term study (concentration as µg Al/L versus time). The grand mean ± one standard error of the mean is plotted for the syringe results and the bottle results (N= 12 for each error bar).

A t-test was performed on the pooled data to determine if the mean Al concentration of the syringe-held samples differed significantly from the mean of the bottle-held samples. For NEx, there was no significant difference between storage containers ($p = 0.05$). For TRx, the t-test result of 2.08 exceeded the decision point ($t_{0.05,\infty} = 1.96$), which indicated that there was a significant difference between storage containers. Of note was that the standard error of the mean for TRx-bottle samples was less than that of the syringe results on six of the seven experimental days.

To test whether the magnitude of the standard deviations were concentration dependent, the variances of a range of synthetic standards were compared using the F-distribution. The variance of a 150 µg Al/L standard, a concentration which approximated the natural audit sample TRx concentration, was compared with the variance of the 75 μg Al/L QC standard. The variance of a 50 μg Al/L standard, which approximated NEx concentration measured in the natural audit sample, was also compared with the variance of the 75 μg Al/L QC standard. In both cases, the result was not significant, indicating that the variances were not concentration-dependent over the range tested. All results were recorded using a single calibration with ten replicates of each standard.

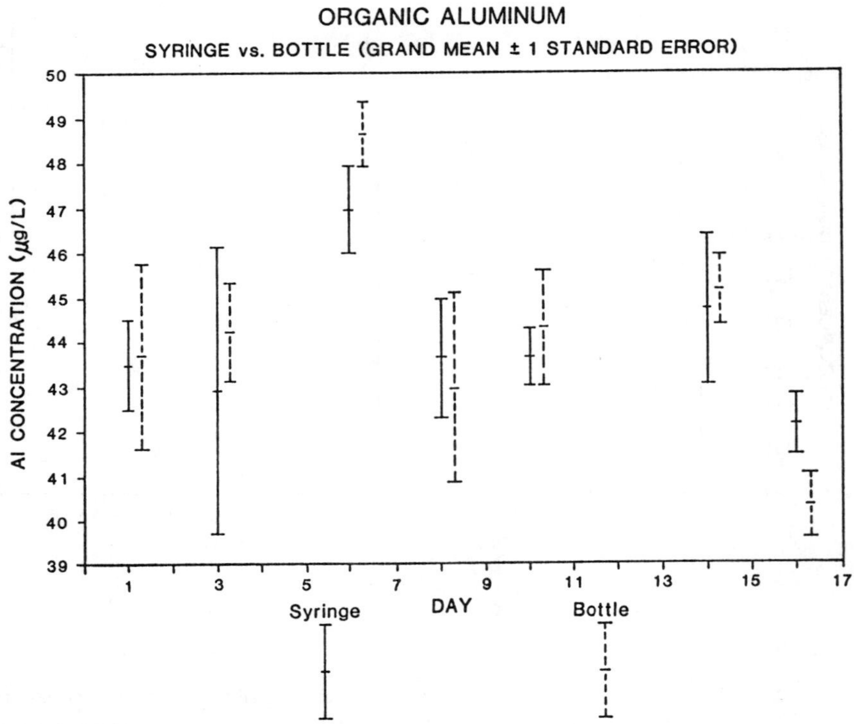

Figure 5. NEx results for the short-term study (concentration as μg Al/L vs. time). The grand mean ± one standard error of the mean is plotted for the syringe results and the bottle results (N = 12 for each error bar).

The experimental results are shown in Table 3 for TRx and NEx. The standard deviation of the synthetic QC solution (75 μg Al/L) represents the method variability and the standard deviation of the Big Moose sample represents the natural audit sample variability. We used an F-test to compare the variances of the synthetic QC standard and the natural audit sample. No significant difference between variances was determined.

Table 3. Comparison of the results of the synthetic quality control standard and the natural audit sample.

Sample	Mean (µg/L)	Standard Deviation (µg/L)	N
Total Monomeric Al			
Synthetic (75 µg Al/L)	79.6	3.6	14
Natural Sample (Syringe)	140	3.4	84
Natural Sample (Bottle)	141	2.8	84
Organic Monomeric Al			
Synthetic (75 µg Al/L)	77.4	4.1	14
Natural Sample (Syringe)	43.9	2.8	84
Natural Sample (Bottle)	44.2	3.4	84

DISCUSSION

The audit sample examined in the present study should not be considered a truly natural sample given the filtration and storage undergone by the sample. However, the stability of Al species in the processed sample was notable. Whether such stability is common among all samples processed in a similar manner is difficult to ascertain given the paucity of the data available in the literature.

When the dissolved concentration of an analyte is of interest, the sample is commonly filtered through a 0.45-µm pore size membrane filter before analysis. Some critical items to evaluate concerning the filtration of samples include: the sample interaction with the filter and/or filtration apparatus,[20] the contamination of the sample by the filter,[21] and the effects of filter load.[22] Kennedy et al.[23] recommend the use of 0.1-µm filters over 0.45-µm filters because fine particles can pass through a 0.45-µm pore size filter and cause major errors in the determination of Al in low concentration samples.[23] The particulates which pass through 0.45-µm filters may affect total Al concentrations, but appeared to have little affect on the stability of dissolved monomeric Al species in the natural audit sample probably because the analytical method was specific for monomeric Al. We believe that removing the particulates (0.45-µm) contributed to the stability of the natural audit sample. The stability of monomeric Al species in an unfiltered portion of the natural audit sample was not examined.

Royset et al.[24] observed minimal change in total Al between filtered (cellulose acetate membrane, 0.45- and 0.22-µm pore size, Millipore) and unfiltered samples. The samples were acidified and stored at 4°C until analysis. A flow injection method utilizing PCV as the complexing agent was the primary analytical method used.[25] Examining over 350 soil leachate and surface water samples, the authors concluded that the effects of filtration on total dissolved Al were insignificant for those samples containing high ("1000 µg/L") levels of Al. For 80% of the samples, the difference between filtered and unfiltered samples was reported as less than 100 µg/L for samples having Al concentrations in the 10-15,000 µg/L range. For samples within this range, the authors noted little difference in measured Al concentrations for samples filtered through 0.45-µm pore size membranes as compared to samples filtered using 0.22-µm pore size membranes. Although the long-term stability of total dissolved Al after filtration was not examined, the necessity for filtration was noted for samples low in dissolved Al.

Goenaga et al.[26] examined the difference in TRx and NEx in filtered and unfiltered stream samples. They used polycarbonate membranes with nominal pore sizes of 0.4-0.1-, 0.05-, and 0.015-µm. Aluminum concentrations were determined colorimetrically using PCV as the chromogenic reagent.[27] Nonexchangeable monomeric Al was measured directly after passage of the sample through a CEC.[11] Samples for TRx and NEx determination were not acidified and were stored at 4°C in the dark until analysis (within one week). For natural samples containing high concentrations of TRx (>100 µg/L), the measured levels of TRx and exchangeable Al decreased significantly with diminishing filter pore size. For samples with less than 100 µg Al/L, TRx values for the filtered natural sample were "slightly below those of the unfiltered sample," but the results indicated that filter pore size had little effect on Al levels measured in these filtrates. Exchangeable Al levels found in unfiltered aliquots of low Al concentration samples were not significantly different from the concentrations determined in the filtered samples.

Natural samples were used to study the common practice of acidifying filtrates prior to trace metal analysis.[23] A filtrate (0.1-µm, Millipore) that was not acidified was extracted immediately after collection using an 8-oxine/methyl isobutyl ketone (MIBK) solution,[28,29,30] and analyzed by atomic absorption spectrophotometry (AAS) eight days later. Also, a set of acidified filtrates was prepared using 0.45-, 0.22-, and 0.10-µm pore size filters, extracted, and stored for 19 days before analysis by the same method. The results from the acidified set of samples were consistently greater than the nonacidified sample results. The acidified filtrates were reanalyzed on Day 19 and Day 94 after collection. Significant increases in extractable Al were measured. Acidification of filtered samples is a valuable preservation technique which is believed to reduce adsorption and/or precipitation losses of constituents in the sample.[23,31] Although filtrates for dissolved Al analysis are commonly acidified, colloids or particulates may pass through the filters and be dissolved upon acidification resulting in an overestimation of dissolved monomeric Al. Colloidal Al in the filtrate may be released over time by the acid used for preservation, and the rate of release has been shown to be inconsistent.[23] Variability in the release of Al in an acidified sample would not be beneficial for a viable QC sample.

Rogeberg and Henriksen[32] conducted sample storage experiments to examine the stability of TRx and NEx in water samples collected in Norway. They used an automated modification of the PCV method of Dougan and Wilson[27] in conjunction with Driscoll's fractionation procedure[11] to measure Al species in natural water samples held at 4°C in darkness. In their initial experiment, they determined no significant changes in TRx or NEx in five stream samples during a storage period of 0-14 days. Their results from a second holding time study involving samples from nine rivers in Norway showed no storage effect for NEx over a period of three days. For TRx, however, the results were pH dependent, with significant storage effects reported for samples with a pH of 5 or greater. They postulate that the decrease in exchangeable Al "may be due to precipitation and/or adsorption to the container walls." For the natural audit sample, our results show a decrease in TRx of 10 μg/L during the 11-month sample holding time period (Table 2). Since NEx was not significantly different between the experimental periods, the apparent decrease in TRx may be attributed to decreases in exchangeable Al. Within each of the two study periods, however, the natural audit sample was stable and was very useful as a QC sample.

Aluminum species were measured by Seip and co-workers in unfiltered stream samples from the Birkenes and Storgama catchments of Norway.[33] Two spectrophotometric methods were used based on the reaction of Al with ferron[34] and with PCV.[27] Aluminum fractions were separated using a CEC.[11] An unfiltered sample was analyzed seven times over a 60-day period for total Al (acidified), TRx and NEx. Monomeric Al determinations were performed on nonacidified samples. The authors reported that the variations in all Al fractions were minimal. For TRx, the mean of the final two results was approximately 6% higher than the mean of the first two results.

Sullivan et al.[35] reported considerable variation in concentrations of TRx and NEx for some samples collected from the Birkenes and Flekkefjord areas of southern Norway. Their samples were stored at 5°C in the dark. The unfiltered samples were fractionated and extracted[11,14] prior to analysis by AAS. Not all samples collected yielded this variation (i.e. greater than the 1% standard deviation reported for replicate analyses). The authors suggest that samples with high total organic carbon (TOC) concentrations may experience greater shifts from monomeric to polymeric forms during storage. Although TOC was not determined, the average concentration of dissolved organic carbon measured in the natural audit sample was 3.25 ± 0.13 mg/L during the synoptic survey. This value is at the low end of the dissolved organic carbon range encountered in the synoptic survey. As suspected by Sullivan et al.[35], the relatively low organic carbon concentration may have contributed to the stability of TRx and NEx in the natural audit sample. Sullivan and co-workers also cite that changes in sample temperature and/or pH may lead to "supersaturation and subsequent precipitation or shifts in speciation." An increase in temperature can cause a decrease in the solubility of Al and the precipitation of synthetic gibbsite.[33] Before processing, the audit sample undoubtedly underwent an increase in temperature and a possible change in pH due to CO_2 re-equilibration. However, the sample was maintained at a constant temperature after the 2-liter aliquots were prepared. We do not purport that the Big Moose audit sample is representative of *in situ* lake chemistry.

Campbell et al.[36] examined the temporal evolution of Al species in stream water collected from the inlet of Lake Laflamme (Quebec, Canada) on three separate occasions. Each sample was held at 0°C (the *in situ* temperature) in a polyethylene container and was stirred constantly until analysis. The samples were passed through polycarbonate membranes (0.4-μm pore size, Nucleopore) prior to the determination of filterable Al and filterable extractable Al[14,37] by AAS. Both of these Al species were found to be stable for up to 24 hours based on seven replicate analyses of each sample collection.

Following the treatment procedure used to process the Big Moose audit sample it appears that Al speciation is remarkably stable for long periods of time. Other scientists have examined the short- and long-term stability of Al species in lake and stream waters using various preservation techniques. The conclusions are not definitive. The stability of Al species seems to be highly site specific based on previous studies.

CONCLUSIONS

We conclude that the natural audit sample prepared as described can serve as a reliable QC sample for Al speciation studies and offers an advantage over the use of a synthetic standard. For Al fractionation studies, no stable, synthetic organic-Al standard is available, whereas the Big Moose natural audit sample contained stable concentrations of organically-bound Al. The Big Moose audit sample was successfully used to determine removal efficiency of exchangeable (inorganic monomeric) Al by the CEC. Another advantage in the use of a natural sample as a QC standard is that it provides both inorganic (by calculation) and organic Al fractions, provided precautions are taken to minimize change in sample temperature and pH. Both storage containers evaluated during the 16-day study were considered adequate for our purposes. Routine water quality measurements for the Big Moose audit sample were presented in order to characterize a natural sample shown to be stable with respect to TRx and NEx during short- and long-term periods. We believe that natural samples with similar matrices may also serve as viable QC samples for the determination of Al species.

NOTICE

Although the research described in this article has been supported by the U.S. Environmental Protection Agency through contract number 68-03-3249 to Lockheed-ESC, it has not been subjected to Agency review and therefore does not reflect the views of the Agency and no official endorsement should be inferred. Mention of trade names or commercial products does not constitute endorsement or recommendation for use.

ACKNOWLEDGMENTS

The authors would like to thank J. Henshaw, D. Hillman, R. Metcalf, M. Miah, D. Peck, and M. Stapanian for the valuable discussions concerning the preparation of this chapter. B. Whitfield is acknowledged for her analytical support during the synoptic survey. The Big Moose Lake sample was first used as an audit sample by D. Chaloud during a previous study.

REFERENCES

1. Kelso, J. R. M., R. J. Love, J. H. Lipsit, and R. Dermott. "Chemical and biological status of headwater lakes in the Sault Ste. Marie District, Ontario," in *Acid Precipitation: Effects on Ecological Systems*. F. D'Itril, Ed. (Ann Arbor: Ann Arbor Science Publ., 1982), pp.165-207.

2. Hall, R. J., G. E. Likens, S. B. Fiance, and G. R. Hendrey. "Experimental acidification of a stream in the Hubbard Brook Experimental Forest, New Hampshire," *Ecology* 61(4):976-989 (1980).

3. Hutchinson T. C. "A historical perspective on the role of aluminum in toxicity of acidic soils and lake waters," in *Proceedings of the International Conference of Heavy Metals in the Environment*, Vol. I, (Heidleberg: 1983), pp. 17-25.

4. Jensen K., and E. Snekvik. "Low pH levels wipe out salmon and trout populations in Southernmost Norway," *Ambio*. 1:223-225 (1972).

5. Leivistad, H., and I. P. Muniz. "Fish kill at low pH in a Norwegian river," *Nature* 259:391-392 (1976).

6. Beamish, R. J., and H. H. Harvey. "Acidification of the La Cloche Mountain Lakes, Ontario, and resulting effects on fishes," *J. Fish. Res. Board Can.* 29:1131-1143 (1972).

7. Henriksen, A., O. K. Skogheim, and B. O. Rosseland. "Episodic changes in pH and aluminum speciation kill fish in a Norweqian salmon river," *Vatten* 40:255-260 (1984).

8. Driscoll, C. T. Jr., J. P. Baker, J. J. Bisogni, and C. L. Schofield. "Effect of aluminum speciation of fish in dilute acidified waters," *Nature* 284:161-164 (1980).

9. Baker, J. P., and C. L. Schofield. "Aluminum toxicity to fish in acidic waters," *Water, Air, and Soil Pollut.* 18:289-309 (1982).

10. Drouse, S. K., D. J. C. Hillman, L. W. Creelman, J. F. Potter, and S. J. Simon. "Eastern lake survey-phase I, Quality assurance plan," U.S. EPA Report 600/4-86-008 (1986).

11. Driscoll, C. T. Jr. "A procedure for the fractionation of aqueous aluminum in dilute acidic waters," *Intern. J. of Environ. Anal. Chem.* 16:267-283 (1984).

12. Royset, O. "Flow-injection spectrophotometric determination of aluminum in water with pyrocatechol violet," *Anal. Chim. Acta* 185:75-81 (1986).

13. Henshaw, J. M., T. E. Lewis, and E. M. Heithmar. "A semi-automated colormetric method for the determination of monomeric aluminum species in natural waters by flow injection analysis," *Intern. J. of Environ. Anal. Chem.* (in press).

14. Barnes, R. B. "The determination of specific forms of aluminum in natural water," *Chem. Geol.* 15:177-191 (1975).

15. *Methods for Chemical Analysis of Water and Wastes*. U.S. EPA Report 600/479-020 (1983).

16. *Annual Book of ASTM Standards*. Vol. 11.01, "Standard test methods for anions in water by ion chromatography," (Philadelphia, PA: 1984), pp. 4327-4384.

17. Hillman, D. J. C., J. F. Potter, and S. J. Simon. "National Surface Water Survey, Eastern Lake Survey-Phase I, Analytical Methods Manual," U.S. EPA Report 600/4-86-009 (1986).

18. Kramer, J. R. "Modified gran analysis for acid and base titrations," Environmental Geochemistry Report No. 1984-2 (1984).

19. Skougstad, M. W., M. J. Fishman, L. C. Friedman, D. E. Erdman, and S. S. Duncan, Eds. *Methods for Determination of Inorganic Substances in Water and Fluvial Sediments: Techniques of Water-Resources Investigation of the U.S. Geological Survey*. Book 5, Chapter A1, (Washington, D.C.: Government Printing Office, 1979).

20. Marvin, K. T., R. R. Proctor Jr., and R. A. Neal. "Some effects of filtration on the determination of copper in freshwater and saltwater," *Limnol. and Oceanogr.* 15:476-479 (1970).

21. Robertson, D. E. "Role of contamination in trace element analysis of seawater," *Anal. Chem.* 40:1067-1072 (1965).

22. Danielsson, L. G. "On the use of filters for distinguishing between dissolved and particulate fractions in natural waters," *Water Res.* 16:179-182 (1982).

23. Kennedy, V. C., G. W. Zellweger, and B. F. Jones. "Filter pore size effects on the analysis of Al, Fe, Mn, and Ti in water," *Water Resour. Res.* 10:785-790 (1974).

24. Royset, O., A. O. Stuanes, G. Ogner, and G. Sjotveit. "Effect of filtration and centrifugation on the determination of aluminum in water," *Intern. J. Environ. Anal. Chem.* 29:141-149 (1987).

25. Royset, O. "Flow-injection spectrophotometric determination of aluminum in water with PCV," *Anal. Chem. Acta* 185:75-81 (1986).

26. Goenaga, X., R. Bryant, and D. J. A. Williams. "Influence of sorption processes on aluminum determinations in acidic waters," *Anal. Chem.* 56:2673-2678 (1987).

27. Dougan, W. K., and A. L. Wilson. "The absorbtiometric determination of aluminum in water: A comparison of some chromogenic reagents and the development of an improved method," *Analyst* 99:413-430 (1974).

28. Gentry, C. H. R., and L. G. Sherrinton. "Extraction and photometric estimation of some metals with 8-hydroxyquinoline," *Analyst* 75:17-21 (1950).

29. Stary, J. "Systematic study of the solvent extraction of metal oxinates," *Anal. Chim. Acta* 28:132-149 (1963).

30. Stary, J. *The Solvent Extraction of Metal Chelates* (New York: Macmillan, 1964), pp. 80-94.

31. Campbell, P. G. C., M. Bisson, R. Bougie, A. Tessier, and J. P. Villeneuve. "Speciation of aluminum in acidic freshwaters," *Anal. Chem.* 55:2246-2252 (1983).

32. Rogeberg, E. J. S., and A. Henriksen. "An automated method for fractionation and determination of aluminum species in fresh-waters," *Vatten* 41:48-53 (1985).

33. Seip, H. M., L. Miller, and A. Naas. "Aluminum speciation: Comparison of two spectrophotometric analytical methods and observed concentrations in some acidic aquatic systems in southern Norway," *Water, Air, Soil Pollut.* 23:81-95 (1984).

34. Rainwater F. H., and L. L. Thatcher. "Methods for collection and analysis of water samples," U.S. Geol. Surv. Water-Supply Pap., 1454 (1960).

35. Sullivan, T. J., I. P. Muniz, and H. M. Seip. "A comparison of frequently used methods for the determination of aqueous aluminum," *Intern. J. Environ. Anal. Chem.* 26:61-75 (1986).

36. Campbell, P. G. C., D. Thomassin, and A. Tessier. "Aluminum speciation in running waters on the Canadian pre-cambrian shield: Kinetic aspects," *Water Air Soil Pollut.* 30:1023-1032 (1986).

37. LaZerte, B. "Forms of aqueous aluminum in acidified catchments of southern Ontario: A methodological analysis," *Can. J. Fish. Aquat. Sci.* 41:766-776 (1984).

CHAPTER 4

MODELING THE INTERACTIONS OF Al SPECIES, PROTONS AND Ca^{2+} WITH HUMIC SUBSTANCES IN ACID WATERS AND SOILS

E. Tipping, C.A. Backes, M.A. Hurley

INTRODUCTION

High concentrations of dissolved aluminum are a major feature of acid environments. The soluble metal exists as a variety of species among which are complexes of Al^{3+} and $AlOH^{2+}$ with humic substances (HS). Often the organically bound Al is the predominant form in surface water or soil solution. Complexation with insoluble HS in the soil itself is also important. The aim of the work reviewed in this chapter was to model the metal-humic interactions so that extents of binding in acid waters and soils can be calculated, given appropriate input data. This would be helpful in predicting the effects of changes in environmental conditions on Al speciation. Because the chosen model explicitly incorporates H^+/Al^{3+} and $H^+/AlOH^{2+}$ competition it is additionally useful in that it can be used to calculate proton buffering by the Al-HS complexes, and thereby to predict solution pH in cases where such complexes dominate. Furthermore, the model involves calculation of the net humic charge, which is useful in understanding the solubility of HS in acid environments.

Much research effort has gone into elucidating and modeling metal-humic interactions, and various models are available, each with its own advantages and drawbacks.[1,2] For this work it was required to describe Al-humic interactions as functions of metal activity in solution, pH, humic concentration, and ionic strength, and to allow for possible competition between Al and other metals such as calcium. The chosen model was based on that of Marinsky and coworkers,[3,4] the HS being treated as macroions, with a limited number of types of ionizable group. In our approach to modeling we differ from Marinsky in that we do not seek a full detailed description of metal-proton-humic equilibria, but rather an approximation with a minimum number of parameters; our model is to be used for predicting complexation in the field, while retaining the essential physicochemical attributes of the system.

The model discussed in this paper is referred to as Model III.[5] It is a direct descendant of Model II[6] which turned out to be too simple for general application. (Model I[6] is a simple empirical method of describing experimental data.)

MODEL DESCRIPTION

Acid-base titrations of HS show them to contain, broadly speaking, two classes of proton-dissociating group.[7] One class ionizes in the pH range 2-7, the other at pH ≥ 8. In Model III the lower-pK groups are taken to be carboxyl groups, the others are referred to simply as "weak acid groups" (WAH). Because of the wide range of pH over which the COOH groups ionize, it is evident that there is considerable heterogeneity with regard to pK.[4,8] We take this into account by postulating two kinds of COOH group, present in equal amounts, and referred to as type I and type II carboxyls. The "weak acid" groups do not ionize at pH ≤ 7, but in Model III they are assumed to be able to exchange a proton with a metal ion at such pH values. The total number of ionizable groups is *Ac* (eq/gHS), consisting of $n/2$ type I and $n/2$ type II COOH and $(Ac\text{-}n)$ WAH. Metal complexation is assumed to take place at $(Ac\text{-}n)/2$ bidentate sites[9] and for simplicity these are restricted to type A sites (consisting of a type I COOH and a type II COOH), type BI (type I COOH, WAH) and type BII (type II COOH, WAH). The reactions considered are summarized in Table 1. Further simplification is made by assuming that

$$K_{A,Al} = K_{BI,Al} = K_{BII,Al} = K_{Al} \tag{1}$$

and

$$K_{A,Ca} = K_{BI,Ca} = K_{BII,Ca} = K_{Ca} \tag{2}$$

Because humic molecules are, in general, macroions,[4] there are purely electrostatic contributions to metal ion and proton complexation. These are incorporated as described by Tanford,[10] by use of the expression e^{-2wzZ} to modify the activities of species making up the equilibrium quotients. Here, z is the charge on the complexed ion, Z (eq/g) the net humic charge, and w (gHS/eq) the electrostatic interaction factor. At a given value of Z, the effective equilibrium constant - $K(Z)$ - is thus a combination of the (truly-constant) intrinsic equilibrium constant and the electrostatic terms for the different ions involved in the particular reaction (Table 1). The electrostatic interaction factor w depends on the ionic strength (I), and for some humic samples examined also on Z. In the absence of an adequate theoretical way of relating w to I or Z for HS, the following empirical equation (based on preliminary analysis of humic titration data) is adopted:

$$w = P \log (I) \exp (Q|Z|) \tag{3}$$

where P and Q are constants and |Z| is the modulus of Z. It is seen that $w = 0$ when I = 1M, i.e., electrostatic influences are taken to be completely abolished in concentrated salt solution, as found experimentally.[4] The parameter Q is found to be < 0, i.e., when

Table 1. Intrinsic and charge-dependent equilibrium constants

K_{I}	$= \dfrac{[\text{(I)-COO}^-]\,\{H^+\}}{[\text{(I)-COOH}]}$	$K_{I}(z) = K_{I}\, e^{2wZ}$
K_{II}	$= \dfrac{[\text{(II)-COO}^-]\,\{H^+\}}{[\text{(II)-COOH}]}$	$K_{II}(Z) = K_{II}\, e^{2wZ}$
$K_{A,Al}$	$= \dfrac{\left[\begin{matrix}\text{(I)-COO}\\ \text{(II)-COO}\end{matrix} > \text{Al}^{+}\right]\{H^+\}^2}{[\text{(I)-COOH}][\text{(II)-COOH}]\{Al^{3+}\}}$	$K_{A,Al}(Z) = K_{A,Al}\, e^{-2wZ}$
$K_{BI,Al}$	$= \dfrac{\left[\begin{matrix}\text{(I)-COO}\\ \text{WA}\end{matrix} > \text{Al}^{+}\right]\{H^+\}^2}{[\text{(I)-COOH}][\text{WAH}]\{Al^{3+}\}}$	$K_{BI,Al}(Z) = K_{BI,Al}\, e^{-2wZ}$
$K_{BII,Al}$	$= \dfrac{\left[\begin{matrix}\text{(II)-COO}\\ \text{WA}\end{matrix} > \text{Al}^{+}\right]\{H^+\}^2}{[\text{(II)-COOH}][\text{WAH}]\{Al^{3+}\}}$	$K_{BII,Al}(Z) = K_{BII,Al}\, e^{-2wZ}$
$K_{Al,H}$	$= \dfrac{[\text{AlOH}]\,\{H^+\}}{[\text{Al}^+]}$	$K_{Al,H}(Z) = K_{Al,H}\, e^{2wZ}$
$K_{A,Ca}$	$= \dfrac{\left[\begin{matrix}\text{(I)-COO}\\ \text{(II)-COO}\end{matrix} > \text{Ca}\right]\{H^+\}^2}{[\text{(I)-COOH}][\text{(II)-COOH}]\{Ca^{2+}\}}$	$K_{A,Ca}(Z) = K_{A,Ca}$
$K_{BI,Ca}$	$= \dfrac{\left[\begin{matrix}\text{(I)-COO}\\ \text{WA}\end{matrix} > \text{Ca}\right]\{H^+\}^2}{[\text{(I)-COOH}][\text{WAH}]\{Ca^{2+}\}}$	$K_{BI,Ca}(Z) = K_{BI,Ca}$
$K_{BII,Ca}$	$= \dfrac{\left[\begin{matrix}\text{(II)-COO}\\ \text{WA}\end{matrix} > \text{Ca}\right]\{H^+\}^2}{[\text{(II)-COOH}][\text{WAH}]\{Ca^{2+}\}}$	$K_{BII,Ca}(Z) = K_{BII,Ca}$

Note: Square brackets represent concentrations, curly brackets activities.

w depends on *Z* it decreases as *Z* increases: this may reflect expansion of the humic molecule as charge builds up.[10]

Because the HS carry a net charge (which is invariably negative) they have associated with them a diffuse layer in which charged ions of opposite sign (i.e., cations) predominate. In other words, as well as interacting chemically with H^+, Al^{3+}, $AlOH^{2+}$, and Ca^{2+}, there are also purely electrostatic interactions giving rise to binding of cationic species, including those such as Na^+ which are hardly complexed at all by the HS. In natural waters the HS are present at sufficiently low concentrations for the diffuse layer cation concentrations to be negligible. However, in organic soils--which may be regarded, for the purposes of describing solid-solution interactions, as concentrated suspensions of HS--the diffuse layer is an important part of the system. To take this into account, we assume a simple Donnan relationship[11] between cations in the diffuse layer and those in the bulk solution. By further assuming that activity coefficients in the two compartments for a given ion are the same we can write

$$\frac{[M^+]_D}{[M^+]_S} = \frac{[M^{2+}]_D^{1/2}}{[M^{2+}]_S} = \frac{[M^{3+}]_D^{1/3}}{[M^{3+}]_S} \qquad (4)$$

where M represents metal or proton, *D* diffuse layer and *S* bulk solution. Implementation of the expressions in (4) requires knowledge of the volumes of diffuse layer and bulk solution. An advantage of this formulation, instead of, for example, one based on the Boltzmann distribution, is that charge balances are readily satisfied, without introducing more parameters. However, it may prove necessary, in the light of data gathered in the future, to modify or replace the present description of the electrical double layer.

When applied to organic soils, the model is referred to as CHAOS (*C*omplexation by *H*umic *A*cids in *O*rganic *S*oils). A schematic illustration is presented in Fig. 1. Further details are given in ref. 12.

DETERMINATION OF PARAMETER VALUES

Parameter values were estimated from laboratory measurements on aquatic HS isolated by adsorption on XAD resins[13,14] and on stream sediment humic acid isolated by extraction with NaOH.[15] Values of total acidity, *Ac*, were measured by titration of hydrogen-saturated HS to pH 11.5. Other parameter values were estimated from the results of acid-base titrations (pH 3-6) carried out at different ionic strengths (0.001 - 0.1 M), total Al concentrations (100 - 1000 μM) and humic concentrations (50 - 500 mg/L) at 25°C. For sets of trial parameter values, pH was calculated for each experimental point and compared with the observed value. The calculations were done by writing mass (total Al, total HS) and charge-balance equations[5] which were solved iteratively. Allowance was made for hydrolysis reactions of Al. The best parameter set was obtained by minimizing the expression $\Sigma\{\log_e(pH_{obs}-3) - \log_e(pH_{calc}-3)\}^2$, the pH values being transformed in this way to achieve an approximately constant variance. Figure 2 shows titration data, together with calculated curves, for one of the humic samples studied. Table 2 shows parameter values for 4 isolated humic samples.

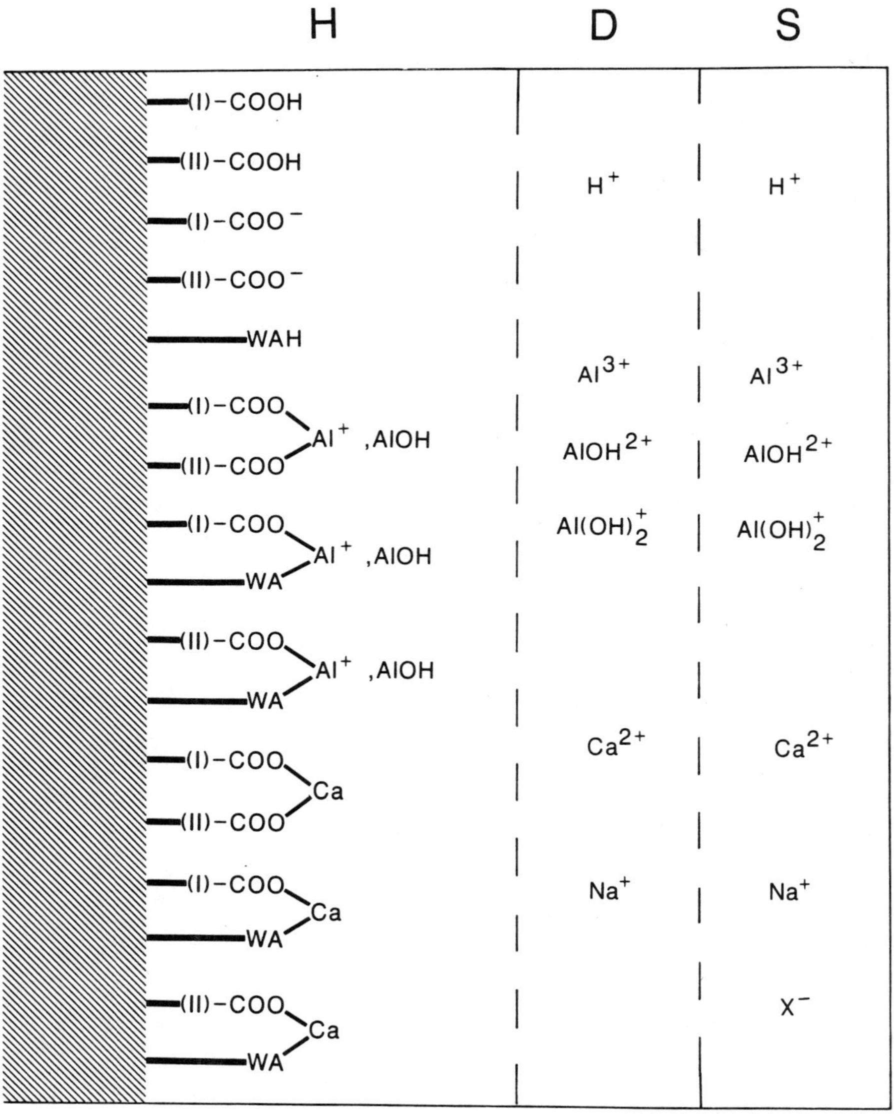

Figure 1. Species considered in CHAOS: those in *H* are complexed by humic binding sites; *D* is the diffuse part of the electrical double layer, *S* is the bulk solution.

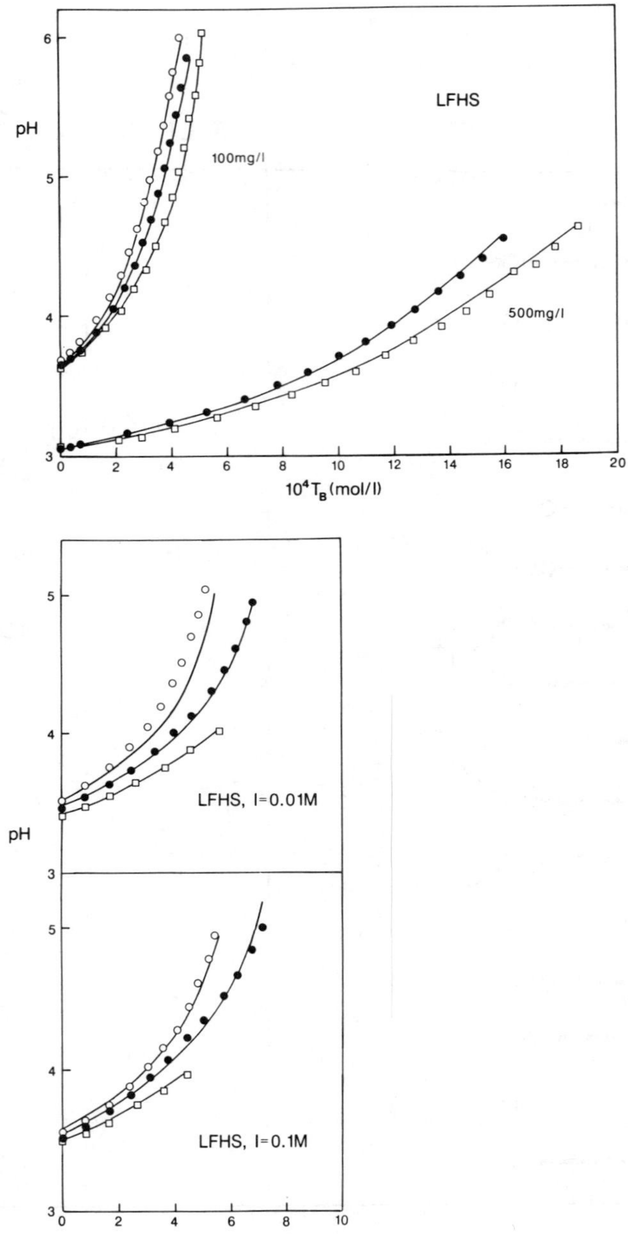

Figure 2. Acid-base titrations of Lochard Forest HS. Top panel: Al absent, humic concentrations as shown, ionic strengths 0.001 M (open circles), 0.01 M (solid circles), 0.1 M (open squares). Bottom panel: total [HS] = 100 mg/L, total Al = 100 μM (open circles), 200 μM (solid circles), 500 μM (open squares). The points are experimental values, the curves are calculated using Model III. T_B = total added base concentration. Temp = 25°C. From Ref. 5.

Table 2. Parameters for Model III, derived from acid-base titrations

	LFHS	*MBHA*	*PRHS-A*	*PRHS-B*
$10^3 Ac$ (eq/gHS)	9.73	4.80	5.85	8.10
$10^3 n$ (eq/gHS)	5.46(0.12)	2.51(0.06)	2.98(0.08)	4.95(0.04)
$10^3 K_I$	4.85(0.54)	2.11(0.36)	2.38(1.16)	5.10(3.78)
$10^5 K_{II}$	6.87(1.84)	3.42(1.56)	4.95(1.38)	2.63(0.62)
$-P$	120(20)	1680(240)	810(220)	12(12)
$-Q$	0	870(120)	410(140)	0
$10^4 K_{Al}$	2.19(0.50)	1.65(0.28)	2.97(1.84)	4.43(0.72)
$10^5 K_{Al,H}$	0.27(0.28)	4.41(1.9)	3.24(2.14)	0.48(0.24)
$10^7 K_{Ca}$	1.46(1.48)	1.63(2.34)	----	----

Note: Total acidity *Ac*) was determined in a separate experiment: n, K_I, K_{II}, P and Q were estimated from metal-free titrations, for which the correlation coefficients between observed and calculated pH values were ≥ 0.995: K_{Al}, $K_{Al,H}$, and K_{Ca} were estimated from titrations in the presence of metal (correlation coefficients ≥ 0.97). Figures in brackets are 2 x asymptotic standard error. The HS studied were from a stream at Lochard Forest (LF), Central Region, Scotland, Mosedale Beck (MB), Cumbria, England, and Penwhirn Reservoir (PR), Galloway, Scotland. PRHS-A and PRHS-B are high- and low-molecular weight fractions, respectively. From ref. 5.

It is evident from Figure 2 and Table 2 that Model III is reasonably successful at fitting the experimental data. The equilibrium constants show less variation than do P and Q among the different humic samples, suggesting that the chemical interactions are much the same for each sample, whereas the electrostatic interactions vary considerably. The pK values for types I and II COOH groups are ~ 2.5 and ~ 4.5, respectively, in the same range as values estimated by other authors using different humic models and different HS.[4,16,17] The value of P increases with the molecular size of the HS; for the smallest HS (PRHS-B) the titration data can be accounted for ignoring electrostatic interactions, whereas such interactions are very important for the HA sample. For the two larger-sized HS (MBHA, PRHS-B), Q is non-zero, possibly because of expansion of the humic molecules due to increasing charge-charge repulsion as Z increases (see above). The estimated values of pK_{Al} are in the range 3.4-3.8; comparison with analogous values for the simple organic diprotic acids salicylic acid (pK = 2.9) and phthalic acid (pK = 4.0) suggests that the model is reasonably chemically realistic. Both $K_{Al,H}$ and K_{Ca} are poorly estimated in the present experiments because neither $AlOH^{2+}$ nor Ca^{2+} was bound to a sufficient extent to cause major displacement of the pH titration curves. Nevertheless, the low value of K_{Ca} indicates that Ca^{2+} will not compete significantly with Al^{3+} in binding to HS at pH < 5.5.[5]

In attempting to model results for natural waters and soils, parameter values determined for isolated HS were retained, except that the capacity factors *Ac* and n were adjusted to fit the data (keeping the ratio n/Ac constant). In other words, the HS were assumed to behave as the isolated material, but at different (lower) effective concentrations. This is discussed further below.

In the case of soils a value for the volume of the diffuse part of the electrical double layer is required (see above). At the present stage of model development, this volume is estimated very crudely by assigning a notional specific surface area of the order of 500 m^2/g to the HS, and combining it with the Debye-Hückel parameter, K.[10] Consider, for example, an organic soil containing 70 g of HS per liter of total soil solution (diffuse layer + bulk solution). At an ionic strength of 0.0003 M--a typical value for the Cumbrian soils we have studied (see below)--the value of K is 5.7×10^7 m^{-1}. The diffuse layer volume is thus 500×70 m^2 (total surface area) x $1/5.7 \times 10^7$ m (diffuse layer thickness) = 6×10^{-4} m^3, or 0.6 liters. Because of probable double-layer overlap, and because the calculation is so approximate, a rounded value of 0.5 liters is adopted.

MODEL TESTING BY DIRECT MEASUREMENTS OF ALUMINUM BINDING BY ISOLATED HS

As explained above, the general approach to determining parameter values is the analysis of acid-base titration data. With some HS--those of sufficient molecular size not to diffuse through a dialysis membrane, or those present in suspension--direct measurements of Al binding can be made, and these can be compared with predicted values in order to test the model. This was done for two humic samples (MBHA and PRHS-A) and reasonably good agreements were obtained for a range of experimental conditions (see Table 3 and Fig. 3). It thus appears that the model is able to approximate Al and H^+ equilibria simultaneously, under laboratory conditions.

Table 3. Summary of experiments in which Al binding by HS was measured directly

Type of HS	pH	T_{Al},μM	T_{HS},mg/L	I	% Al bound
MBHA	3.2-5.1	30-500	170	0.01,0.1	13-94
PRHS-A	3.1-4.8	20-160	80	0.01	0-91

Note: For MBHA, binding was measured by equilibrium-dialysis and by centrifugation (depletion method): for PHRS-A, only equilibrium-dialysis was used. Plots of observed and calculated concentrations of bound Al are shown in Figure 3. From ref. 5.

ALUMINUM BINDING BY HS IN NATURAL WATERS: GENERAL PREDICTIONS

The Lochard Forest HS are probably the most representative studied, and therefore the parameter set for these HS is the most appropriate for performing calculations to explore the extent of Al binding by HS in natural waters. Figure 4 shows plots of % Al bound, (bound Al/unit weight of HS), Al^{3+} concentration, and net humic charge (Z) as functions of pH for different concentrations of LFHS at a constant, typical, total

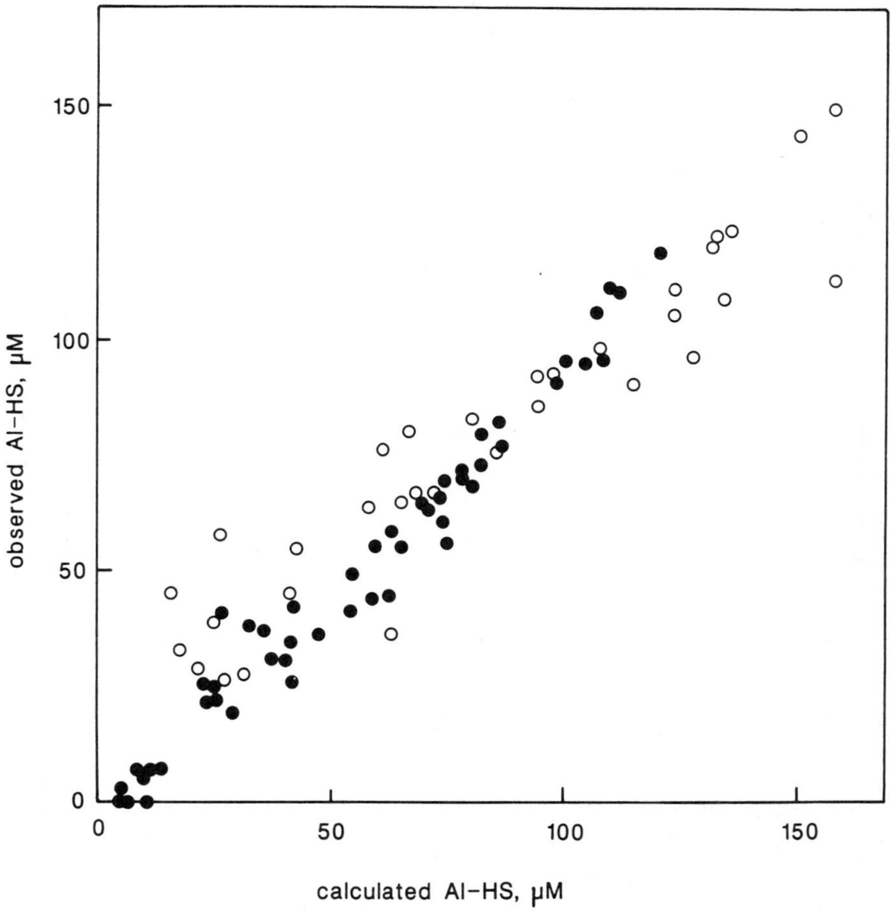

Figure 3. Plot of observed vs calculated concentrations of organically complexed Al for MBHA (open circles) and PRHS-A (solid circles). See Table 3 for details of experimental conditions.

dissolved monomeric Al concentration of 10 μ*M*. It is seen that the binding of Al by HS in natural waters is expected to increase substantially between pH 3 and pH 5, levelling off thereafter. The binding has a marked effect on the concentration of Al^{3+}, reducing it by several orders of magnitude under some circumstances (pH ≥ 4.5, T_{HS} ≥ 10 mg/L). The interactions also affect the HS, the net charge of which can be decreased considerably as a result of the complexation of Al, tending to make them less soluble.

It is also interesting to consider the H^+ buffering capabilities of HS and Al, alone and in combination. Figure 5 shows calculated plots of pH against added base for some solutions that are representative of acid natural waters. In each case the starting pH is

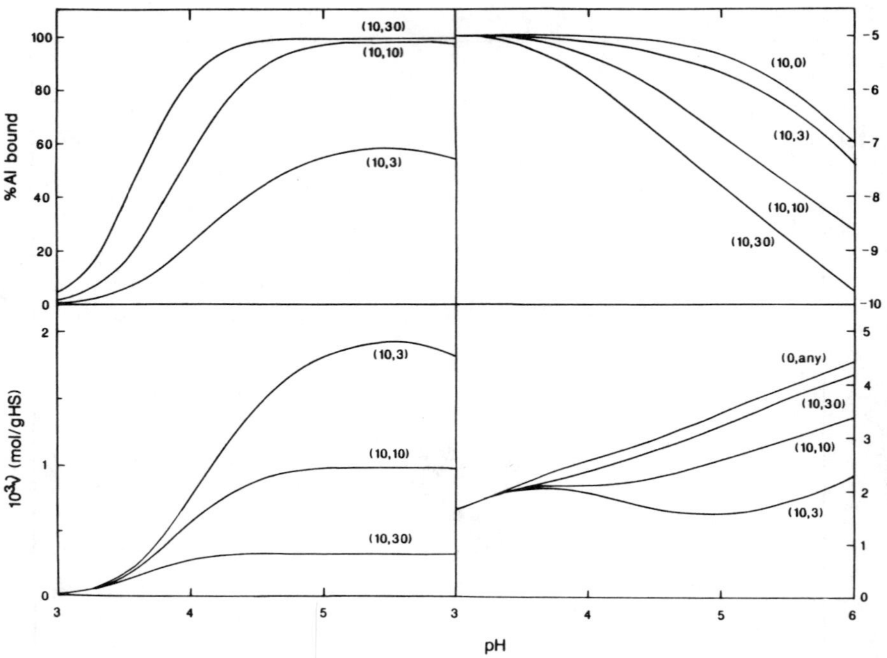

Figure 4. Calculations for LFHS based on Model III. Values in brackets: (T_{Al}, T_{HS}) in μM and mg/L, respectively. I = 0.001 M.

4.5. The calculations show that, acting separately, 10 μM Al and 10 mg/L LFHS have very similar buffer capacities. When both are present, however, there is no additivity in the buffer capacity; this is due to the Al-humic interactions diminishing both the binding of protons by the HS and the hydrolysis reactions of Al^{3+} and $AlOH^{2+}$. A similar effect is found for 30 mg/L HS in the absence and presence of 10 μM Al.

ALUMINUM BINDING BY HS IN NATURAL WATERS: COMPARISON OF PREDICTED AND OBSERVED VALUES

Organically-complexed monomeric Al (Al_m-org) in natural waters can be estimated by the cation-exchange method of Driscoll[18]--with corrections applied if necessary[19]--or by equilibrium-dialysis.[20] If sufficient ancillary data are available, the concentration of Al_m-org can be calculated by using Model III together with equilibrium constants for Al^{3+} hydrolysis and complexation by F^- and SO_4^{2-}. The data required are pH, ionic strength, and total concentrations of monomeric Al, fluoride, sulfate, and HS. Competition by Ca^{2+} is, on the basis of the K_{Ca} values given in Table 2, negligible. A survey of the literature published up to 1986, together with our own measurements on field samples,[21,22,23] provide a set of 114 water analyses (from 8 different sites) suitable

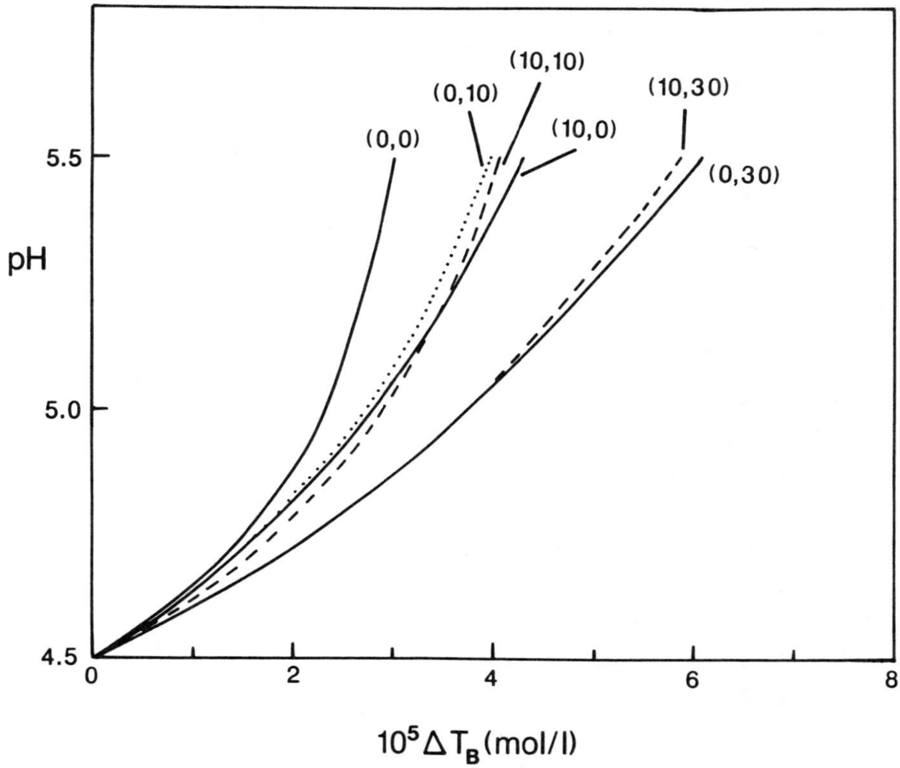

Figure 5. Buffering of H^+ by Al, HS and combinations of the two, calculated for LFHS with Model III. Values in brackets: (T_{Al}, T_{HS}) in μM and mg/L, respectively. I = 0.001 M.

for model-testing. The data are summarized in Table 4. In calculating [Al_m-org] it was assumed that the natural water samples all had a temperature of 10°C. Equilibrium constants for the complexation of Al^{3+} by OH^-, F^- and SO_4^{2-} were temperature-corrected using published enthalpy data (compiled in ref. 21). Except for K_{AlH}, Model III parameters were assumed to be temperature-independent. Clearly, *Ac* and *n* do not vary with temperature, while theoretical expressions for the electrostatic free energies of macroions[10] indicate that *P* and *Q* would tend to be temperature-independent. Enthalpies of proton-metal exchange at COO^- groups are close to zero,[24] and in the absence of literature values we assume this also to be true of the "weak acid" groups. Thus, K_I, K_{II}, K_{Al} and K_{Ca} are likely to be approximately temperature independent. Values of K_{AlH} *at* 10°C were estimated from 25°C values using the enthalpy of hydrolysis of Al^{3+}, i.e., 12.8 kcal mol^{-1}.[25]

Because the HS isolated from the Lochard Forest stream are likely to be the most representative of HS in solution in natural waters, calculations were performed first with the LFHS Model III parameters (Table 2). The calculated values of [Al_m-org] were found to follow the same trends as the observed values, but were consistently high by

Table 4. Summary of analytical data on Al speciation in natural waters. For R. Duddon, Falls Brook, L. Hövvatn, R. Swale and Whitray Beck, the cation-exchange method of Driscoll[18] was used to estimate [Al_m-org], and published values[22,23] were corrected as described by Backes and Tipping.[19]

Location	Ref.	No. of samples	I, μ*M*	pH	[Al_m], μ*M*	[T_F], μ*M*	[SO_4^{2-}], μ*M*	[HS], mg/L	[Al*m*-org]
R. Duddon	21	9	500	4.86 - 5.69	1.4 - 20.3	1.2 - 2.0	22 - 48	0.2 - 0.6	0.1 - 1.7
Falls Brook	22	19	300	4.82 - 5.79	2.3 - 13.6	0.1 - 3.4	48 - 60	4.1 - 8.5	0.7 - 9.0
L. Hövvatn	23	11	500	4.48 - 5.46	4.3 - 6.6	1.1	36	8.0	2.1 - 2.7
Paint L. inflow	20	14	500	5.24 - 5.83	2.7 - 7.9	2.8 - 3.5	81 - 98	3.8 - 6.0	2.0 - 4.3
Plastic L. inflow	20	21	500	4.30 - 4.84	4.6 - 11.6	3.3 - 3.9	29 - 88	9.8 - 33.0	1.3 - 9.7
Plastic L. outflow	20	20	500	4.88 - 5.92	0.4 - 4.3	2.2 - 2.8	66 - 82	4.0 - 7.2	0.4 - 2.7
R. Swale	21	10	500	4.18 - 5.57	1.3 - 7.5	1.1 - 4.1	64 - 111	4.8 - 27.5	0.8 - 5.8
Whitray Beck	21	10	300 - 1800	3.87 - 5.98	2.5 - 10.1	0.9 - 3.9	34 - 290	5.0 - 55.0	1.6 - 5.1
All Data	-	114	300 - 1800	3.87 - 5.98	0.4 - 20.3	0.1 - 4.1	22 - 290	0.2 - 55.0	0.1 - 9.7

approximately 1 μM Al. Better agreement could be obtained by decreasing the values of *Ac* and *n* (in proportion), i.e., by lowering the apparent concentration of the HS in the field samples. For the data as a whole, optimized values of *Ac* and *n* of 5.7 x 10^{-3} eq/gHS and 3.2 x 10^{-3} eq/gHS, respectively, gave the best least-squares agreement. The mean residual was then found to be 0.1 μM Al and the root-mean-square residual 0.8 μM Al. Optimized values of *Ac* were also calculated for results from individual sites (Table 5). An appreciable part of the root-mean-square residual is due to analytical errors in the input values and in the observed value of [Al_m-org].

Table 5. Application of Model III to field data (cf. Table 4)

Location	Optimum Value of Ac (eq/gHS x 10^3)	Residual Mean	Residual RMS
R. Duddon	7.8	-0.1	0.7
Falls Brook	6.4	0.4	1.1
L. Hovvatn	3.2	0.1	0.3
Paint Lake inflow	5.3	0	0.4
Plastic Lake inflow	7.1	0.1	0.5
Plastic Lake outflow	5.5	0	0.1
R. Swale	4.1	0	0.5
Whitray Beck	5.3	-0.1	1.1
All	5.7	0.1	0.8

Note: The residual referred to is the measured concentration of Al_m-org subtracted from the calculated value, in μM. From ref. 5.

Further model testing was carried out[26] by comparing calculated and measured values of [Al_m-org] in a series of solutions obtained by adding acid and/or $Al(NO_3)_3$ to aliquots of a water sample from the stream at Lochard Forest, the sample being taken at the same time as the much larger sample used to isolate the LFHS. This allowed the complexation properties of the isolated HS to be compared with those of the same HS close to their natural state. The results of the experiments are shown in Figure 6. Again it is found that the values of *Ac* and *n* for the isolated HS give calculated values of [Al_m-org] greater than those observed, the optimal value of *Ac* being 6.5 x 10^{-3} eq/gHS. Some bias is evident, the residuals changing from positive to negative as the pH increases (Figure 6).

Adjusting *Ac* in order to improve the fit of the field data is appropriate because of problems in estimating the concentration of HS in the samples. For the data used in our analysis, values of [HS] were obtained either from dissolved organic carbon con-

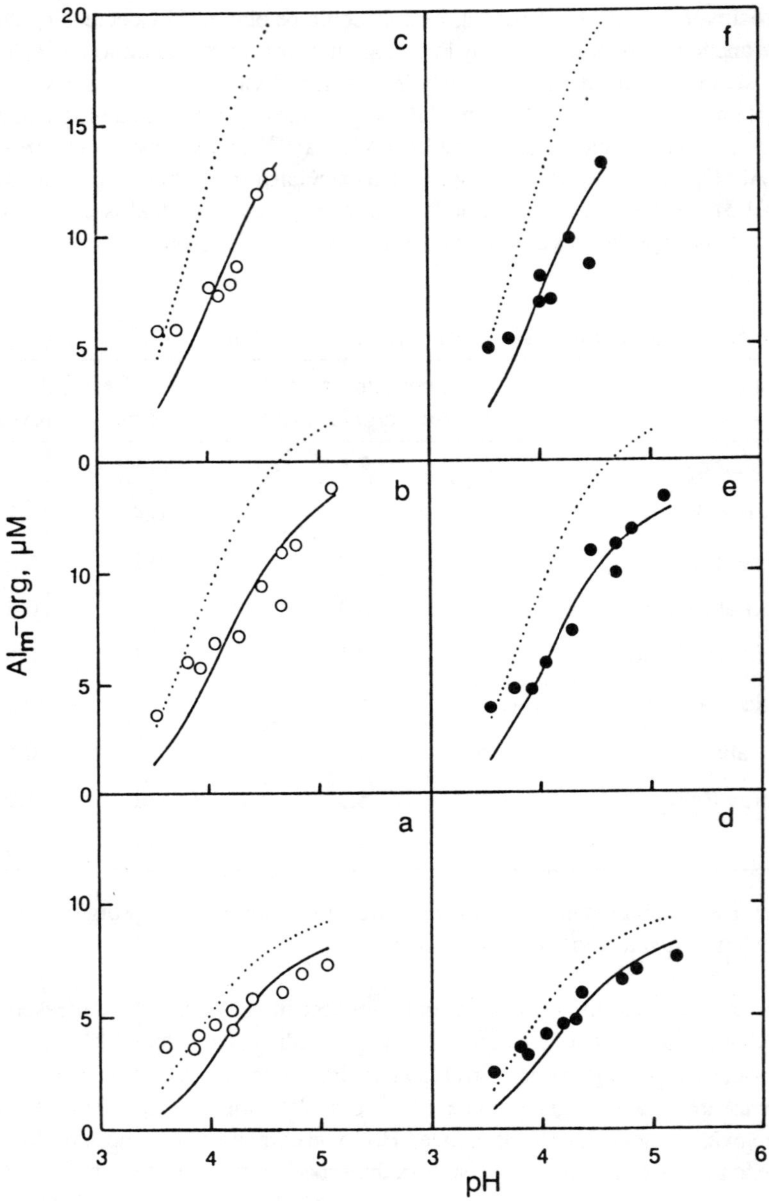

Figure 6. Dependence of [Al$_m$-org] on pH, total monomeric Al concentration, and temperature, for a streamwater sample from Lochard Forest. Left-hand panel 10°C; right-hand panel 25°C. Total monomeric Al = 10 μ*M* (a, d), 18.5 μ*M* (b, e), 28 μ*M* (c, f). Points represent measured values. Lines calculated with Model III: .... *Ac* = 9.7 x 10^{-3} eq/gHS, ——— *Ac* = 6.5 x 10^{-3} eq/gHS. See text. From ref. 26.

centrations, by multiplying by 2, or from the optical densities of the solutions.[21] By either means, only approximate values of [HS] can be expected. Another problemis the possibility that some of the humic binding sites may be occupied by competing metals, notably Fe (e.g., ref. 27). This would explain not only the general decrease in the number of Al binding sites, but also, if the competitive effect increased with pH, the bias observed in the titration experiments of Figure 6, at low pH (~3.5) the data for the natural sample coincide with predictions based on the *Ac* value for isolated LFHS, whereas at pH~5 a lower *Ac* is required to fit the data.

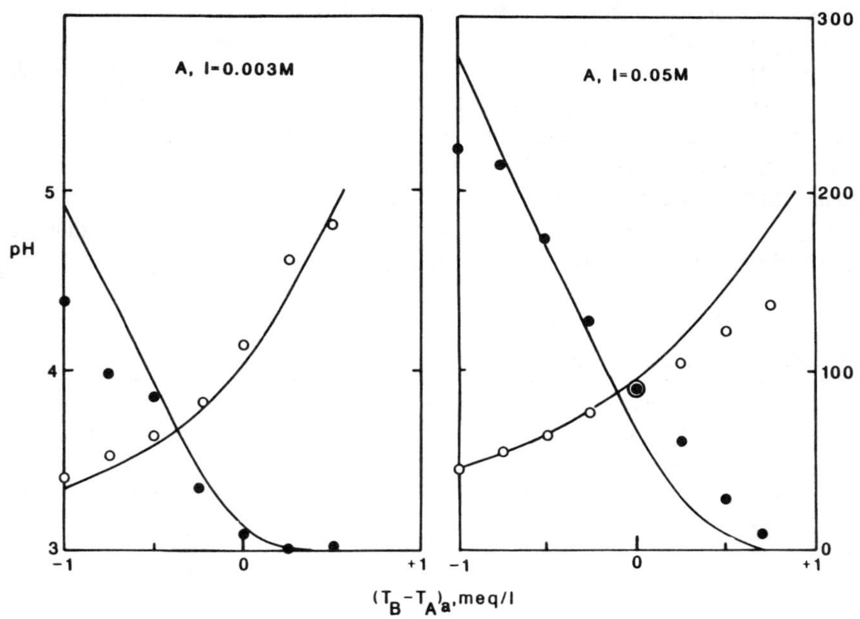

Figure 7. Batch titrations of soil sample A (see Table 6). Supernatant pH and concentration of monomeric inorganic Al are plotted against *added* base-acid. The concentration of soil was 5.8 g dry weight (equivalent)/L. Points are experimental, lines are predicted by CHAOS on the basis of total Al and HS concentrations, with $Ac = 3.57 \times 10^{-3}$ eq/gHS (Table 7), and taking the diffuse layer volumes to be 0.01 L/L at $I = 0.003$ M and 0.001 L/L at $I = 0.05$ M. In order to match the observed and calculated values, it was necessary to assume that when no base or acid had been added, the *actual* value of $(T_B - T_A)$ was -0.02 meq/L. Temp = 16°C. From ref. 12.

SOLID-SOLUTION INTERACTIONS IN ORGANIC SOILS

As already explained, combination of Donnan equilibrium expressions with the basic humic complexation model allows us to construct a model of organic soils, CHAOS.[12] Preliminary laboratory experiments have been carried out in order to explore the validity of this approach. The five soils samples taken for study were from Mosedale, a glaciated valley with acid soils in Cumbria, U.K. Analytical data on the samples are given in Table 6. The Al contents are assumed to represent metal complexed by the HS, metal present in the diffuse part of the electrical double layer and in the bulk soil solution, but not metal present in aluminosilicate or oxide mineral phases. In modeling the solid-solution interactions, organic matter not extractable by NaOH is assumed to be inert. Batch titrations were carried out by adding known amounts of acid or base to aqueous suspensions (I = 0.003 or 0.05 M) of the soils, and equilibrating for 24 hr. Supernatant pH values (range 3-5) and concentrations of monomeric Al were determined (Figure 7), and Al speciations calculated using Model III (with parameter values for MBHA; see Table 2) and equilibrium constants for Al hydrolysis. The parameters *Ac* and *n* were then adjusted (keeping the ratio *n*/*Ac* constant) to fit the data (pH, Al^{3+} activity) to the CHAOS equations. The values of *Ac* estimated for the different soils are shown in Table 7. It is seen that the optimum *Ac* was similar for each of the soils studied, and that the values are ~60% of that for isolated MBHA, similar to optimized values of *Ac* for HS in waters.

Table 6. Analytical data on soil samples.

Sample	pH	Water % by weight	Organic matter % by dry weight	HS % dry weight dry weight	Al µmol/g
A	3.7	77	80	23	119
B	4.0	72	61	28	154
C	3.7	86	87	29	54
D	4.0	78	72	19	105
E	3.3	80	85	24	20

Note: The samples were from the top 10 cm of the soil, and were dark brown or black. Measurements of pH were made on the supernatant after suspending 50 g of wet soil in 50 cm^3 of 10^{-3}M NaCl. Humic substances were estimated by assuming 0.01 M-NaOH-extractable organic carbon to be 50% by weight of the HS. The Al contents were determined by extracting for 24 hours with 0.01M HNO_3/0.1M $NaNO_3$. From ref. 12.

Table 7. Optimized values of *Ac* for the soil samples (cf. Table 6)

Soil	No. of data points	Mean *Ac* eq/gHS x 10^3
A	12	3.57 (0.34)
B	14	2.98 (0.22)
C	12	2.31 (0.28)
D	13	3.01 (0.15)
E	14	3.15 (0.45)
All	65	3.01 (0.50)

Note: Values in brackets are standard deviations. From ref. 12.

During the batch titration experiments it was observed that the concentration of HS in the supernatant solution increased with pH, but to a different extent for each soil (Figure 8). However, if the concentration of HS in solution is plotted against *Z* (the net humic charge) it is found that the points for the different soils approximately coincide (Figure 9), indicating that there is in CHAOS the capability to predict quantitatively concentrations of HS in soil water.

CHAOS can be used to calculate steady state ionic concentrations for soils, given the necessary inputs and outputs. First let us consider an organic soil with a bulk composition representative of samples taken from Mosedale (Table 6), with inputs of a simplified "acid rain" (see Table 8 for composition) and aluminum (added at a sufficient rate to keep the total Al concentration constant). For simplicity, we assume that humic production in the soil is zero, and that no humics are lost from the soil by transfer to the (mobile) aqueous phase. The calculated steady state concentrations of ionic species are given in Table 8, column 1. The composition of compartment *S* is found to be similar to that of the rain, with a small replacement of H^+ by Al^{3+} and $AlOH^{2+}$. The Al is largely in compartment *H*, i.e., humic-complexed, whereas the Ca and Na reside chiefly in the diffuse part of the electrical double layer, compartment *D*. The net humic charge, -263 μeq/g, is probably too low for much humic material to dissolve. It is neutralized mainly by Ca^{2+} (69%) and Al^{3+} (24%). If the cation-exchange capacity (usually estimated by titration to an arbitrary, near-neutral pH) is taken to be equal to *n*, the content of carboxyl groups, then the base saturation of the soil is 12%. The concentration of Al^{3+} in the soil solution is calculated to be ~ 1 μ*M*, of similar magnitude to field values,[28,29,30] and much lower than values of ≥ 100 μ*M* that would be

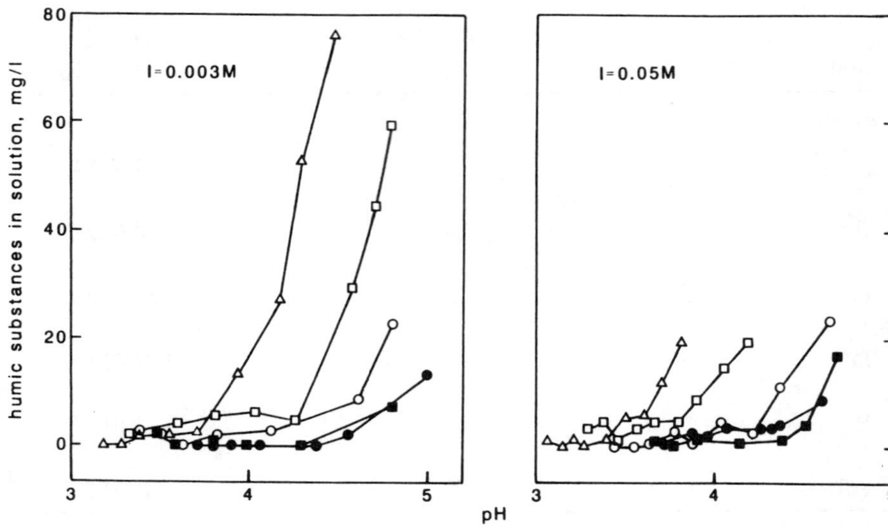

Figure 8. Concentrations of HS in supernatant solutions in batch titrations of soils A (open circles), B (solid circles), C (open squares), D (solid squares) and E (open triangles). Soil concentrations (g dry weight/L) were A 5.8, B 9.8, C 7.8, D 8.4, E 11.3. From ref. 12.

expected for equilibrium with an aluminum trihydroxide phase [cf. ref. 30]. When high concentrations of monomeric Al are found in organic soil solutions, this is due to organically-complexed metal.[31]

Now consider the same soil, but a less-acid rain (pH 5.01: see Table 8), again in steady-state. Table 8, column 2 shows that under these circumstances the Al concentration in compartment *S* is calculated to be very small and that the humic charge, now neutralized almost entirely by Ca^{2+} (base saturation 23%), is significantly more negative, which would mean that HS are more likely to pass into solution under these conditions. Thus, one consequence of reducing rain acidity might be to increase the concentration of dissolved HS in soil, and presumably stream, water. There is evidence from palaeolimnological studies that the reverse process, i.e., a decrease in dissolved organic matter concentrations, occurred on acidification of soil and water.[32]

If simplifying assumptions about the soil hydrology are made, CHAOS can be used to estimate how long it would take to pass from one steady state to the other. The rain-soil interaction is assumed to take place in a series of steps, during each of which the water in compartment *S* is replaced by rainwater and re-equilibration is allowed to occur. The results of the calculation are shown in Figure 10. The change in rain composition is supposed to take place over 50 steps. It is found that the soil solution requires more than 5000 steps to achieve the new steady state concentration values, because of the buffering capacity of the soil HS. The steps in the calculations can be

Table 8. Steady state ionic distributions for an organic soil

		1	2
Soil			
T_{HS} g/L		70	70
T_{Al}	μ*M*	25000	25000
T_{Ca}	μ*M*	6520	14400
T_{Na}	μ*M*	850	1150
Compartment H			
Al	μmol/g HS	336	355
Ca	μmol/g HS	2	33
Compartment D			
Al^{3+}	μ*M*	3000	235
$AlOH^{2+}$	μ*M*	13	4
Ca^{2+}	μ*M*	12700	24200
Na^{+}	μ*M*	1590	2200
H^{+}	μ*M*	760	220
Compartment S			
Al^{3+}	μ*M*	0.74	0.02
$AlOH^{2+}$	μ*M*	0.05	0.01
Ca^{2+}	μ*M*	50.0	50.0
Na^{+}	μ*M*	100.0	100.0
H^{+}	μ*M*	47.7	10.0
pH		4.33	5.01
Net humic Charge μeq/g		-263	-368
Neutralizing Cations (%)			
Al^{3+}		24	1
$AlOH^{2+}$		0	0
Ca^{2+}		69	94
Na^{+}		4	4
H^{+}		2	0
Aluminum, μeq/g soil			
H		252	267
D		16	1
S		0	0
total		268	268
Calcium, μeq/g soil			
H		1.0	16.5
D		45.3	86.4
S		0.2	0.2
total		46.5	103.1
Sodium, μeq/g soil			
H		0.0	0.0
D		2.8	3.9
S		0.2	0.2
total		3.0	4.1
Apparent base saturation		12	23

Note: Column 1 is for a "rain" composition of 50 μ*M* Ca^{2+}, 100 μ*M* Na^{+}, 50 μ*M* H^{+}, 250 μ*M* X^{-}; Column 2 is for 50 μ*M* Ca^{2+}, 100 μ*M* Na^{+}, 10 μ*M* H^{+}, 210 μ*M* X^{-}. It was assumed that one-half of the soil water was in compartment *D*, the other half in compartment *S* (Figure 1). Total concentrations are denoted by T. To calculate cation equivalents per g of soil, it was assumed that HS comprise 25% of the soil dry weight (cf. Table 6). The apparent base saturation was calculated by assuming that the conventionally-estimated cation exchange capacity would be equivalent to *n* (see text). From ref. 12.

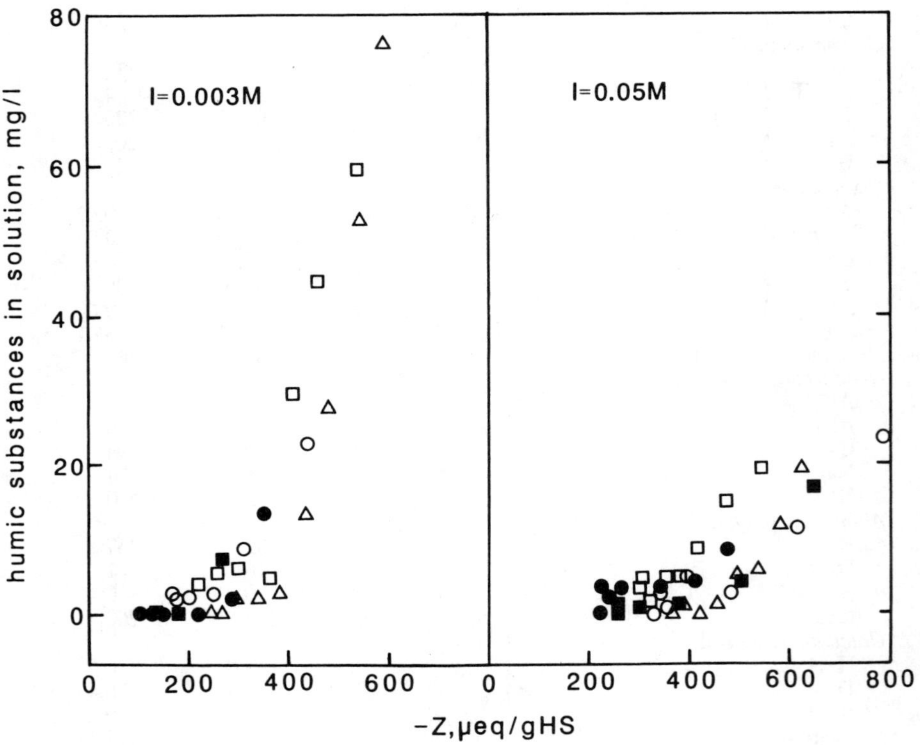

Figure 9. Data in Figure 8 plotted against calculated values of Z, the net humic charge. Symbols as for Figure 8. From ref. 12.

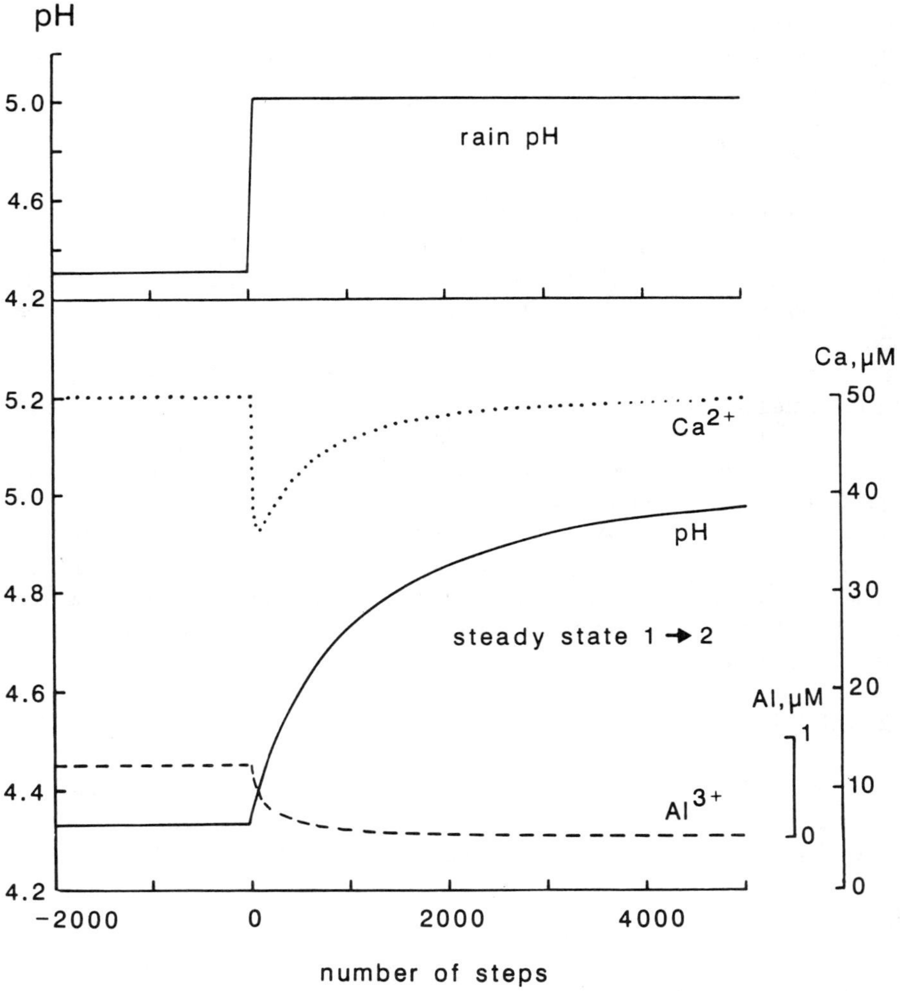

Figure 10. Predictions of soil solution composition (compartment *S*) during conversion from steady state 1 to steady state 2 (Table 8). For a 10 cm depth of soil and rainfall of 2 $m \cdot yr^{-1}$, 1000 steps represent ca. 20 years. From ref. 12.

related to real time by calculating the residence time of water in compartment *S*. For example, given a soil depth of 10 cm and a soil water content of 80% by volume, of which half is replaceable, the total volume of replaceable water per m^2 of soil area is 0.04 m^3. For an annual rainfall of 2 m the residence time is therefore 0.02 years, so that 50 calculation steps represent 1 year of rain-soil interactions. In this case, therefore, more than 100 years would be required to reach the new steady state pH (Figure 10), although most of the change in Al^{3+} concentration would occur within 10 years. Of course, this calculation is approximate and relies on a number of assumptions. It does suggest, however, that acid organic soils have a substantial buffering capacity, toward both H^+ and Al^{3+} (cf. ref. 33).

CONCLUSIONS

The model we have developed appears to describe the equilibria involving HS, Al species, Ca^{2+}, and H^+ under acid conditions (pH < 6) to at least a good first approximation. Further developments could be made, in general to account for possible competition by Fe, and in the case of CHAOS, the soil version of the model, to improve upon the highly simplified formulation of bulk solution/diffuse layer concentration relationships. It may also be worthwhile to increase the number of permitted types of metal-binding site (with a concomitant increase in the number of equilibrium constants) in order to improve data-fitting. It is intended to make CHAOS more realistic by (a) incorporating humic dissolution and export, using expressions based on charge-solubility relationships (cf. Figures 8, 9), (b) specifying rates of production of HS in theorganic soil, and (c) introducing water passage through macro- and micropores. Testing of the model in more detail is also desirable, by field measurements and by further laboratory experiments with soil samples. In the meantime the model should be useful in predicting approximately how the chemistry of acid environments responds to changing conditions.

ACKNOWLEDGMENTS

We thank H. Gardberg for typing the manuscript. This work was financially supported by the Commission of European Communities, the Natural Environment Research Council and the Royal Society Surface Waters Acidification Programme.

REFERENCES

1. Buffle, J. "Natural organic matter and metal-organic reactions in aquatic systems," in *Metal Ions in Biological Systems*, Vol. 18, H. Sigel, Ed. (New York and Basel: Marcel Dekker, 1984), pp. 166-221.

2. Sposito, G. "Sorption of trace metals by humic materials in soils and natural waters," *C.R.C. Rev. Environ. Control* 16:193-229.

3. Marinsky, J. A. and J. Ephraim. "A unified physicochemical description of the protonation and metal ion complexation equilibria of natural organic acids (humic and fulvic acids). 1. Analysis of the influence of polyelectrolyte properties on protonation equilibria in ionic media: fundamental concepts," *Environ. Sci. Technol.* 20:349-354 (1986).

4. Ephraim, J., S. Alegnet, A. Mathuthu, M. Bicking, R. L. Malcolm, and J. A. Marinsky. "A unified physicochemical description of the protonation and metal ion complexation equilibria of natural organic acids (humic and fulvic acids). 2. Influence of polyelectrolyte properties and functional group heterogeneity on the protonation equilibria of fulvic acid," *Environ. Sci. Technol.* 20:354-366 (1986).

5. Tipping, E., C. A. Backes, and M. A. Hurley. "The complexation of protons, aluminum and calcium by aquatic humic substances: a model incorporating binding-site heterogeneity and macroionic effects," *Water Research* 22:597-611 (1988).

6. Backes, C. A. and E. Tipping. "Aluminum complexation by an aquatic humic fraction under acidic conditions," *Water Research* 21:211-216 (1987).

7. Thurman, E. M. *Organic Geochemistry of Natural Waters* (Dordrecht: Nijhoff/Junk, 1985).

8. Perdue, E. M. "Acidic functional groups of humic substances," in *Humic Substances in Soil, Sediment, and Water*, G. R. Aiken, D. M. McKnight, R. L. Wershaw and P. MacCarthy, Eds. (New York: John Wiley & Sons, Inc., 1985).

9. Gamble, D. S., A. W. Underdown, and C. H. Langford. "Copper (II) titration of fulvic ligand sites with theoretical, potentiometric and spectrophotometric analysis." *Anal. Chem.* 52:1901-1908 (1980).

10. Tanford, C. *Physical Chemistry of Macromolecules* (New York: Wiley, 1961).

11. Arnold, P. W. "Surface-electrolyte interactions," in *The Chemistry of Soil Constituents*, Greenland, D. J. and M.H.B. Hayes, Eds. (Chichester: John Wiley & Sons, 1978) pp. 355-404.

12. Tipping, E. and M. A. Hurley. "A model of solid-solution interactions in acid organic soils, based on the complexation properties of humic substances," *J. Soil Sci.*, 39:505-519 (1988).

13. Mantoura, R.F.C. and J. P. Riley. "The analytical concentration of humic substances from natural waters," *Anal. Chim. Acta* 76:97-106 (1975).

14. Thurman, E. M. and R. L. Malcolm. "Preparative isolation of aquatic humic substances," *Environ. Sci. Technol.* 15:463-466 (1981).

15. Schnitzer, M. and S. U. Khan. *Humic Substances in the Environment* (New York: Marcel Dekker 1972).

16. Lövgren, L., T. Hedlund L.-O. Öhman, and S. Sjöberg. "Equilibrium approaches to natural water systems. 6. Acid-base properties of a bog-water and its complexation reactions with aluminum (III)," *Water Research*, 21:1401-1407 (1987).

17. Perdue, E. M. and C. R. Lytle. "A distribution model for binding of protons and metal ions by humic substances," *Environ. Sci. Technol.* 17:654-660 (1983).

18. Driscoll, C. T. "A procedure for the fractionation of aqueous aluminum in dilute acidic waters," *Int. J. Environ. Anal. Chem.* 16:267-283 (1984).

19. Backes, C. A. and E. Tipping. "An evaluation of the use of cation-exchange resin for the determination of organically-complexed Al in natural acid waters," *Intern. J. Environ. Anal. Chem.* 30:135-143 (1987).

20. LaZerte, B. D. "Forms of aqueous aluminum in acidified catchments of central Ontario: A methodological analysis," *Can. J. Fish. Aquat. Sci.* 41:766-776 (1984).

21. Tipping, E., C. Woof, C. A. Backes and M. Ohnstad. "Aluminum speciation in acidic natural waters: testing of a model for Al-humic complexation," *Water Research* 22:321-326 (1988).

22. Johnson, N. M., C. T. Driscoll, J. S. Eaton, G. E. Likens and W. H. McDowell "'Acid rain,' dissolved aluminum and chemical weathering at the Hubbard Brook Experimental Forest, New Hampshire," *Geochim. Cosmochim. Acta* 45:1421-1437 (1981).

23. Wright, R. F. and 0. Skogheim. "Aluminum speciation at the interface of an acid stream and a limed lake," *Vatten* 39:301-304 (1983).

24. Martell, A. E. and R. M. Smith. *Critical Stability Constants* Vol. 3 (New York: Plenum, 1977).

25. Couturier, Y., G. Michard and G. Sarazin. "Constantes de formation des complexes hydroxydés de l'aluminium en solution aqueuse de 20 à 70°C", *Geochim. Cosmochim. Acta* 48:649-659 (1984).

26. Tipping, E. and C. A. Backes. "Organic complexation of Al in acid waters: model-testing by titration of a streamwater sample," *Water Research* 22:593-595 (1988).

27. Perdue, E. M., K. C. Beck, and J. H. Reuter. "Organic complexes of iron and aluminum in natural waters," *Nature* 260:418-420 (1976).

28. Nilsson, S. I. and B. Bergkvist. "Aluminum chemistry and acidification processes in a shallow podzol on the Swedish westcoast," *Water Air Soil Pollut.* 20:311-329 (1983).

29. Driscoll, C. T., N. Van Breemen and J. Mulder. "Aluminum chemistry in a forested spodosol," *Soil Sci. Soc. Am. J.* 49:437-444 (1985).

30. Cronan, C. S., W. J. Walker and P. R. Bloom. "Predicting aqueous aluminum concentrations in natural waters," *Nature (Lond.)* 324:140-143 (1986).

31. David, M. B. and C. T. Driscoll. "Aluminum speciation and equilibria in soil solutions of a Haplorthod in the Adirondack Mountains (New York, U.S.A.,") *Geoderma* 33:297-318 (1984).

32. Davis, R. B., D. S. Anderson and F. Berge. "Palaeolimnological evidence that lake acidification is accompanied by loss of organic matter," *Nature (Lond.)* 316:436-438 (1985).

33. Rosenqvist, I. Th. "Alternative sources for acidification of river water in Norway," *Sci. Tot. Environ.* 10:39-49 (1978).

CHAPTER 5

CHEMISTRY AND TRANSFER OF ALUMINUM IN A FORESTED WATERSHED IN THE ADIRONDACK REGION OF NEW YORK, USA

C.T. Driscoll, B.J. Wyskowski, P. DeStaffan, R.M. Newton

INTRODUCTION

The transport of Al is an important aspect of the development of soils in northern temperate regions. For example, the formation of Spodosols is generally thought to involve the mobilization of Al (and Fe) from the upper (E) to lower (B) mineral soil horizons by organic acids leached from foliage as well as from decomposition of the forest floor.[1] Reuss and Johnson[2] have also demonstrated that production of CO_2 may facilitate the dissolution of Al in acidic soils. However, soil retention of organic acids and/or degassing of CO_2 restrict the mobility of Al, so little transport is evident from soil to surface water.[3,4]

In recent years, it has been hypothesized that strong acids from atmospheric deposition remobilize soil Al, causing elevated concentrations in surface waters.[3] Many investigators have reported an exponential increase in Al concentrations with decreasing pH in acid-impacted surface waters.[5] The speciation of aqueous Al depends on a variety of factors, including solution contact with solid phases, pH, concentrations of complexing ligands (F^-, SO_4^{2-}, organic anions) and temperature.[6] In regions exhibiting surface water acidification, organic forms of Al predominate in circumneutral waters, while concentrations of inorganic Al become significant at pH values below 5.5.[7]

The ecological effects of Al in drainage waters are a function of its speciation as well as concentration. Elevated concentrations of Al in natural solutions may influence water quality through 1) pH buffering,[8] 2) interactions with critical element cycles (e.g. phosphorus, dissolved organic carbon),[9] 3) coagulation of particulates and organic carbon, which increases the transparency and reduces the thermal stability of acidic lakes,[10] and 4) potential toxicity of inorganic species to vegetation[11] and aquatic biota.[12]

The objective of this study was to monitor variations in soil water and stream composition as well as selected stream sediment parameters in a small acid-sensitive watershed in the Adirondack region of New York, U.S.A. in order to evaluate processes regulating the chemistry and transport of Al. This analysis was facilitated by calibration and application of the Integrated Lake/Watershed Acidification Study (ILWAS) model to the watershed.

METHODS

The study was conducted in a forested watershed, Pancake-Hall Creek (74 ha), located within the west-central Adirondack region of New York, U.S.A. (43°51'N, 74°52'W; Figure 1). Vegetation within the catchment is primarily deciduous (*Acer saccharum, Fagus grandifolia, Betula alleghaniensis*) with stands of coniferous species interspersed (*Picea rubens, Abies balsamea, Tsuga canadensis*). The watershed is predominately covered with shallow deposits (< 3 m depth) of glacial till, with thicker deposits (> 3 m depth) at lower elevations.[13] The mineralogy of the soils is largely quartz, plagioclase and potassium feldspar, with minor quantities of hornblende, ilmenite and magnetite.[14] The soils are well drained Spodosols (Typic Fragiorthod, Haplorthod), which are acidic with low concentrations of exchangeable basic cations (13% base saturation in the E horizon decreasing to 7% in the BC horizon;).[15] A relatively uniform grain size distribution occurs throughout the soil profile, sand being the predominant fraction, with minor amounts of silt.[13] The clay fraction is small (< 2%), consisting largely of vermiculite and kaolinite. A summary of soil chemical parameters for the study site is provided in Table 1.

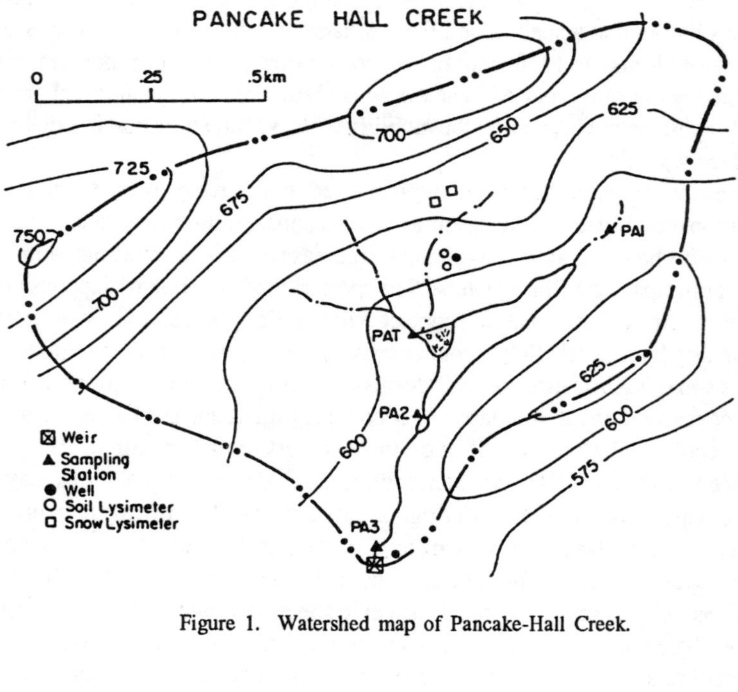

Figure 1. Watershed map of Pancake-Hall Creek.

Table 1. A summary of soil chemical parameters for the Pancake-Hall Creek watershed

Soil Horizon	pH Water	Cation Exchange Capacity ($meq \cdot 100\ g^{-1}$)	% base Saturation	% Organic Matter
Oi	5.0	39.0	84	86
Oa	3.6	32.0	38	76
E	3.9	3.7	13	5
Bh	3.9	12.0	7	16
Bhs	4.4	7.0	5	14
Bs	4.6	5.3	8	10
BC	4.9	1.2	7	4
C	5.1	0.61	17	2

Note: Data are based on collections from five soil pits (Cronan 1986).

The watershed was instrumented for hydrologic and water quality monitoring during the fall 1985. Replicate zero tension lysimeters (Figure 2) and soil gas collectors were placed immediately beneath the Oa horizon and within the Bs2 horizon (at a depth of 0.5 m) in two soil pits. In addition, aliquots of soil were collected from the Oa, Bh and Bs2 horizons for analysis of free (non-silicate bound) Al fractions. After lysimeter installation, pits were backfilled to eliminate freezing of soil/soil solutions and to allow for winter collection of solutions. Sampling of soil solutions was accomplished by evacuation of the lysimeter reservoir with a hand vacuum pump. Soil gas collections were made using a syringe connected to the soil gas tubing by a luer lock fitting.

Groundwater solutions were monitored from a PVC well placed in the thin till adjacent to the lysimeter plots (Figure 1). Solutions were collected by a hand vacuum pump through a length of tubing placed within the well. Streamwater has been collected since January 1984 at the confluence of two first order tributaries, downslope from the soil solution monitoring site and above a beaver impoundment (PAT; Figure 1).

Sediments were collected for analysis at the stream sampling site on 4 June and 21 September 1985. Collection of replicate sediment samples was made by sampling sediments using a polyethylene scoop and filtration between 250-µm and 56-µm mesh Nitex screening. The size fraction contained within the screens was transported to the laboratory at Syracuse University and stored at 4°C until extracted for analysis.

Groundwater stage was measured from a 14-cm diameter well, located near the watershed outlet, using a data logger wired to a Stevens water level recorder (locations given in Figure 1). To monitor stream discharge, a 1.4 $m^3 \cdot s^{-1}$ Parshall flume with a The removable 90° V-notch weir was installed in the stream channel. The data logger was again used to monitor water level for continuous stream discharge measurements. data logger measured stream and groundwater stage every 10 seconds, with average values recorded at 10-minute intervals. Stream temperature was measured continuously using a thermistor and monitored by the data logger.

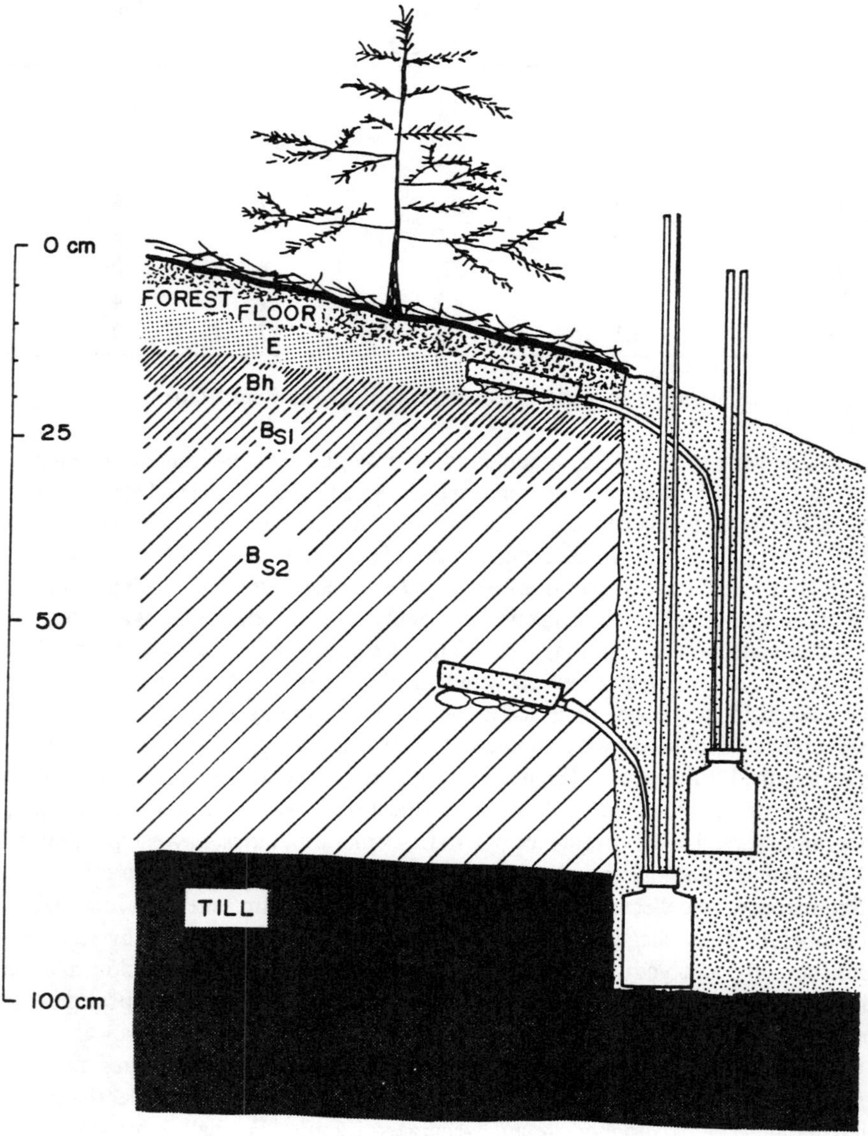

Figure 2. Schematic illustration of zero-tension lysimeter.

Atmospheric deposition was not monitored as part of this study, however, data were obtained from the Utilities Acid Precipitation Sampling Program (UAPSP) station at Big Moose Lake, approximately 2 km from the study site. A summary of the concentration and deposition of major solutes in precipitation for the study period is provided in Table 2.

Table 2. Summary of the precipitation chemistry and wet deposition for the study year at the Utilities Acid Precipitation Program Station at Big Moose Lake. Wet deposition for the study year was 128 cm.

parameter	annual concentration ($\mu eq \cdot l^{-1}$)	annual wet deposition ($eq \cdot ha^{-1} \cdot yr^{-1}$)
H^+	55	700
C_B	13	139
NH_4^+	17	217
SO_4^{2-}	46	590
NO_3^-	32	411
Cl^-	4.2	54

C_B is the sum of basic cations (Ca^{2+}, Mg^{2+}, Na^+, K^+)

Solution and soil gas samples were collected nominally at monthly intervals, with more frequent collections during the high flow snowmelt period. Shortly after collection, samples were measured for pH and processed for the determination of monomeric Al.[16] After field processing, samples were transported to Syracuse University for analyses. Analyses of soil gas for CO_2 and soil solutions for dissolved organic carbon (DOC), dissolved inorganic carbon (DIC), NH_4^+, SO_4^{2-}, NO_3^- and Cl^- were conducted immediately upon receipt at Syracuse University, generally within 48 hours of collection. Samples were stored at 4°C until the remainder of the analyses were completed, generally within one month of collection. The water chemistry parameters monitored and the analytical methods used in this study are summarized in Table 3.

To facilitate an evaluation of Al chemistry, Al was fractionated into non-labile monomeric Al, labile monomeric Al and acid soluble Al using the method described by Driscoll.[16] Non-labile monomeric Al is considered to represent organic complexes of Al. Labile monomeric Al is monomeric Al less nonlabile monomeric Al and is an estimate of inorganic species of monomeric Al, including aquo, OH^-, F^- and SO_4^{2-} complexes. Acid soluble Al is total Al less monomeric Al. This fraction represents particulate Al and very strongly bound or occluded alumino-organic substances. Although operationally defined, we refer to labile monomeric Al as inorganic monomeric Al and nonlabile monomeric Al as organic monomeric Al in this report.

Table 3. Analytical methods used in analysis of Pancake-Hall Watershed samples

SO_4^{2-}, NO_3^-, Cl^-	Ion chromatography	Small et al.[17]
NH_4^+	Phenate colorimetry; autoanalyzer	US EPA[18]
Ca^{2+}, Mg^{2+}, K^+, Na^+	Atomic Absorption Spectrophotometry (AAS)	Slavin[19]
pH	Potentiometrically with glass electrode, field measurement	Standard Methods[20]
Acid Neutralizing Capacity (ANC)	Strong acid titration with Gran plot analysis	Gran[21]
Dissolved Inorganic Carbon (DIC)	CO_2 detection by infrared (IR) spectrophotometry	Dohrmann[22]
Dissolved Organic Carbon (DOC)	UV enhanced persulfate oxidation, detection of CO_2 by IR spectrophotometry	Dohrmann[22]
Dissolved Silica	Heteropoly blue complex colorimetry; autoanalyzer	US EPA[18]
Total F	Potentiometrically with ion selective electrode after TISAB addition	Orion[23]
Total Al	Acid digestion (lhr,pH 1), 8-hydroxyquinoline extraction into methyl isobutyl ketone (MIBK), analysis by AAS, graphite furnace.	Barnes;[24] Driscoll[16]
Monomeric Al	Field extraction by 8-hydroxy-quinoline into MIBK, analysis by AAS, graphite furnace	Barnes[24]
Nonlabile monomeric Al	Fractionation by ion exchange column, analysis for monomeric Al	Driscoll[16]
CO_2	Gas chromatography	

Free (non-silicate bound) Al fractions in soil and stream sediments were determined by separate extractions with (1) 1 M KCl for 10 min., (2) 0.1 M sodium pyrophosphate for 18 hr and (3) 0.2 M ammonium oxalate adjusted to pH 2 for 4 hr.[25] Potassium chloride is thought to displace exchangeable Al, pyrophosphate extractable Al is an estimate of soil/sediment organic Al and acid oxalate is an effective extractant for both organic and amorphous forms of free Al.[26] In addition, organic carbon was determined on the pyrophosphate extracts to estimate the organic carbon associated with free Al and Fe.

Using water chemistry data collected from this study, the distribution of inorganic carbon (partial pressure of CO_2 (pCO_2), H_2CO_3, HCO_3^-, CO_3^{2-} and Al were calculated using the chemical equilibrium model, ALCHEMI.[27] The solubility of Al with respect to mineral phase equilibrium was evaluated using saturation indices (SI).

$$SI = \log (Qp/Kp)$$

where, Qp is the ion activity product of the solution with respect to the mineral phase of interest and Kp is the thermodynamic solubility product. Driscoll et al[7], have reported that the pH dependence of aquo Al (Al^{3+}) activity in surface waters of the west-central Adirondacks is similar to the solubility predicted from the dissolution of natural $Al(OH)_3$ (log *Kso = 8.77; May et al.[28]). As a result, natural $Al(OH)_3$ was used as a reference mineral in our analysis. The thermodynamic data used in aqueous and solid phase calculations are summarized in Schecher and Driscoll[27].

ALCHEMI was also used to help assess the role of CO_2 in the mobilization of Al within soils. To quantify CO_2 mediated leaching of Al, measured solution solute concentrations and pCO_2 were entered in ALCHEMI and solution equilibrium determined. Following this calculation, ALCHEMI was again used to reequilibrate the solution chemistry with atmospheric CO_2 ($10^{-3.5}$ atm), to simulate soil, CO_2 degassing. If the CO_2-degassed system exceeded the solubility of natural $Al(OH)_3$, Al was allowed to precipitate. The amount of Al precipitated from solution following equilibration with atmospheric CO_2 provides an estimate of Al solubilized by the production of CO_2 within the soil.

To help evaluate pathways of Al transport, the ILWAS model was applied to Pancake-Hall Watershed. ILWAS is a process-oriented research model that was developed to predict the effects of atmospheric deposition on surface water quality. To accurately represent the acid-base status of waters, Gherini et al.[29] found it necessary to simulate the routing of water through lake/watershed systems as well as all the major processes consuming and producing acidity along the hydrologic cycle. We calibrated the hydrologic subcomponent of the ILWAS model using precipitation, groundwater stage and stream discharge data collected at Pancake-Hall Watershed for the period November 1985 to November 1986. A detailed description of the hydrologic submodel and calibration procedure is available elsewhere.[29,30] We coupled model simulations of water movement through soil horizons with soil solution, groundwater and streamwater monitoring data to calculate solute fluxes at the study site.

RESULTS

Drainage waters in the Pancake-Hall Watershed were highly acidic. In all solutions, SO_4^{2-} was the dominant anion and concentrations of strong acid anions (SO_4^{2-}, NO_3^-, Cl^-) exceeded concentrations of basic cations (C_B: Ca^{2+}, Mg^{2+}, Na^+, K^+) resulting in negative values for the acid neutralizing capacity (ANC). Moreover, spatial and temporal variations in ANC were evident that may be attributed to a variety of processes regulating the concentrations of major solutes in solution.

Organic horizon leachates were characterized by low pH and ANC due to elevated concentrations of SO_4^{2-}, NO_3^- and organic acids (DOC) relative to concentrations of basic cations (Figure 3). As solutions migrated through the B horizon, concentrations of NO_3^- and DOC were reduced due to soil retention, and Al concentrations increased due to release from soil, which served to increase ANC (Figure 4). This increase in ANC was restricted, to some extent, by increases in SO_4^{2-} concentrations within mineral soil solutions (Bs2 horizon and groundwater). Little change in the sum basic cation concentrations was evident between Oa and Bs horizon leachates.

Groundwater was generally less acidic than Bs2 horizon solutions due to additional release of basic cations and retention of DOC within the lower mineral soil and glacial till. However, SO_4^{2-} concentrations were enriched in groundwater, relative to soil solutions, which served to attenuate the increase in ANC. The higher pH of groundwater relative to Bs2 horizon leachate resulted in lower Al concentrations (Table 4).

The chemical characteristics of the tributary upstream of the beaver impoundment in Pancake-Hall Watershed (PAT; Figure 1) were somewhat different than either Bs2 horizon leachate or groundwater (Figure 5). Values of pH and DOC were intermediate between Bs2 solutions and groundwater, while ANC, SO_4^{2-} and basic cation concentrations were similar to values in the Bs2 horizon. Both NO_3^- and Al concentrations in streamwater were significantly lower than in mineral soil solutions and the groundwater.

Temporal variations in chemical composition were particularly evident in the Oa horizon leachates (Figure 3). Here, the pH remained relatively constant in Oa soil solutions during the fall and early winter period. However, decreases in basic cations (ΔC_B = -80 $\mu eq \cdot L^{-1}$) and DOC concentration coupled with increases in NO_3^- (ΔNO_3^- = +60 $\mu eq \cdot L^{-1}$) and SO_4^{2-} (ΔSO_4^{2-} = +50 $\mu eq \cdot L^{-1}$) resulted in a net decrease in ANC (ΔANC = -30 $\mu eq \cdot L^{-1}$) during this time. In February and early March, water movement through the catchment was minimal due to cold conditions and snowpack accumulation. This period coincided with an extremely acidic Oa horizon leachate (pH = 4.0; ANC = -120 $\mu eq \cdot L^{-1}$) due to the release of very high NO_3^- concentrations (200 $\mu eq \cdot L^{-1}$). In mid March/early April, snowmelt produced a large hydrologic input to the forest floor, which apparently diluted the concentrations of solutes in the Oa horizon leachate causing a marked increase in pH and ANC (ANC = +100 $\mu eq \cdot L^{-1}$). Following snowmelt, basic cation concentrations remained essentially constant, SO_4^{2-} concentrations increased (ΔSO_4^{2-} = +30 $\mu eq \cdot L^{-1}$), and NO_3^- decreased (ΔNO_3^- = -60 - $\mu eq \cdot L^{-1}$), coinciding with an increase in ANC through the summer/fall season (ΔANC = +50 $\mu eq \cdot L^{-1}$). Dissolved organic carbon concentrations in forest floor leachate increased after snowmelt and remained high throughout summer and fall.

High concentrations of Al were evident in the forest floor leachate (mean value 42 $\mu mol \cdot L^{-1}$), with significant fractions of organic monomeric Al, inorganic monomeric Al as well as acid-soluble Al (Figure 3; Table 4). The distribution of Al was relatively constant throughout the study period, except for a large increase in inorganic monomeric Al during the winter season.

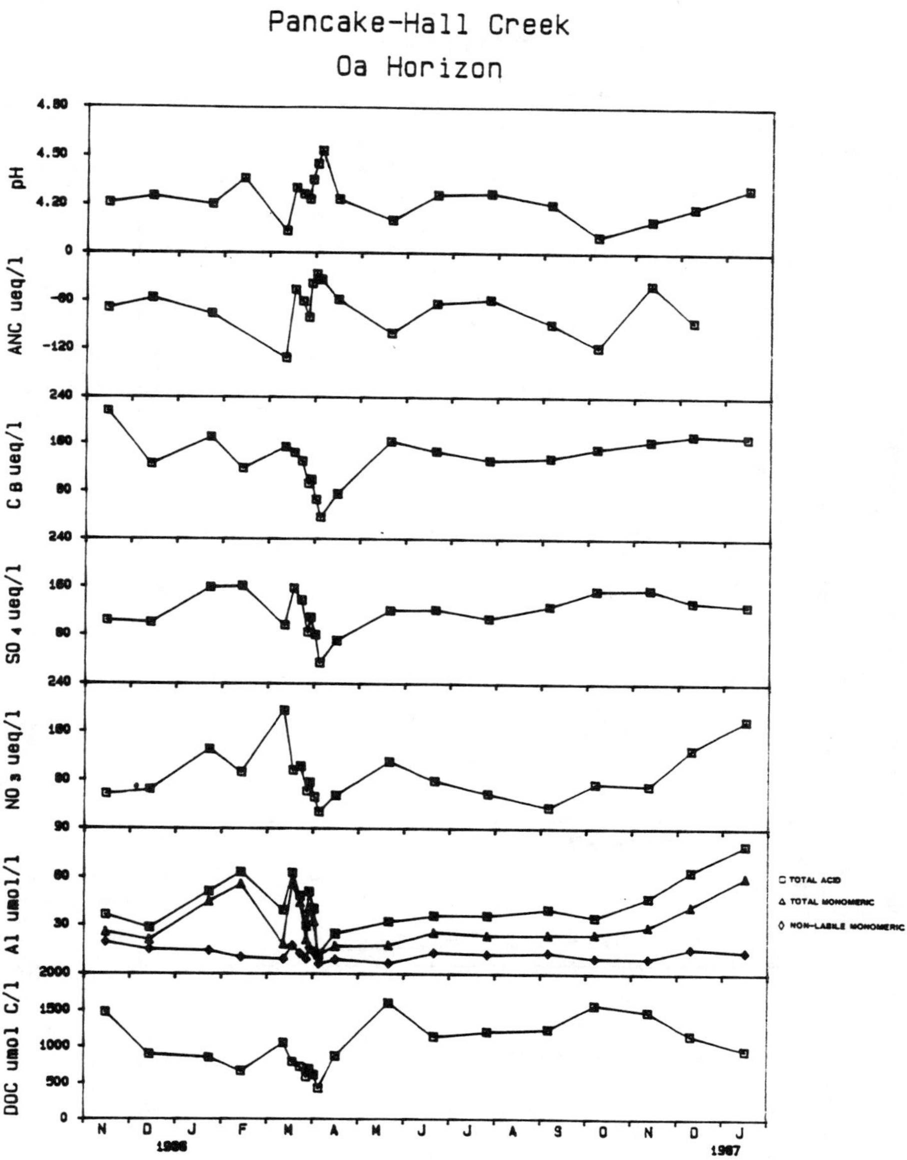

Figure 3. Temporal trends in the chemistry of Oa horizon leachate at Pancake-Hall Creek. C_B is the sum of basic cations (Ca^{2+}, Mg^{2+}, Na^+, K^+). DOC is dissolved organic carbon concentration.

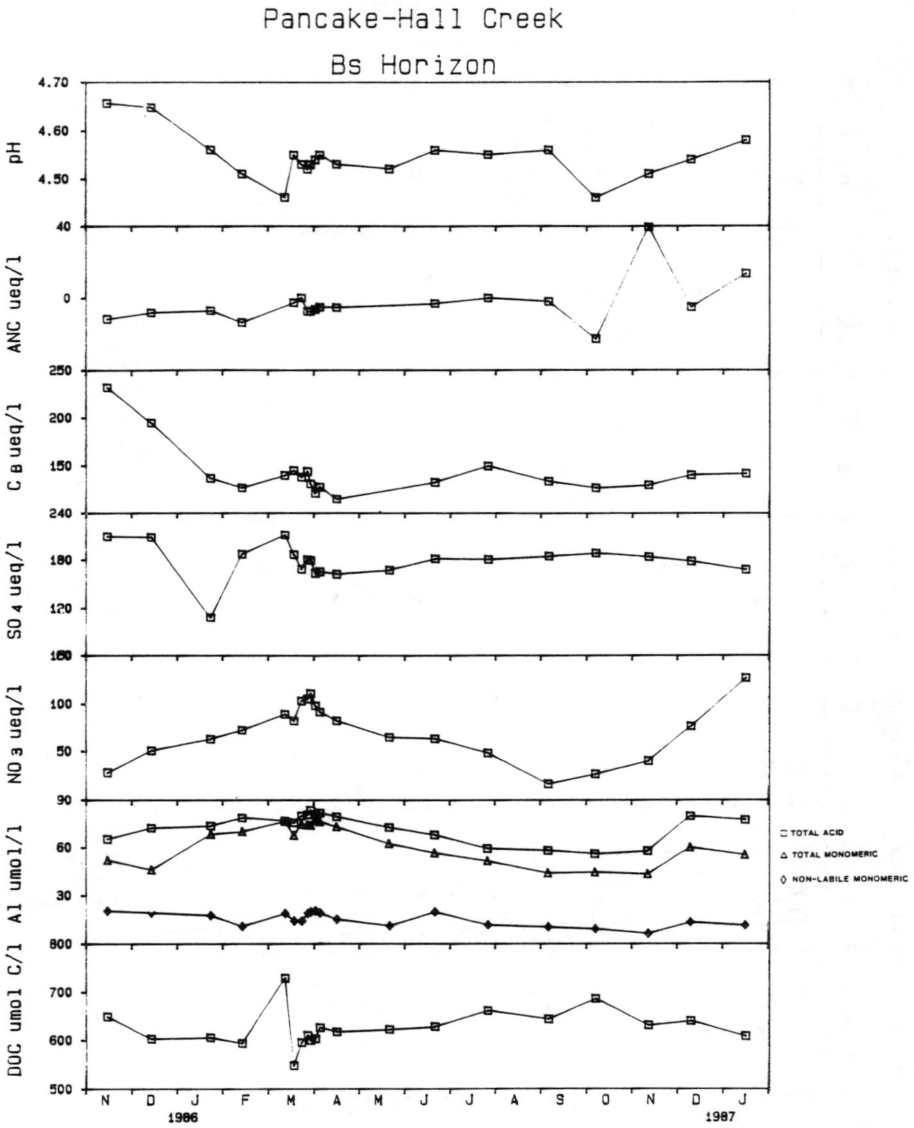

Figure 4. Temporal trends in the chemistry of Bs2 horizon leachate at Pancake-Hall Creek.

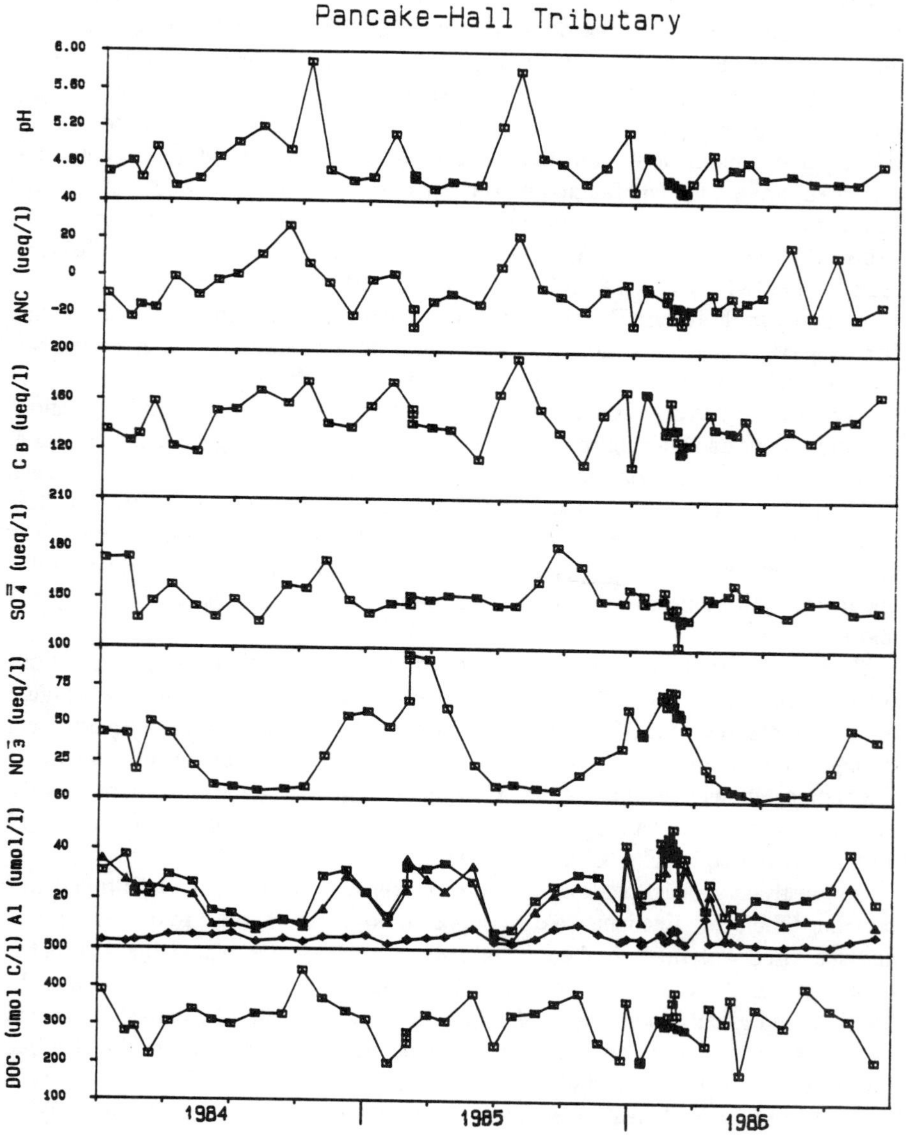

Figure 5. Temporal trends in the chemistry of streamwater at Pancake-Hall Creek (PAT).

Most water chemistry parameters exhibited little temporal variation in the Bs2 horizon or in groundwater. However, NO_3^- concentrations followed distinct seasonal trends in the mineral soil and groundwater solutions. Nitrate values increased from the fall through the winter season peaking at values of 100 μeq·L^{-1} during snowmelt and decreased through the summer season (Figure 4). Concentrations of Al were extremely high in Bs2 soil solutions (mean value 72 μmol·L^{-1}) and this Al was largely present as inorganic monomeric Al (Figure 4; Table 4). Changes in the concentration of inorganic monomeric Al coincided with seasonal variations in NO_3^-. Concentrations increased through the late fall and winter, peaking in the spring and decreased in the summer/fall. Aluminum concentrations were strongly correlated with variations in NO_3^- in both Bs2 horizon leachate ($Al^{n+} = 70 + 0.83\ NO_3^-$; in μeq·L^{-1}, $r^2 = 0.63$, $P < 0.001$) and groundwater ($Al^{n+} = 47 + 0.86\ NO_3^-$; in μeq·L^{-1}, $r^2 = 0.35$, $P < 0.001$).

A variety of short-term fluctuations in water chemistry were observed at the stream sampling site. Values of pH and ANC were lowest during the winter/spring (December-June) period, and increased to maximum values during the low flow summer period (July-September). Acidic conditions in winter/spring coincided with low concentrations of basic cations and high concentrations of NO_3^- (Figure 5). During the summer period, basic cation concentrations were elevated and NO_3^- concentrations were reduced, resulting in less acidic conditions. Although seasonal fluctuations in SO_4^{2-} concentrations occurred, they did not coincide with changes in ANC. For example, SO_4^{2-} decreased in February 1984 (from 171 to 134 μeq·L^{-1}), a period of low pH and ANC, and increased in August 1984 (from 135 to 158 μeq·L^{-1}), a period of relatively high pH and ANC.

Concentrations of monomeric Al exhibited seasonal variations; these fluctuations were almost entirely due to changes in the inorganic fraction of monomeric Al (Figure 5). Concentrations of DOC and organic monomeric Al in streamwater were constant throughout the study period. Changes in the concentration of inorganic monomeric Al were coincident with pH fluctuations. Concentrations of inorganic monomeric Al were also strongly correlated and apparently stoichiometric with changes in NO_3^- ($Al^{n+} = 19 + 1.00 \cdot NO_3^-$; in μeq·L^{-1}, $r^2 = 0.77$, $P < 0.001$).

Soil analysis showed patterns of free (non-silicate bound) Al in Oa, Bh and Bs2 horizons (Table 5). Exchangeable Al concentrations were highest in the Oa horizon and decreased with depth in the mineral soil. Pyrophosphate extractable Al was the dominant free Al form in all horizons. Concentrations were similar in Oa and Bh horizons and greatly increased in the Bs2 horizon. Aluminum obtained from acid oxalate extractions were comparable to pyrophosphate extractions, suggesting that free soil Al was largely associated with organic matter.[26]

Stream sediments also exhibited significant trends in concentrations of free Al (Table 6). Sediment values were similar to soil and pyrophosphate extractable Al was again the dominant fraction. No differences in KCl extractable Al were evident between the June and September samplings. However, the other Al fractions and pyrophosphate extractable organic carbon apparently exhibited temporal variations, with lower concentrations occurring in June following the spring high flow period and higher concentrations in September after the low flow period ($P < 0.01$; t-test).

Table 4. Speciation of Al (in μmol·L^{-1}) and saturation indices with respect to the solubility of natural $Al(OH)_3$ for drainage waters in Pancake-Hall Watershed

Parameter	Oa horizon	Bs2 horizon	groundwater	tributary
total Al	42 ± 15	72 ± 11	61 ± 21	33 ± 10
labile (inorganic) monomeric Al	17 ± 14	46 ± 12	34 ± 10	21 ± 10
non-labile (organic) monomeric Al	13 ± 7	15 ± 6	5 ± 2	6 ± 3
acid soluble Al	11 ± 6	11 ± 7	20 ± 14	6 ± 3
aquo Al	14 ± 12	38 ± 11	23 ± 7	16 ± 9
OH- complexed Al	1 ± 0.7	3 ± 0.9	4 ± 1.4	2 ± 0.5
F- complexed Al	2 ± 0.6	3 ± 0.5	5 ± 0.5	3 ± 0.5
SO_4^{2-} complexed Al	1 ± 0.7	2 ± 0.6	1 ± 0.5	1 ± 0.4
SI natural $Al(OH)_3$	-2.29 ± 0.64	-0.79 ± 0.28	-0.27 ± 0.28	-0.91 ± 0.25

Note: Values represent mean ± standard deviation.

To better evaluate the chemistry of Al, we determined the distribution of Al species and mineral phase saturation indices (SI) using ALCHEMI (Table 4). Results of the thermodynamic calculations indicate that aquo Al(Al^{3+}) was the predominant form of inorganic monomeric Al in the highly acidic drainage waters of Pancake-Hall Watershed. Temporal fluctuations in pH and inorganic monomeric Al resulted in relative changes in the speciation of Al, with F^- and OH^- complexed forms becoming more significant during the higher pH summer conditions. However, seasonal changes in the concentrations of inorganic monomeric Al occurred largely through variations in aquo Al, as concentrations of Al greatly exceeded concentrations of available ligands.

No clear pattern of mineral phase solubility control was evident in Pancake-Hall drainage waters. Organic horizon solutions were highly undersaturated with respect to natural $Al(OH)_3$ solubility (Table 4). Natural $Al(OH)_3$ SI values increased in mineral soil solutions, and approached saturation in groundwater samples. Streamwater natural $Al(OH)_3$ SI values were also somewhat undersaturated with respect to mineral phase equilibrium and similar to values determined for Bs2 soil solutions.

Table 5. Soil concentrations of free Al in $\mu mol \cdot g^{-1}$. Values represent mean ± standard deviation.

Horizon	KCl extractable Al	pyrophosphate extractable Al	oxalate extractable Al
Oa	10 ± 0.1	110 ± 14	83 ± 6
Bh	9 ± 0.8	120 ± 21	100 ± 1
Bs2	3 ± 0.6	360 ± 24	350 ± 4

Table 6. Chemical characteristics of stream sediments. Values represent mean ± standard deviation.

Date	KCl extractable Al	pyrophosphate extractable Al*	oxalate extractable Al*	pyrophosphate extractable C*
6/85	2.4 ± 0.2	110 ± 18	73 ± 12	890 ± 140
9/85	1.9 ± 0.8	240 ± 12	240 ± 20	1790 ±180

*indicates that concentrations are different ($P<0.01$) between the two dates as determined by t-test.

The production of CO_2 and subsequent reactions involving inorganic carbon may affect the acid-base status and regulate Al transfer within forest soil systems.[2] In the mineral soil at Pancake-Hall Watershed good agreement between measured partial pressures of CO_2 (pCO_2) and values calculated from solution pH, DIC and temperature using ALCHEMI were obtained (measured $pCO_2 = -0.2 \pm 1.07 \cdot$calculated pCO_2; $r^2 = 0.75$, $P < 0.001$). Although a strong correlation was also evident between measured and calculated gaseous CO_2 from the Oa horizon (measured $pCO_2 = -1.2 + 0.80 \cdot$calculated CO_2; $r^2 = 0.77$, $P < 0.001$), measured values were systematically lower than concentrations calculated from solution composition. This discrepancy was probably due to the difficulty in sampling soil gas in the Oa horizon. Apparently atmospheric gas, low in CO_2 ($pCO_2 = 10^{-3.5}$ atm), became entrained during the collection of Oa horizon gas, diluting the actual concentrations. Therefore, CO_2 calculated from solution measurements using Henry's law equilibrium appears to be more appropriate than measured values for the Oa horizon. Moreover, measurements of soil gas were not available for some sampling dates due to saturated conditions within the soil. As a result, calculated values of pCO_2 were used in the analysis of data from this study.

Concentrations of calculated gaseous CO_2 were similar in Oa (log $pCO_2 = -2.66 \pm 0.26$) and Bs2 (log $pCO_2 = -2.69 \pm 0.23$) horizons and slightly higher in groundwater (log $pCO_2 = -2.33 \pm 0.10$): with all values substantially in excess of atmospheric concentration (log $pCO_2 = -3.5$). Streamwater pCO_2 was considerably lower than soil and groundwater values (log $pCO_2 = -3.01 \pm 0.10$), suggesting that degassing of CO_2 occurs as water emerges from the terrestrial environment.

Temporal trends in drainage water inorganic carbon chemistry were also evident. Measured dissolved inorganic carbon (DIC) concentrations, as well as values of pCO_2 and HCO_3^- calculated from ALCHEMI all exhibited similar patterns. During the fall, concentrations were relatively high and decreased through the winter and spring, reaching a minimum after snowmelt. In the summer DIC, pCO_2 and HCO_3^- values increased, reaching maximum values during the late summer, while declining again during the fall.

Although soil and groundwater pCO_2 were greatly enriched relative to atmospheric values, inputs of this weak acid were not effective in mobilizing soil Al in Pancake-Hall Watershed. Calculated values of CO_2 mobilized Al were only evident in the Bs2 horizon, and those concentrations were extremely low (0.2 ± 0.9 $\mu mol \cdot L^{-1}$). At no time over the annual cycle did this process significantly contribute to soil Al release.

Using measurements of groundwater stage and stream discharge, we calibrated values of hydraulic conductivity for individual soil horizons for application of the ILWAS model. Hydraulic conductivity was very high in the Oa horizon (1.4 $cm \cdot sec^{-1}$) and the mineral soil (0.12 $cm \cdot sec^{-1}$), with this latter value representing the sum of matric and macropore hydraulic conductivity. The upper soil in Pancake-Hall Watershed is characterized by an abundance of root channels and stream hydrology is dominated by macropore flow. The hydraulic conductivity of the glacial till was much lower, estimated at 2.9×10^{-4} $cm \cdot sec^{-1}$.

Hydrologic analysis of the upper reach of Pancake-Hall Watershed using the ILWAS model indicates that most of the drainage water migrated laterally via macropores through the B horizon prior to entering the stream channel (Table 7). The upper reach of the watershed contains very shallow deposits of glacial till.[13] As a result, groundwater inputs from the glacial till contributed only a small fraction of the total stream discharge.

Variations in hydrologic flowpaths were evident over the annual cycle. For example, during spring 1986 (March, April, May), elevated flow associated with snowmelt contributed 43% of the total annual discharge (45.2 cm) prior to entering the stream channel. The predominant component of this flow was routed through the B horizon (43.6 cm) prior to entering the stream channel. In contrast, during June and July 1986 only 3.3 cm of water was exported from the catchment, with groundwater from the shallow glacial till contributing 27% (0.9 cm) of this discharge. At no time over the annual cycle was water derived from the glacial till an important component of the hydrologic budget of Pancake-Hall Watershed.

Solute flux calculations provided additional detail on the acid-base transformations occurring within the catchment (Table 7; Figure 6). The movement of water through the Oa horizon was characterized by elevated transport of SO_4^{2-}, NO_3^- and DOC relative to basic cations, causing a large negative flux of ANC to the mineral soil. Within the mineral soil, very little release of basic cations was evident (0.13 $keq \cdot ha^{-1} \cdot yr^{-1}$), while some release of SO_4^{2-} (0.68 $keq \cdot ha^{-1} \cdot yr^{-1}$) and retention of NO_3^- (0.14 $keq \cdot ha^{-1} \cdot yr^{-1}$) occurred. Although release of SO_4^{2-} exceeded the sum of basic cation release and NO_3^- retention (0.27 $keq \cdot ha^{-1} \cdot yr^{-1}$), there was considerable production of ANC within the mineral soil. The neutralization of negative ANC could largely be attributed to the retention of organic acids (DOC retention 5.7 $kmol \cdot ha^{-1} \cdot yr^{-1}$) coupled with the release of inorganic monomeric Al (0.24 $kmol \cdot ha^{-1} \cdot yr^{-1}$). Organic monomeric Al was apparently largely derived from the Oa horizon as little deposition/release occurred in the mineral soil (Figure 6). Because the movement of water was predominantly associated with lateral flow from the B horizon, transformations within the glacial till were not quantitatively important, although some retention of SO_4^{2-}, NO_3^- and DOC was evident.

A comparison of solute fluxes leaving the terrestrial environment (lateral flow from the Oa, Bs horizons and glacial till) with streamwater efflux yielded important information on processes regulating surface water Al concentrations (Table 7; Figure 6). Good agreement was obtained between terrestrial influx and stream efflux of basic cations, indicating relatively conservative transport through the near-stream/in-stream environment. However, moderate retention of SO_4^{2-} (19% of inputs), and marked retention of NO_3^- (50% of inputs) and DOC (48% of inputs) occurred in this region. These transformations all served to increase ANC. Nevertheless, a slight loss of ANC (0.07 $keq \cdot ha^{-1} \cdot yr^{-1}$) was evident when comparing terrestrial inputs (-0.06 $kmol \cdot ha^{-1} \cdot yr^{-1}$) with stream outputs (-0.14 $kmol \cdot ha^{-1} \cdot yr^{-1}$; Table 7). This destruction of ANC is qualitatively consistent with the retention of labile (inorganic) monomeric Al in the near-stream/in-stream region (0.26 $kmol \cdot ha^{-1} \cdot yr^{-1}$). Hydrolysis/retention of Al as $Al(OH)_3$ would consume 0.76 $keq \cdot ha^{-1} \cdot yr^{-1}$ of ANC. Apparently the acidity associated with Al retention exceeded the production of ANC due to anion (SO_4^{2-} + NO_3^-, totaling 0.60 $keq \cdot ha^{-1} \cdot yr^{-1}$; organic anions; unmeasured) retention, resulting in a slight net destruction of ANC. In addition, a marked retention of organic monomeric Al occurred in the near-stream/in-stream environment which was qualitatively consistent with the retention of DOC.

Table 7. Solute fluxes through compartments in Pancake-Hall Watershed (in keq·ha^{-1}·yr^{-1} except where indicated) for November 1985 to November 1986

	Water (cm)	ANC	CB	SO_4^{2-}	NO_{3-}	DOC (kmol·ha^{-1}·yr^{-1})
Oa Horizon						
percolation	112.0	-0.67	1.54	1.26	0.77	12.4
lateral outflow	0.1	0.00	0.00	0.0	0.0	0.0
B Horizon						
percolation	7.2	0.00	0.10	0.12	0.03	0.4
lateral outflow	100.2	-0.06	1.57	1.81	0.56	6.3
retention	0	-0.60	-0.13	-0.68	0.14	5.7
Till						
lateral outflow	5.5	0.00	0.11	0.11	0.02	0.1
retention	0.0	0.00	0.00	0.01	0.01	0.3
Stream						
outflow	105.9	-0.14	1.59	1.56	0.33	3.3
retention	0.0	0.07	0.09	0.36	0.29	3.1

Note: Fluxes include vertical percolation, lateral outflow and compartment retention. A negative value of retention within a compartment indicates solute release.

DISCUSSION

Results of this analysis suggest that a variety of processes influenced transport of Al within Pancake-Hall Watershed. Drainage waters were highly acidic due to elevated inputs of strong acids (H_2SO_4, HNO_3, organic acids) relative to the release of basic cations. Moreover, the organic horizon was an important source of solutes to drainage waters, including basic cations, DOC and Al. In particular, inputs of Al to the mineral soil and streamwater appeared to be predominantly derived from the organic horizon. Similar patterns have been reported for forest soils in the northeastern U.S.[31, 32] Soil analysis indicated elevated concentrations of Al in the Oa horizon, and this Al appears to be largely associated with organic matter (pyrophosphate extractable; Table 5). Aluminum may enter the organic horizon by a number of processes, including throughfall[31] and root and leaf litter, as well as mixing of mineral soil associated with physical disturbance (e.g. windthrown trees). Organic monomeric Al and acid soluble Al leached from the Oa horizon were probably largely organic forms of Al derived as a by-product of microbial degradation of organic matter.[31]

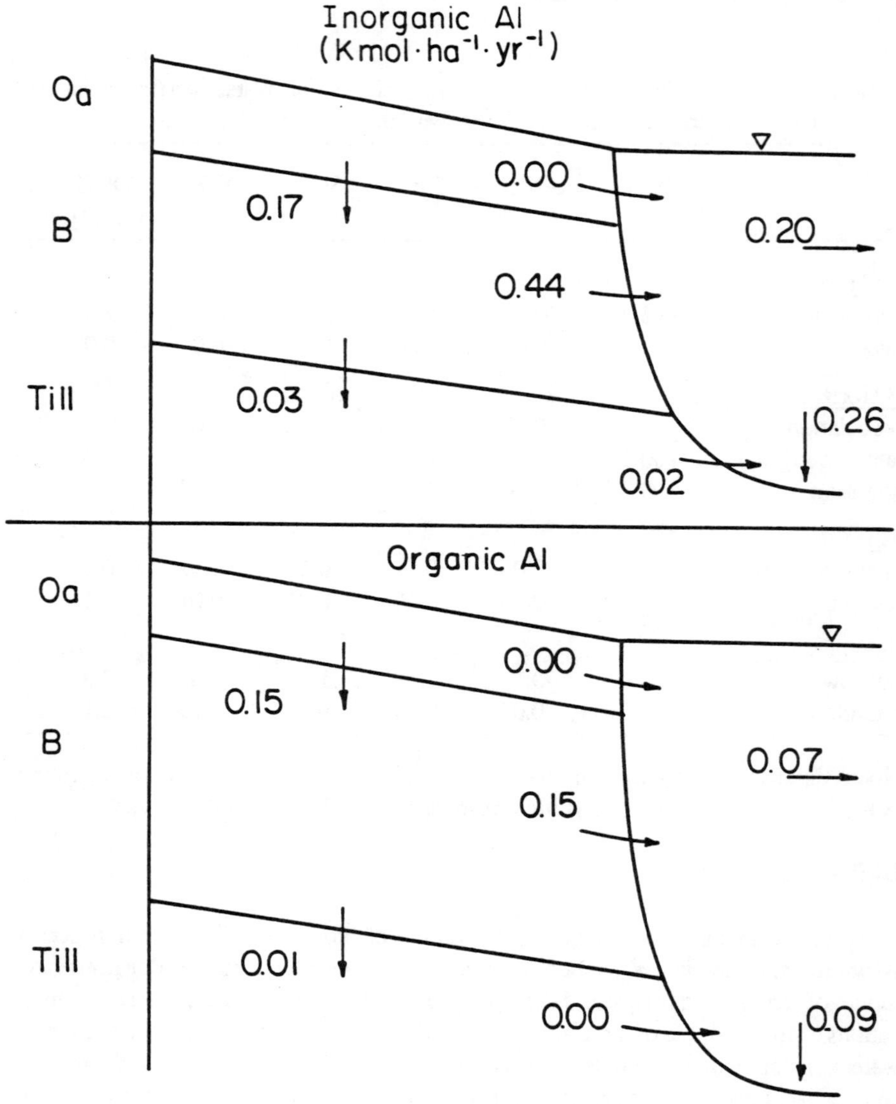

Figure 6. Fluxes of inorganic monomeric Al and organic monomeric Al within the watershed at Pancake-Hall Creek (in $kmol \cdot ha^{-1} \cdot yr^{-1}$).

Elevated concentrations of inorganic monomeric Al in the Oa horizon leachate have not been previously reported and are difficult to explain. Concentrations of this Al fraction were particularly high during the winter/spring period (Figure 3). Since other investigations of Al in forest soil solutions have not conducted detailed monitoring through the winter period,[31,32,33] this apparent discrepancy may simply reflect differences in sampling programs. The highly acidic conditions during winter/spring period would be conducive to the release of inorganic Al from organic horizon minerals.

The mineral soil was the predominant source of inorganic monomeric Al to Pancake-Hall drainage waters. Again, highly acidic soil solutions apparently facilitated the release of soil Al. Solutions were apparently not in equilibrium with any thermodynamically defined solid phase minerals. In general, mineral soil solutions were undersaturated with respect to the solubility of natural $Al(OH)_3$, although values approached saturation under low flow, higher pH summer conditions. It would appear that the high input of acids to the mineral soil exceeded the rate of Al release, so thermodynamic equilibrium with respect to mineral phase solubility was not attained. Our results suggest that solution concentrations of Al in mineral soil may be regulated by kinetically controlled dissolution.[29]

Mass transport analysis of organic monomeric Al indicates that this fraction of Al was relatively conservative within the mineral soil (Figure 6). This trend would appear to be inconsistent with the marked deposition of DOC (Table 7) and the apparent accumulation of soil organic Al (pyrophosphate extractable Al; Table 5). Unfortunately, the behavior of alumino-organic solutes within the mineral soil is not well understood. While its generally assumed that organically complexed Al is retained in the mineral soil during podzolization, Driscoll,[3,32] reported limited retention of these solutes in an acid-impacted catchment at the Hubbard Brook Experimental Forest in New Hampshire. The Al fractionation procedure used in this study is operationally defined, and a detailed characterization of alumino-organic solute is not available. However, given the contrasting patterns of Al leaching between pristine sites[3] and regions impacted by acidic deposition,[34] one might hypothesize that acidic conditions not only facilitate leaching losses of inorganic Al, but may also contribute to the mobile behavior of alumino-organic solutes through the soil profile to surface waters. Mobile transport of organic Al might be enhanced under acidic conditions if alumino-organic solutes exhibited a cationic charge. Regardless of the mechanism of transport, better analytical tools are needed to characterize Al-organic interactions if we are to improve our understanding of the biogeochemistry of this element.

Probably the most interesting aspect of our study was the contrasting patterns of soil and surface water Al. A comparison of solution concentrations (Table 4) and mass transport calculations (Figure 6) between terrestrial and surface waters strongly suggests retention of both inorganic monomeric Al and organic monomeric Al within the near-stream/in-stream environment. The marked loss of organic monomeric Al was consistent with the retention of DOC, and may be due to microbial or chemical immobilization. Similarly, the retention of inorganic monomeric Al could be due to a variety of biogeochemical processes. However, our assessment of CO_2-mediated mobilization of Al suggests that this process is not an important mechanism of Al transfer. Drainage waters in Pancake-Hall Watershed were highly acidic, restricting the hydration/deprotonation of CO_2 to H^+ and HCO_3^-. As a result, acidity derived from the dissociation of CO_2 was limited and did not significantly contribute to soil Al release.

Mass transport calculations suggest that marked retention of NO_3^- coincided with the deposition of inorganic monomeric Al in the near-stream/in-stream region. Note that seasonal fluctuations in NO_3^- were strongly correlated with concentrations of inorganic monomeric Al in Bs2 horizon solutions, groundwater and streamwater. Although concentrations of basic cations and SO_4^{2-} generally exceeded NO_3^- in drainage waters, concentrations of basic cations and SO_4^{2-} were relatively invariant while NO_3^- concentrations exhibited marked temporal and spatial fluctuations. As a result, variations in drainage water acidity and Al were closely coupled to the processes regulating NO_3^- concentrations. During the summer/fall period, NO_3^- concentrations were low due to biological retention. In contrast, limited biological activity during the winter/spring period[35] coupled with elevated snow inputs[36] resulted in high drainage water concentrations of NO_3^-. The apparent retention of NO_3^- in the hydrologic zone linking terrestrial waters and streamwater served to increase ANC and pH, which in turn facilitated the hydrolysis/retention of Al.

Unfortunately the exact location of Al retention cannot be determined from our study; although it appears to occur somewhere along the near-stream/in-stream region. The process of Al deposition in this zone has potentially important implications for stream ecosystems. Hydrolysis of Al facilitates the coagulation and removal of particulate matter and DOC,[10] and may scavenge phosphorus from water through direct precipitation or adsorption reactions.[9] Therefore, in-stream retention of Al may alter important nutrient cycles for stream ecosystems. Evidence for these processes may be available from stream sediment data (Table 6). Free Al fractions and pyrophosphate extractable carbon were elevated during the September sampling. This collection followed a low flow period when stream water was close to saturation with respect to natural $Al(OH)_3$ solubility (SI for June-September 1985 = -0.21 ± 0.19). High values of Al and carbon fractions in stream sediments might be the result of hydrolysis/coagulation/deposition processes.

Deposition of freshly precipitated Al within the stream bed may be an important source of mobile Al during episodic events.[37,38] During fall rains and spring snowmelt, Pancake-Hall Creek was highly acidic and undersaturated with respect to natural $Al(OH)_3$ solubility (e.g. SI for March-June 1985 = -1.04 ± 0.24). These conditions could allow stream-bed release of Al to the water column providing an important source of bioavailable Al. Low concentrations of stream sediment Al (Table 6) during the June sampling following snowmelt would appear to be consistent with this process.

CONCLUSIONS

Drainage waters from Pancake-Hall Watershed were highly acidic due to elevated inputs of H_2SO_4 and HNO_3 relative to the limited release of basic cations. These conditions facilitated the mobilization of soil Al. Alumino-organic solutes were largely derived from the soil organic horizon, while inorganic monomeric Al was released from the mineral soil and to a lesser extent from the organic horizon. Drainage water Al was largely in an inorganic monomeric form, with aquo Al (Al^{3+}) predominating. Mass transport calculations suggest that the near-stream/in-stream environment is a zone of Al retention. Organic monomeric Al deposition coincided with DOC retention, while NO_3^- retention appeared to facilitate the deposition of inorganic monomeric Al. Hydrolysis/deposition of Al along the terrestrial/aquatic interface may influence nutrient cycling and could serve as an important source of mobile Al during episodic acidification.

ACKNOWLEDGMENTS

This research was supported as part of the ALBIOS and ALSS studies by the Electric Power Research Institute and with a Presidential Young Investigator (PYI) award to CTD. Funding for the PYI award is provided by the National Science Foundation with matching funds from the Empire State Electric Energy Research Corporation and the Electric Power Research Institute. We gratefully acknowledge the precipitation data provided by UAPSP which was essential for our analysis. Finally, we would like to thank C. Cronan, C. Cosentini, M. Hanna, R. Munson, S. Gherini, and Dr. and Mrs. Steinhauser for their support of this study.

REFERENCES

1. Bloomfield, C. "The possible significance of polyphenols in soil formation," *J. Sci. Food Agric.* 8:389-392 (1957).

2. Reuss, J.O. and D.W. Johnson. "Effect of soil processes on the acidification of water by acid deposition," *J. Environ. Qual.* 14:26-31 (1985).

3. Ugolini, F.C., R. Minden, H. Dawson and J. Zachara. "An example of soil processes in the *Abies amabilis* zone of Central Cascades, Washington," *Soil Sci.*, 124:291-302 (1977).

4. Cronan, C.S. and C.L. Schofield. "Aluminum leachate response to acid precipitation, effects on high-elevation watersheds in the Northeast," *Science*, 204:304-306 (1979).

5. Baker, J.P. "Effects on fish of metals associated with acidification," in *Acid Rain/Fisheries*, R.E. Johnson, Ed. (Bethesda, MD: American Fisheries Soc., 1982), p. 165.

6. Driscoll, C.T. "Aluminum in acidic surface waters, chemistry transport and effects," *Environ. Health Perspectives* 63:93-104 (1985).

7. Driscoll, C.T., J.P. Baker, J.J. Bisogni and C.L. Schofield. "Aluminum speciation in dilute acidified surface waters in the Adirondack region of New York State," in *Acid Precipitation, Geological Aspects*, O.R. Bricker, Ed. (Boston, Butterworth, 1984), p. 55.

8. Driscoll, C.T. and J.J. Bisogni. "Weak acid/base systems in dilute acidified lakes and streams in the Adirondack region of New York State," in *Modeling of Total Acid Precipitation Impacts*, J.L. Schnoor, ed. (Boston, Butterworth, 1984), p. 53.

9. Dickson, W. "Some effects of the acidification of Swedish lakes," *Verh. Ver. Limnol.* 20:851-856 (1978).

10. Effler, S.W., G.C. Schafran and C.T. Driscoll. "Partitioning light attenuation in an acidic lake," *Can. J. Fish. Aquat. Sci.* 42:1707-1711 (1985).

11. Foy, C.D. "Effects of aluminum on plant growth," in *The Plant Root and Its Environment*, E.W. Carson, Ed. (Charlottesville, VA, Univ. Press of Virginia, 1974) p. 601.

12. Driscoll, C.T., J.P. Baker, J.J. Bisogni and C.L. Schofield."Effect of aluminum speciation on fish in dilute acidified waters," *Nature* 284:161-164 (1980).

13. Newton, R.M. "Surficial geology and hydrology of the ALBIOS watersheds," in *ALBIOS Annual Report for 1986*, C.S. Cronan, Ed. (Orono, ME, University of Maine, 1986).

14. April, R. "Summary of results Colgate University," in *ALBIOS Annual Report for 1986*, C.S. Cronan, Ed. (Orono, ME, University of Maine, 1986).

15. Cronan, C.S. "Biogeochemical influence of vegetation and soil in the ILWAS watersheds and factors controlling solution aluminum," in *ALBIOS Annual report for 1986*, C.S. Cronan, Ed. (Orono, ME: University of Maine, 1986).

16. Driscoll, C.T. "A procedure for the fractionation of aqueous aluminum in dilute acidic waters," *Int. J. Environ. Anal. Chem.* 16:267-283 (1984).

17. Small, H., T.S. Stevens, W.C. Bauman. "Novel ion exchange chromatographic method using conductimetric detection," *Anal. Chem.* 47:1801-1809 (1975).

18. U.S. Environmental Protection Agency. *Methods for Chemical Analysis of Water and Wastes*, EPA-600/4-79-020 (Cincinnati OH: Environmental Monitoring and Support Laboratory, 1983)

19. Slavin, W. *Atomic Absorption Spectroscopy* (New York: John Wiley Interscience, 1968.

20. *Standard Methods for the Examination of Water and Wastewater*, 16th Ed. (New York: American Public Health Association, 1985).

21. Gran, G. "Determination of the equivalence point in potentiometric titrations," *International Congress on Analytical Chemistry*, 77:661-671 (1952).

22. Dorhmann. *Total Organic Carbon Systems Manual*, (Santa Clara, CA, 1984).

23. Orion Instruction Manual. *Fluoride Electrodes*, (Cambridgs, MA: Orion Research Inc., 1976).

24. Barnes, R.B. "The determination of specific forms of aluminum in natural water," *Chem. Geol.* 15:177-191 (1976).

25. USDA. *Soil Survey Laboratory Methods and Procedures for Collecting Soil Samples*, Soil Survey Investigations Report No. 1 (Washington D.C., U.S. Gov. Printing Office, 1972).

26. Farmer, V.C., J.D. Russell and B.F.L. Smith. "Extraction of inorganic forms of translocated Al, Fe and Si from a podzol B horizon," *J. Soil Sci.* 34:571-576 (1983).

27. Schecher, W.D. and C.T. Driscoll. "An evaluation of uncertainity associated with aluminum equilibrium calculations," *Water Resour. Res.* 23:525-534 (1987).

28. May, H.M., P.A. Helmke and M.L. Jackson. "Gibbsite solubility and thermodynamic properties of hydroxy-aluminum ions in aqueous solution at 25 C." *Geochim. Cosmochim. Acta*, 43:861-868 (1979).

29. Gherini, S.A., L. Mok, R.J.M. Hudson, G.F. Davis, C.W. Chen and R.A.Goldstein. "The ILWAS model: formulation and application," *Water Air Soil Pollut.* 26:425-459 (1985).

30. Chen, C.W., J.D. Dean, S.A. Gherini and R.A. Goldstein, "Acid rain model, hydrological module." *J. Environ. Engr. Div. ASCE*, 108:455-472 (1982).

31. David, M.B. and C.T. Driscoll. "Aluminum speciation and equilibria in soil solutions of a Haplorthod in the Adirondack Mountains, New York," *Geoderma* 33:297-318 (1984).

32. Driscoll, C.T., N. van Breemen and J. Mulder. "Aluminum chemistry in a forested Spodosol." *Soil Sci. Soc. Am. J.* 49:437-44 (1985).

33. Cronan, C.S. "Biogeochemical influence of vegetation and soils in the ILWAS watersheds," *Water Air Soil Pollut.* 26:355-371 (1985).

34. Driscoll, C.T., N.M. Johnson, G.E. Likens and M.C. Feller. "The effects of acidic deposition on stream water chemistry: a comparison between Hubbard Brook, New Hampshire and Jamieson Creek, British Columbia," *Water Resour. Res.* 24:195-200 (1988).

35. Vitousek, P.M. "The regulation of element concentrations in mountain streams in the northeastern United States," *Ecol. Monogr.* 47:65-87 (1977).

36. Galloway, J.N., C.L. Schofield, G.R. Hendrey, N.E. Peters and A.H. Johannes. "Sources of acidity in three lakes acidified during snowmelt," in *Ecological Impact of Acid Precipitation*, D. Drablos and A. Tollan, Eds. (Oslo, SNSF Project, 1980) p. 264.

37. Hall, R.J., G.E. Likens, S.B. Fiance and G.R. Hendrey. "Experimental Acidification of a stream in the Hubbard Brook Experimental Forest, New Hampshire," *Ecology* 61:976-989 (1980).

38. Norton, S.A. and A. Henriksen. "The importance of CO_2 in evaluation of effects of acidic deposition," *Vatten.* 4:346-354 (1983).

CHAPTER 6

HOW ALUMINUM LEVELS IN SUBSURFACE DRINKING WATER SUPPLIES IN CANADA CAN BE USED TO PREDICT POSSIBLE IMPACT BY ACIDIC DEPOSITION

J. C. Méranger

INTRODUCTION

The impact of acid rain on the chemical content of surface waters is well documented, and studies have shown that reduction in pH, increased levels of nitrates and sulfates and aluminum have been observed.[1-6] However, very little information is available on the impact of acid rain on groundwater supplies in Canada.

Recently, some 500 surface and groundwater samples selected from three areas of Canada were analyzed for major and trace elements. The sampling areas were chosen to represent a range of acidic deposition and a model was proposed to study the impact of this acidic deposition on the quality of surface and groundwater supplies in these areas.[7-8] A field study on Chelex-100 extractable aluminum was also conducted in Ontario and Nova Scotia.[9]

MATERIALS AND METHODS

Description of the Model[6]

Three modes of water flow were identified in this model: surface water flow (SWF), interfacial water flow (IWF), and groundwater flow (GWF). It was assumed that the water chemistry in each flow mode was controlled by the water pH and by the minerals present in rocks, glacial till and soils. The model also assumed a quasi-thermodynamic equilibrium between the boundary minerals and the water. As a result of increases in SO_4^{2-} and NO_3^- and decreasing pH, the equilibrium between the three flow modes was displaced in the order SWF, IWF, and GWF, and over a long period of time the GWF will re-equilibrate with the contaminated SWF and IWF.

Description of Site Areas

The site areas chosen were La Ronge, Saskatchewan, Sioux Lookout, Ontario and Shelburne, Nova Scotia, having resident populations of approximately 4000, 3200 and 2500, respectively. The areas had similar geologies characterized by sandy gravel overburden and granitic bedrock which is known to have a low buffering capacity and is therefore sensitive to acid rain precipitation. They were well-characterized geochemically so that well-defined samples of each water flow regime (SWF, IWF, GWF) could be obtained from each sampling. They were also chosen to represent a gradient of acidic deposition, with La Ronge, Saskatchewan being the least impacted and chosen as the control area and Sioux Lookout, Ontario and Shelburne, Nova Scotia being the most heavily impacted (Figure 1).

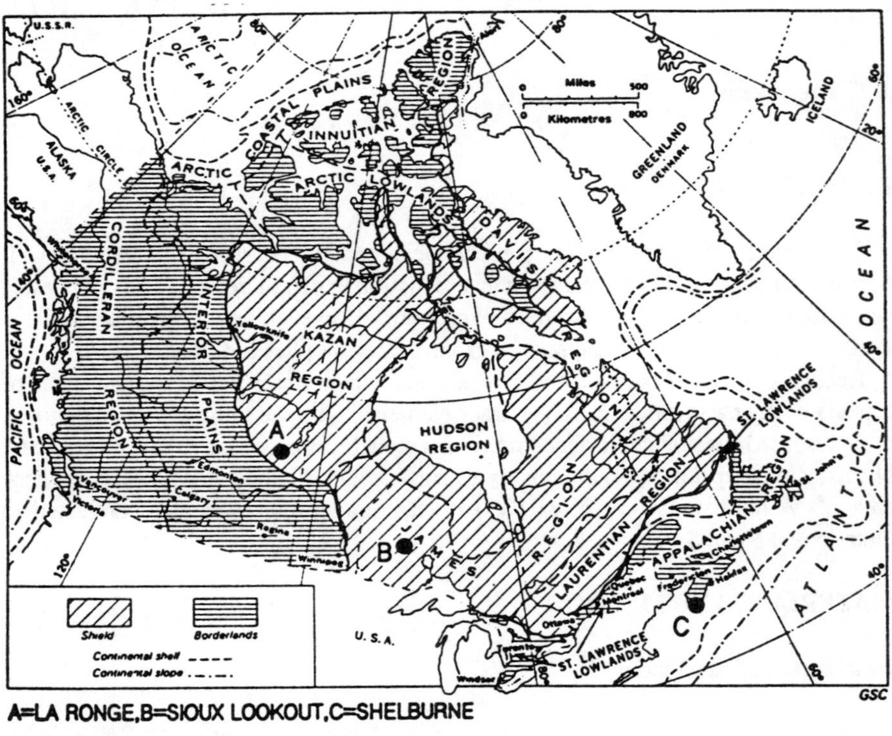

Figure 1. Location of three sampling sites within the Canadian Shield.

Description of Thermodynamic Equilibria

The model chosen for this study, WATEQF, had originally been used for hydrogeochemical mineral exploration.[8,10] Models for these types of studies have been constructed based on an ion association theory, or on the mixed electrolyte theory.[7] Within the ion association framework, the species distribution can be solved by using either the equilibrium approach or the Gibbs free energy approach. Using either approach there is a requirement that the computed sum of free and derived species be equal to the analyzed total concentration. The equilibrium constant approach requires simultaneous solution of a number of non-linear equations, derived by substituting the mass action expressions into the mass balance conditions. Since there are more reliable, available equilibrium constants than free energy values, this approach was used in the proposed model. The mathematical problem for both approaches involves the solution of a large set of non-linear equations. In general this problem can be solved using an iteration procedure since it is not feasible to solve simultaneously a large number of such equations in exact form. The program used as the basic algorithm for this study included the following desirable features: major anion and cation analyses of the water as input data, a representative number (approximately 100) of aqueous species in the equilibrium thermodynamic calculations and major redox species in the equilibrium calculations. Since water samples were collected from a wide variety of environments (surface to deep groundwater) and in different seasons, temperature correction is also a prerequisite. From a computational point of view, convergence by the method of continued fraction was preferred. Since WATEQF satisfied the above requirements, had been used for a number of hydrogeochemical studies[8,10] and had also been used by our contractor for previous hydrogeochemical exploration purposes, it was chosen as the basis for the hydrogeochemical equilibrium thermodynamic calculations in this study.

The program used consists of a main program and five subroutines as follows:

a. Recalculates all temperature dependent data to the temperature of the water sample.
b. Initializes values of individual species for the iterative Mass-action/Mass-balance loop.
c. Calculates activity coefficients and solves the Mass-action/Mass-balance equations for the species concerned.
d. Prints the results calculated from the aqueous model and stores relevant information for statistics and computer plotting.
e. Prints and stores the thermodynamic saturation state of the water (IAP ion activity products) with respect to the 101 minerals considered by the program.

In order to compare the apparent IAP with the thermodynamic equilibrium constant, the saturation index (SI) can be used,[11] where a negative value of SI indicates that the observed product of the dissolved ions in water is less than that which should prevail at equilibrium, and the water is thus thermodynamically capable of dissolving the mineral in question. Since kinetics are not taken into consideration in the calculations, a negative value only indicates that the water is capable and not that this dissolution is actually taking place. Similarly a mineral with an SI value of zero indicates that the water and solid are thermodynamically at equilibrium, and a positive value suggests that the water is supersaturated with respect to the mineral of interest. The SI is of critical importance to this study since it allows comparison between thermodynamic water/mineral equilibria for each water flow mode (SWF, IWF, GWF) and therefore allows direct testing and evaluation of the proposed model.

When selecting areas of study to determine the possible impact of acid deposition on water supplies the importance of having similar geology (bedrock and overburden) in each of the areas cannot be emphasized enough. As indicated earlier, the areas chosen for study must be well characterized geochemically so that well-defined samples of SWF, IWF, and GWF can be obtained for each of the areas studied. The field water samples were obtained by experienced geochemists who had previously conducted hydrogeological surveys in the three areas chosen for the modeling studies.

Sample Preparation

Samples of water were obtained in the fall and spring from each flow mode (25 from each of SWF, IWF, and GWF) in the three site areas and analyzed for major elements (Al, Ca, Fe, K, Mg, Na, Mn, Si), minor elements (Ag, Ba, Be, Cd, Co, Cr, Cu, Mo, Ni, P, Pb, Sr, Th, Ti, V, Zn, Zr), anions (F^-, NO_2^-, NO_3^-, SO_4^{2-}), pH, alkalinity and hardness. Samples were collected in one litre acid-washed PVC bottles and separated into two 500-mL aliquots. One aliquot was filtered through a 0.4-μM polycarbonate filter acidified with 0.1% high purity grade nitric acid and used for major and trace element determinations; the other unfiltered portion was used for anion determinations. All field samples were stored at 4°C prior to shipping and analysis.

Analytical Procedures

Major cations and trace elements were determined by inductively coupled plasma emission spectrometry using a 32-channel Applied Research Laboratory model QA-137. The induced current supplied by a 1500 watt radio frequency source ionizes the argon gas stream to provide a plasma with temperatures in excess of 7500K. Sample solutions were aspirated at an uptake rate of 2-3 mL/min following preconcentration by evaporating to dryness on a medium temperature sand bath and dilution with 0.1% nitric acid to obtain a 20-fold concentration factor.

Arsenic and selenium were determined using heated quartz atomization following borohydride reduction. Cadmium, zinc, lead and copper were determined by heated graphite atomization using a Perkin-Elmer HGA 2100 graphite furnace, and mercury was determined by cold vapor atomic absorption.

Anion concentrations were determined by ion chromatography using a Dionex model 2010 equipped with an AS4 anion column, an AG4 guard column, and an anion fiber suppressor using 0.0028 M $NaHCO_3$/0.0022 M Na_2CO_3 as eluent.

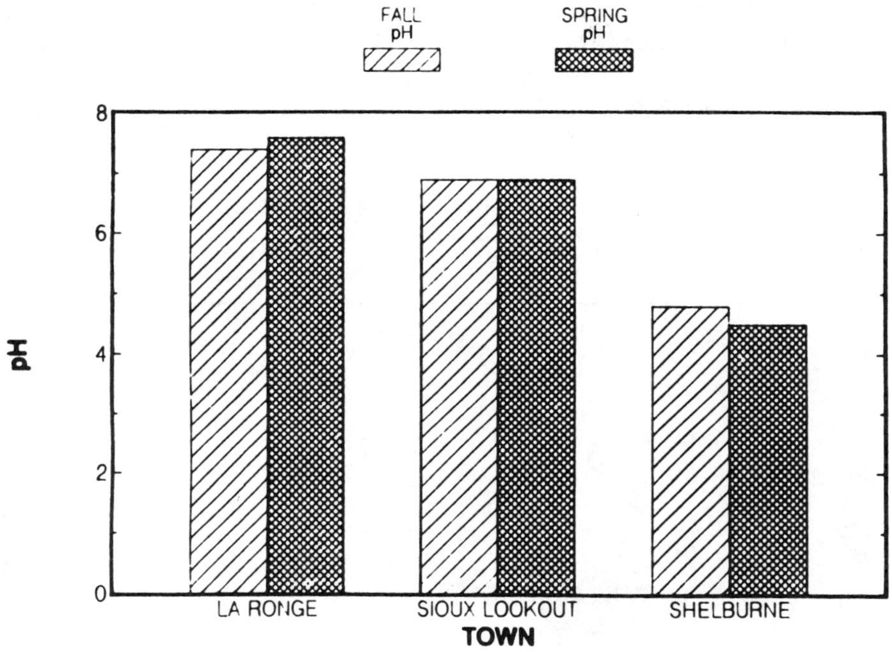

Figure 2. Average pH levels of SWF at three Canadian sampling sites.

Alkalinity and pH measurements were determined using an Orion model 407 pH meter.

Quality Control

Blanks, repeat samples and certified samples were run with a frequency of 10% of the total number of samples analyzed. The overall analytical precision was below 15% for most elements and the blank values were typically less than 0.01 mg/L. The analytical precision for aluminum determinations was generally less than 5% and blank values were 0.5 μg Al/L, which is at the detection limit of the pre-concentration inductively coupled plasma optical emission analytical method employed.[6]

Statistical Analysis and Interpretation

The statistical method used in this study to compare the means and variances is one-way analysis of variance (ANOVA).[6] The ANOVA null hypothesis assumes that the two subpopulations were drawn from parent populations. This hypothesis was tested using the F statistic at both the 0.01 and 0.05 uncertainty level to verify whether the

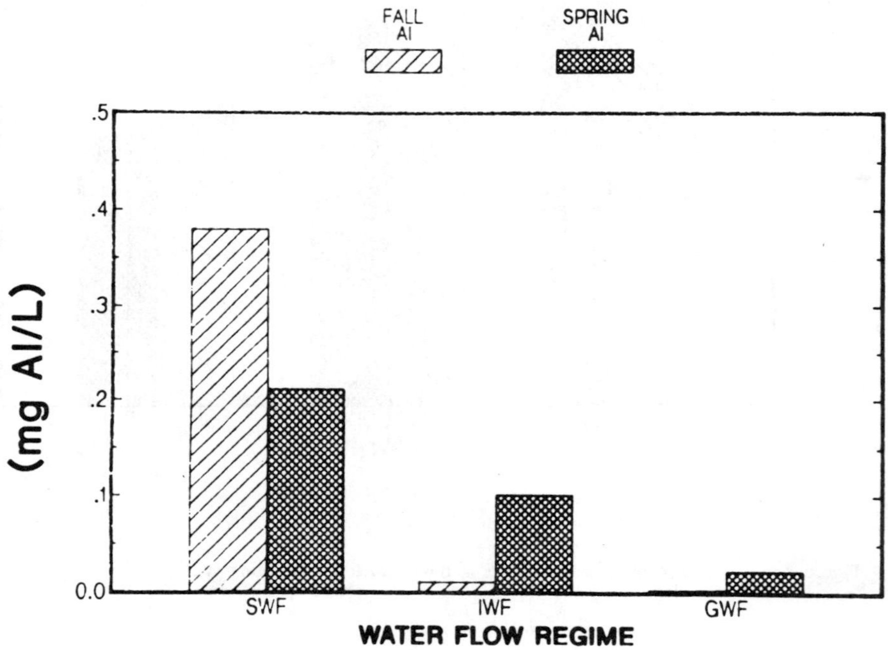

Figure 3. Average aluminum levels in Shelburne, Nova Scotia water flow regimes.

within population variance is sufficient to explain differences between two subpopulations. In order to approach a gaussian distribution, histograms of anion and cation subpopulations were drawn using untransformed and logarithmically (base 10) transformed data. Data values of 2.5 standard deviations below and above the mean were iteratively removed until the iterative estimate of the mean differed by less than 0.001. This technique also tends to normalize otherwise highly skewed or high variance sample populations. In practice, the number of points excluded from each subpopulation varied from 0 to 5. In order to allow for differences in subpopulation numbers the sum of squares were corrected for differences in degrees of freedom between the subpopulations.

Using the ANOVA statistical analysis method outlined above comparisons were then done between water flows, sampling sites and spring and fall sampling periods.

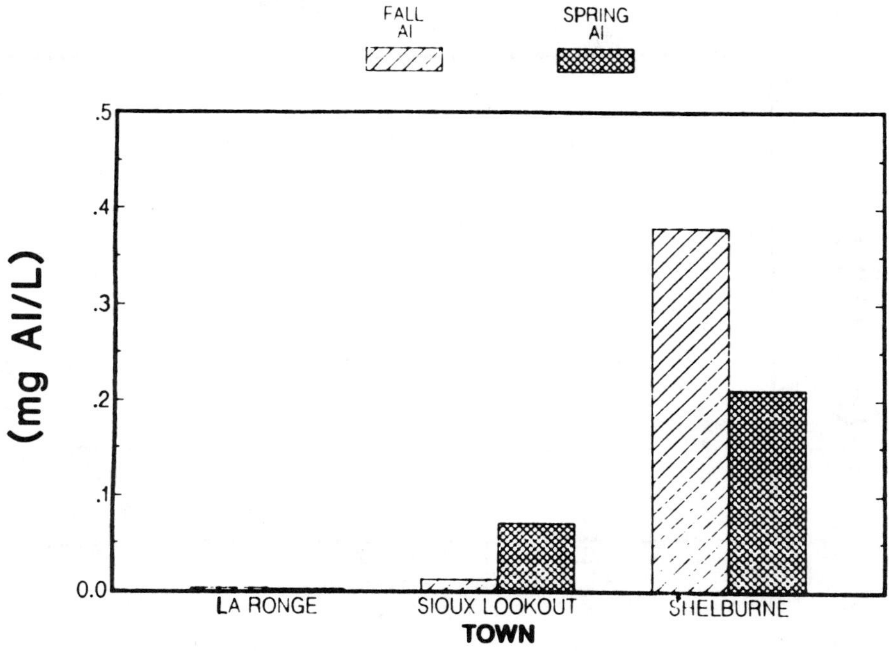

Figure 4. Average aluminum levels in SWF at three Canadian sampling sites.

DISCUSSION

Generally, the degree of nonequilibrium increased in the order: La Ronge < Sioux Lookout < Shelburne. This observation was consistent with the pH measurements with the lower pH values recorded in Shelburne SWF as shown in Figure 2. There was no significant difference in the pH of the GWF between site areas and only slightly lower pH values for Shelburne IWF. Interfacial waters IWF at Shelburne also showed the highest levels of Cl^-, NO_3^-, SO_4^{2-}, Al, Cu, B, and Na.

It was previously reported that high acidity was responsible for the increased leaching of copper, lead, and zinc from private wells.[8] Further analysis of the data showed that aluminum may be used as an index to predict the potential impact of acid rain on subsurface drinking water supplies. The aluminum levels found in Shelburne, the area most affected by acid rain,[12] showed a dramatic increase in subsurface IWF waters between the fall sampling period and the post-snowmelt period, where maximum levels up to 0.10 mg/L were recorded for IWF (Figure 3) compared with maximum levels of up to 0.009 mg/L for the control area La Ronge (Figure 5).

Although the fall SWF water showed aluminum levels up to 0.38 mg/L, only traces were observed in IWF and GWF, and maximum levels of 0.01 and 0.002 mg/L,

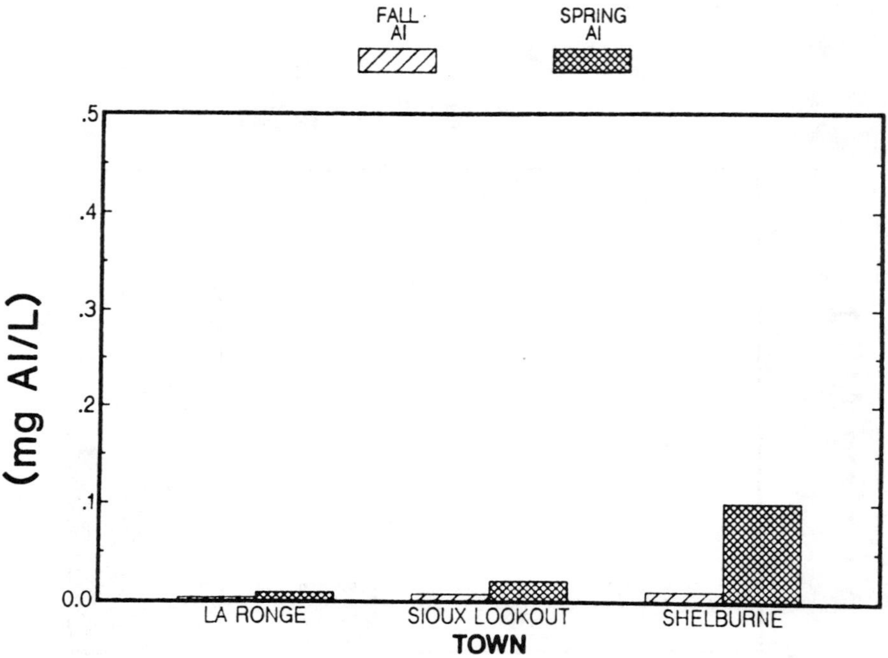

Figure 5. Average aluminum levels in IWF at three Canadian sampling sites.

respectively, were recorded. By contrast spring IWF and GWF samples had levels up to 0.01 and 0.02 mg/L, respectively. It is well known that the spring post-snowmelt period is the period with the maximum short-term acidic deposition impact on surface water supplies.[13,14] Therefore the duplicate sampling of shallow wells in the spring and fall presents the ideal situation for the model since each area concerned can serve as its own control because overburden and bedrock geology remain constant during the sampling period. Furthermore, comparisons of the SWF and IWF from each site area (Figures 4 and 5) showed decreasing levels of aluminum in the order Shelburne, Sioux Lookout and La Ronge, the control area with minimal impact from acidic deposition.[12] Similar trends were not observed in the GWF from each of these site areas.

CONCLUSION

The release of aluminum into subsurface water supplies as a result of mobilization by continued acidic deposition impact is perhaps an early warning to the eventual contamination of water supplies by other toxic elements which could eventually reach the lower water table levels. More recently, a field study using an on-site Chelex-100 resin extraction technique has been completed in the Shelburne area to further refine this approach.[9] The preliminary results of this study indicate that monomeric aluminum accounts for a significant portion of the total aluminum present in shallow overburden wells whose pH values fall below 6.0, while greater amounts of insoluble aluminum are present in more alkaline waters. The higher levels of aluminum in low pH, shallow groundwater most probably reflect more rapid infiltration of acid surface water and acidic precipitation through porous overburden. A more extensive field survey should be carried out to confirm the relationship between monomeric aluminum, pH, and amount of acidic deposition in shallow overburden wells.

In conclusion, based on total aluminum levels and the preliminary results of Chelex-100 extractable levels, there is increasing evidence linking acidic deposition to changes in the water quality of subsurface water supplies in Canada. It is further postulated that aluminum may be used in the future as an index to predict changes in the overall drinking water quality of areas affected by acidic deposition.

REFERENCES

1. Acidification in the Canadian Aquatic Environment, *National Research Council NRCC 18475* (1981).

2. The Case Against Acid Rain, *Saskatchewan Environment Advisory Council* (1980).

3. Johnson, N. M., C. T. Driscoll, J. S. Eaton, G. E. Likens, and W. H. McDowell. *Geochim. et Cosmochim. Acta*; 45 (9):1421-1437 (1981).

4. Andelman, J. B. and J. R. Miller. *WHO Water Quality Bulletin*, Part 1; 11 No. 1, 19-25 (1986).

5. Henriksen, A. and L. A. Kirkhusmo. *WHO Water Quality Bulletin*, Part 1; 11 No. 1, 34-36 (1986).

6. Méranger, J. C. and D. R. Gladwell. *WHO Water Quality Bulletin*, Part 1; 11 No. 1, 26-33 (1986).

7. Méranger, J. C., D. R. Gladwell, and R. E. Lett. *WHO Water Quality Bulletin*, Part 2; 11 No. 3, 152-159 (1986).

8. Méranger, J. C., D. R. Gladwell, and R. E. Lett. *WHO Water Quality Bulletin*, Part 3; 11 No. 4, 179-186 (1986).

9. Méranger, J. C. and R. E. Lett. "Heavy Metals in the Hydrological Cycle," Publications Division, Selper, Ltd, London W4 IBN, UK 47-50 (1988).

10. Plummer, L. N., B. F. Jones, and A. H. Truesdell. *U.S. Geological Survey Resources Inventory 76-13* (1978).

11. Runnells, D. B. and R. D. Lindberg. *Geochem. Explor. 15* (1-3):37-50 (1980).

12. United States-Canada Memorandum of Intent on Transboundary Air Pollution, Impact Assessment Work Group 1, 2-19 (1983).

13. Jeffries, D. S., C. M. Cox, and P. J. Dillon. *J. Fish. Res. Board Canada* 38, 640 (1979).

14. Méranger, J. C., T. R. Khan, C. Vairo, R. Jackson, and W. C. Li. *Intern. J. Environ. Anal. Chem.* 15:185-212 (1983).

CHAPTER 7

CHEMICAL SPECIATION AND PLANT TOXICITY OF AQUEOUS ALUMINUM

D. R. Parker, L. W. Zelazny, T. B. Kinraide

INTRODUCTION

Although it has been some 70 years since Hartwell and Pember[1] first postulated that soluble aluminum (Al) is a major limitation to plant growth in acid soils, the relationships between the aqueous chemistry of Al and phytotoxicity are still not fully understood. That efforts to relate toxicity to levels of Al in soil solution have met with variable degrees of success[2,3,4,5,6,7,8] may be partly ascribable to an insufficient understanding of the importance of speciation to biological toxicity. Recent concerns about the possible effects of acidic deposition on the biogeochemistry of soil Al has broadened the scope of interest in Al toxicity to both terrestrial and aquatic organisms.[9,10,11]

The chemistry of Al is quite complex, its high ionic charge and small crystalline radius combining to yield a level of reactivity unmatched by any other soluble metal found in soil solution, and a variety of hydrolysis, complexation, and chelation reactions have been reported. As solution pH is raised above 4.0, Al^{3+} gives way to the mononuclear species $AlOH^{2+}$, $Al(OH)_2^+$, $Al(OH)_3^0$, and $Al(OH)_4^-$ (waters of hydration are omitted here and elsewhere for clarity).[12] In addition, larger polynuclear hydroxy-Al species may form as metastable intermediates in the precipitation of solid phase $Al(OH)_3$.[13,14] Inorganic ligands such as sulfate and fluoride, as well as a host of organic compounds, may also form soluble complexes with Al.[15,16,17] In general, polynuclear and chelated forms of Al have been considered nontoxic to plants, but agreement is not uniform on this point. At the initiation of our studies, information on the relative phytotoxicities of inorganic mononuclear complexes, including the hydrolysis products, was either sparse or conflicting.

While the exact molecular basis for Al toxicity remains elusive,[18,19] the symptoms are well known. Roots of Al-affected plants are short, stubby, distorted, and discolored.[18,20] The overall stunting of the root system greatly limits the plants' ability to explore the soil for needed water and nutrients.[18] Consequently, Al phytotoxicity is often studied in the laboratory using solution culture techniques wherein root elongation is used as the primary measure of Al stress.[20] In addition to the types of studies reported here, this approach is useful for examining the basic physiology of Al toxicity and for rapidly screening genetic material for differential Al tolerance.[21,22]

In nutrient solution studies, amelioration of Al toxicity has been achieved by the addition of various solutes to the media. The effective ameliorative treatments can, in our view, be divided into two broad categories. The first is the addition of Al-complexing ligands, including hydroxyl ions, which may reduce the concentration of toxic Al species. The second is the addition of simple salts of the alkaline or alkaline earth metals (e.g., $CaCl_2$), wherein the anion is not a significant ligand for Al. In such cases, the concentrations of toxic Al are not reduced, but activities are, and there are both theoretical and empirical bases for using the latter in studies of ionic toxicities.[23,24] Alternatively, salt additions may relieve Al toxicity through a direct physiological effect of the cation, perhaps, as an example, through competition for external binding sites on root cells.[25]

Prior to our investigations, calcium was a well-known ameliorant for Al toxicity, and was demonstrated to be effective for several crops.[26,27,28,29] The implication of these studies was that Ca amelioration resulted from direct physiological effects, and reductions in the activity of toxic Al via increases in ionic strength were not always considered by the authors. In contrast, several studies have attempted to demonstrate that, upon varying solution ionic strength, phytotoxicity could be explained solely by variations in the activities of the toxic Al species.[3,30,31] Recent studies have suggested that both mechanisms may be operative, at least in the case of Ca.[32,33] The relative efficacy of other cations in ameliorating Al toxicity has been little studied. Magnesium was shown to effectively ameliorate Al toxicity and was qualitatively as effective as Ca.[21,26,29] Limited data available for monovalent cations suggests that K and Na can ameliorate toxicity, but much less so than Ca and Mg,[21,26] and it is uncertain whether this entails more than ionic-strength effects.

Comprehensive studies of the toxicities of inorganic Al complexes are few. The $AlSO_4^+$ species has been assumed to be both nontoxic[2,4] and toxic,[8,16,30,32,34] but no justification has been offered for either view. Our studies,[35] summarized here, and a corroborating study[36] both indicate that $AlSO_4^+$ is nontoxic. Similarly, few reports concerning the toxicity of Al-fluoride complexes are available. Konishi and Miyamoto[37] noted that F could overcome Al inhibition of pollen tube growth in tea (*Camellia sinensis*), and Kinraide et al.[21] reported that F reversed Al inhibition in red clover (*Trifolium pratense* L.). Recently, Cameron et al.[36] reported that Al-F complexes were nontoxic to barley (*Hordeum vulgare* L.).

That raising the solution pH will eventually relieve Al-toxicity is well documented; it remains unclear whether this should be ascribed to (i) differential toxicity of the mononuclear hydrolysis products, (ii) formation of nontoxic polynuclear hydroxy-Al complexes or solid phase $Al(OH)_3$, or (iii) a physiological response whereby increasing pH reduces sensitivity to Al. Among the mononuclear species, toxicity has variously been ascribed to Al^{3+},[3,5,6,31] to $AlOH^{2+}$,[20,34] to $Al(OH)_2^+$,[34] and to the sum (expressed as activities) of all mononuclear species.[30,32,34] Polynuclear hydroxy-Al has generally been viewed as nontoxic,[30,34] but several reports have suggested that these species can be highly toxic.[38,39,40] That these conflicting views exist may, in part, be due to difficulties in complete and accurate speciation of the growth media, especially where chemically complex nutrient solutions have been used. Moreover, the polynuclear species can be elusive, their composition and extent depending on Al concentration, pH, ionic strength, ligands present, temperature, and the age and method of preparation of the solution, as well as other factors.[13,14,41,42,43]

Although not under direct study here, other ligands have been reported to ameliorate Al toxicity. Phosphate can be effective, although conditions favoring precipitation of Al-phosphates may be required.[28,38] White[44] reported that formation of soluble polynuclear hydroxy-phosphate-Al species can reduce toxicity. It remains unclear whether the formation of simple, mononuclear Al-phosphate complexes will alleviate toxicity, and the problem is confounded by the uncertain nature of these species. Low molecular-weight organic ligands have generally been found to reduce toxicity,[16,28,45,46,47] but exceptions include findings that citrate and tartrate actually exacerbated Al toxicity to alfalfa (*Medicago sativa*).[48] Even when toxicity has been relieved, additions of organic ligands have yielded variable effects on the uptake of Al and its distribution between plant shoots and roots.[28,46,47]

Clearly, it is presently not possible to make definitive generalizations concerning the relationships between Al speciation and biological responses. In order for such relationships to be evaluated under "field" conditions -- i.e., soil solutions and natural waters -- they must first be established under controlled chemical environments in the laboratory. Whenever possible, Al speciation is most readily accomplished computationally, using known formation and solubility constants. To that end, we have employed a bioassay that offers distinct advantages in speed, simplicity, and sensitivity to Al.[21] Moreover, the chemical simplicity of the basal medium (0.4 m*M* $CaCl_2$) permits more precise computations of Al speciation than do more complete nutrient solutions.[25] However, metastable polynuclear complexes render a computational approach untenable, and we have adopted a kinetically-based empirical method[49,50] to characterize and quantify these species when present. The objectives of this research were (i) to evaluate the ameliorative effectiveness of mono- and divalent cations while differentiating between physiological and ionic-strength effects, (ii) to elucidate the toxic or nontoxic nature of Al complexed by inorganic ligands, especially sulfate, and (iii) to clarify the relationships between speciation and toxicity in hydroxy-Al solutions where a variety of mononuclear, polynuclear, and solid-phase species may form. In addition, some inferences are drawn concerning physiological mechanisms for Al toxicity. Many of the experimental findings summarized here have been extracted from published articles by the authors.[25,35,51]

EXPERIMENTAL

Seedling Culture

Wheat (*Triticum aestivum* L.) seeds were germinated on moist filter paper in the dark for 2 d. Five seedlings were then selected for uniform initial root length and transferred to each 500-mL test solution for 2 d of additional growth at 298 ± 1°K in the dark. The seedlings selected had a primary root that was 15 ± 2 mm long, and two slightly shorter seminal roots (typically 10-12 mm) at transfer. The seedlings were floated in the solutions on slabs of closed-cell polyethylene foam such that the roots extended down into solution. The solutions were generally aerated to prevent buildups of dissolved CO_2 and to reduce pH gradients next to the root via agitation. At termination, the lengths of the two longest roots of each seedling were recorded, and the mean of the 10 values for each beaker used as the measure of Al toxicity. For selected experiments, root growths relative to the control (0 Al) treatments were computed by subtracting the initial 15 mm from all final root lengths:

$$\text{Relative Root Growth (\%)} = \frac{RL - 15}{RL_c - 15} \times 100$$

where RL and RL_c are the final root lengths of the Al treatment and control solutions, respectively. All test solutions contained a basal level of at least 0.4 m*M* $CaCl_2$, but this concentration was varied as needed to manipulate sensitivity to Al.[25] For most experiments the Al-sensitive variety "Tyler" was used, but the more tolerant "Seneca" was used in selected experiments as well.

A few experiments were also conducted with red clover (*Trifolium pratense* L. cv. "Kenland"), but the seeds were germinated for 1 d by floating them on 0.2 m*M* $CaCl_2$. Ten seedlings were then transferred to each test solution, and primary root length measured after 1 d of additional growth.[21] Previous reports have detailed the culture method for both wheat and red clover, and demonstrated its utility as a sensitive assay for Al toxicity.[21,25,35]

Solution Preparation and Speciation

Details varied among experiments and will be provided under "Results", but the following procedures were common to all experiments. All reagents were of analytical grade or higher purity. Double deionized or double distilled water was used throughout, and all glassware was acid-washed to prevent contamination. All growth solutions were prepared from stock solutions (0.5 or 1.0 m*M*) of $AlCl_3{\cdot}6H_2O$ that in some cases contained dilute HCl as well. Where it was necessary to add base to Al solutions, 0.5 m*M* Al stocks were preadjusted by slow injection (generally 0.5 mL min^{-1}) of standardized 0.05 *M* NaOH using an autoburette. The delivery tip was submerged and the solutions vigorously stirred to minimize locally high concentrations of base during neutralization.[41] Otherwise, base was never added, although growth solutions were often supplemented with HCl to achieve a desired pH. The pH of all solutions was measured at the beginning and end of each growth experiment. In selected experiments, a modified version of the spectrophotometric ferron procedure[49,50] was used to estimate the fractions of mono- and polynuclear Al, or to check for the presence of the latter. One-milliliter samples and from 0.4 to 0.8 mL (depending on total Al) of the mixed ferron reagent were used, and absorbance at 363 nm was monitored over time with a Beckman DU-7 UV-VIS spectrophotometer. The data were collected by a microcomputer and fitted to kinetics functions as described previously.[50] Distributions and activities of mononuclear species were computed with a corrected version of the program GEOCHEM,[52,53, and errata in Soil Sci. Soc. Am. J. 51:1680] or with a program written by the third author. Comparisons of computed results from the two programs showed exact agreement. Except where noted in the text, the thermodynamic constants for mononuclear complexes given in Table 1 were used for all computations. In the absence of experimental evidence to the contrary, Cl was considered to be an indifferent ligand towards Al, i.e., no Al-Cl complexes were considered.

For the discussion to follow, we will denote total Al as Al_t and use braces for activities and brackets for concentrations of individual species. The ferron-based estimate of mononuclear Al will be indicated by $[Al_a]$ and the computed sum of the mononuclear species activities will be denoted by Σa_{mono}. Vertical bars in all figures indicate standard error of the mean where it exceeds symbol size.

Table 1. Thermodynamic Constants Used for Computed Speciation of Growth Solutions

Reaction	log K	Reference
$Ca^{2+} + Cl^- = CaCl^+$	-1.00	17
$Ca^{2+} + 2Cl^- = CaCl_2^0$	0.00	17
$Ca^{2+} + SO_4^{2-} = CaSO_4^0$	2.31	17
$Al^{3+} + H_2O = AlOH^{2+} + H^+$	-5.02	17
$Al^{3+} + 2H_20 = Al(OH)_2^+ + 2H^+$	-9.30	17
$Al^{3+} + 3H_2O = Al(OH)_3^0 + 3H^+$	-14.99	17
$Al^{3+} + 4H_2O = Al(OH)_4^- + 4H^+$	-23.33	17
$Al^{3+} + F^- = AlF^{2+}$	6.98	54
$Al^{3+} + 2F^- = AlF_2^+$	12.60	54
$Al^{3+} + 3F^- = AlF_3^0$	16.65	54
$Al^{3+} + 4F^- = AlF_4$	19.03	54
$Al^{3+} + SO_4^2- = AlSO_4^+$	3.20	17
$La^{3+} + H_2O = LaOH^{2+} + H^+$	-8.5	12

RESULTS

Cation Amelioration

For these experiments, the basal medium (0.4 m*M* $CaCl_2$) at pH 4.3 was used throughout. Total Al ranged from 0 to 35 μ*M*, while the added salts ranged from 0 to 3.6 m*M* for $CaCl_2$ and $MgCl_2$, 0 to 3.0 m*M* for $SrCl_2$, and 0 to 40 m*M* for NaCl and KCl. Based on several tests, including assays with the 8-hydroxyquinoline[55] and aluminon[30] reagents (see 25 for details), we found no evidence of polynuclear Al in the test solutions, and computed speciation accordingly. The choice of $\{Al^{3+}\}$ as an independent variable was somewhat arbitrary -- its ratio to that of total Al activity was always 0.72 due to the constancy of pH at 4.3.

In the absence of Al, additions of Ca, Mg, K, or Na improved root growth slightly, presumably due to alleviation of a slight inhibition by H^+ at this pH.[20] Strontium may have inhibited growth slightly at concentrations > 2 m*M*. Other monovalent cations tested -- Rb, Li, Cs -- were toxic at potentially ameliorative levels, while the trivalent metals La and Sc were even more toxic than Al (data not shown).

Attempts to account for differences in toxicity based solely on the activity of Al^{3+} proved unsuccessful when cation levels were varied. Figure 1 shows the results from a typical experiment using Ca. A family of curves can be seen wherein root length is correlated with $\{Al^{3+}\}$ only within a given Ca concentration, or with Ca level only within a given level of Al_t. These results suggested that multiple regression analyses of root length vs $\{Al^{3+}\}$ and cation activity might prove useful. We had accumulated a large amount of data for Ca, Mg, and Na from a number of diverse experiments, some of which were orthogonal with respect to Al and cation concentrations, while others employed only a single concentration of Al or salt. However, none of these early experiments was designed to be orthogonal with respect to $\{Al^{3+}\}$ and $\{C\}$ (cation activity), so that the accrued data were unevenly distributed. Points where $\{Al^{3+}\}$ and $\{C\}$ were mutually high were especially absent, so new experiments were designed that were orthogonal for activities over the range previously employed. Analyses of both original and "improved" data sets yielded virtually identical models, so the two sets were pooled for analysis and presentation. Linear models for root length vs. $\{Al^{3+}\}$ and $\{C\}$ yielded R^2 values that were rather low, not an unexpected result given the curvilinear responses depicted in Figure 1. Transformations using an exponential function for $\{Al^{3+}\}$ or quadratic terms (including interactions) yielded significantly better fits to the data. Two such models are presented in Table 2 for Ca, Mg, and Na. The ranges of experimental data used to fit the models were 0 to 6 μ*M* for $\{Al^{3+}\}$, 0 to 2.2 m*M* for $\{Ca^{2+}\}$ and $\{Mg^{2+}\}$, and 0 to 32.6 m*M* for Na^+. The exponential model offers the advantage of fewer coefficients that require evaluation. However, the quadratic model probably describes the response best in that it accounts for the transition from a concave-upward to convex-upward response to $\{Al^{3+}\}$ as $\{C\}$ increases (Figure 2), a trend we have also observed with Seneca wheat (data not shown).

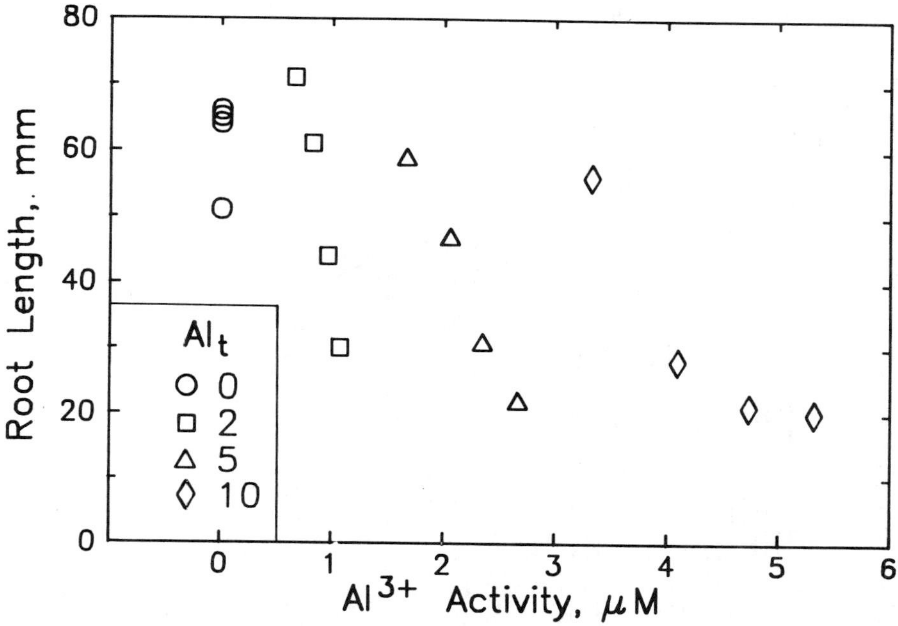

Figure 1. Root length of Tyler wheat vs. $\{Al^{3+}\}$ at three levels of Al_t and various concentrations of $CaCl_2$ at pH 4.3. Within a given Al_t, increasing root length corresponds to increasing $[CaCl_2]$ of 0.4, 0.8, 1.6, or 3.2 m*M*. Data are from a single unreplicated experiment, but representative of many similar experiments.

The interaction terms are significant and are indicative of the steeper decline in growth with increasing $\{Al^{3+}\}$ at low $\{C\}$, and the greater response to $\{C\}$ at high $\{Al^{3+}\}$. Without the interaction terms, opposite edges of the response surface would be parallel.

With respect to relative effectiveness, Ca was approximately 1.5 times more ameliorative than Mg and some 15 times more effective than Na as indicated by the coefficients in Table 2. Additional experiments indicated that Sr was about as effective as Mg, while K was equivalent to Na (data not shown).

Sulfate and Fluoride Effects

These experiments were also conducted at pH 4.3 with a minimum of 0.4 m*M* $CaCl_2$. Al was added as a 1.0 m*M* stock that was also 1 m*M* in HCl to further protect against polymer formation. Test solutions were supplemented with NaCl, Na_2SO_4, or NaF to achieve controlled activities of all species.

Four initial experiments were conducted with Tyler wheat wherein $\{Na^+\}$ was 8.9, 17.3, 25.4, or 33.2 m*M* and the basal level of 0.4 m*M* $CaCl_2$ was used. Within each level of $\{Na^+\}$, $\{Al^{3+}\}$ was 1.5, 3.0, 4.5, or 6.5, µ*M*, while $\{AlSO_4^+\}$ was 0, 2, 4, 6, or 8 m*M*, in a fully factorial combination. Multiple regression analyses demonstrated a

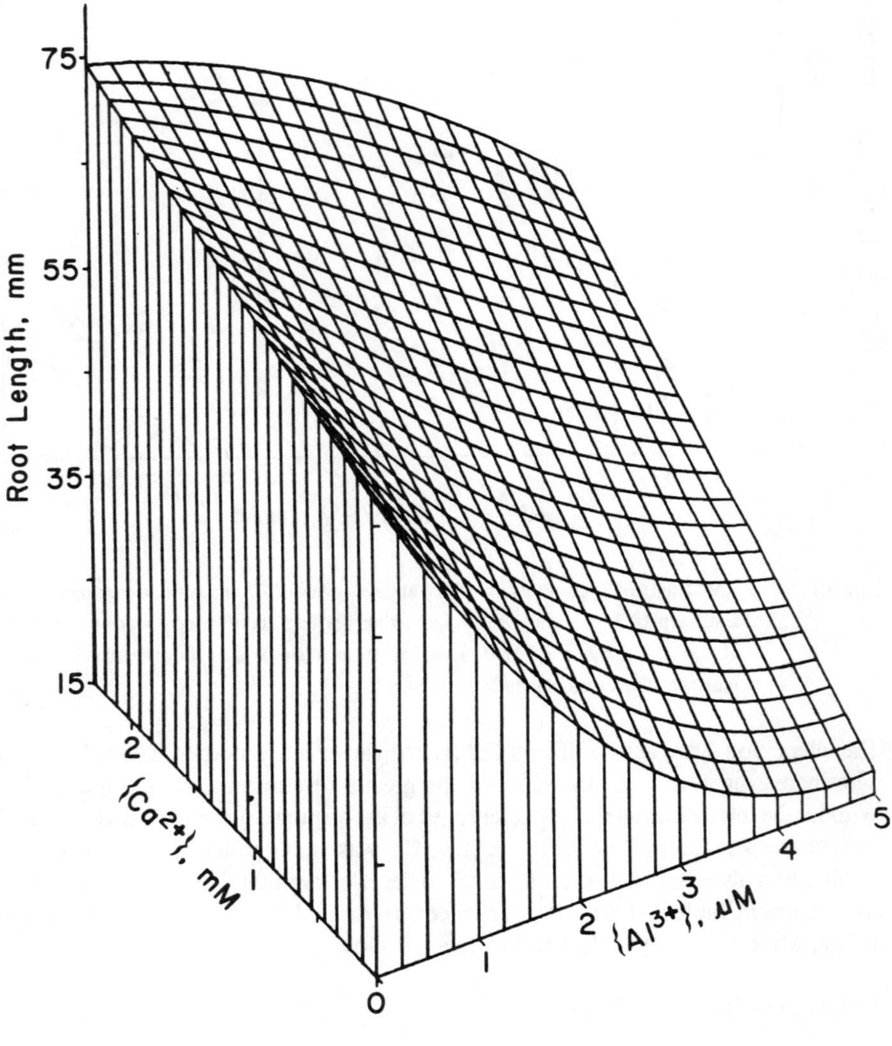

Figure 2. Root length of Tyler wheat vs. $\{Al^{3+}\}$ and $\{Ca^{2+}\}$ (above the basal level) at pH 4.3. Response surface represents the quadratic model given in Table 2. Based on 169 observations, $R^2 = 0.842$.

Table 2. Two models for root length of Tyler wheat as a function of Al^{3+} (in μ*M*) and cation activities (in m*M*) at pH 4.3

	Coefficients						
Salt	β_0	β_1	β_2	β_3	β_4	β_5	R^2
$RL = \beta_0 + \beta_1 e^{-0.4\{Al^{3+}\}} + \beta_2\{C\} + \beta_3\{C\}e^{-0.4\{Al^{3+}\}}$							
$CaCl_2$	11.5	45.4	17.1	-8.07			0.812
$MgCl_2$	13.5	47.1	13.3	-7.89			0.838
NaCl	11.1	43.9	1.02	-0.571			0.873
$RL = \beta_0 + \beta_1\{Al^{3+}\} + \beta_2\{Al^{3+}\}^2 + \beta_3\{C\} + \beta_4\{C\}\{Al^{3+}\} + \beta_5\{C\}\{Al^{3+}\}^2$							
$CaCl_2$	57.5	-17.0	1.79	7.33	6.54	-1.02	0.842
$MgCl_2$	61.2	-17.8	1.89	4.12	5.93	-0.911	0.853
NaCl	55.7	-17.0	1.90	0.397	0.323	-0.046	0.880

Note: {C} indicates the cation activity of the added salt (in m*M*), but does not include the Ca activity of the basal medium. All coefficients given are significantly different from zero ($p < 0.05$).

significant inhibition by $\{Al^{3+}\}$ and an ameliorative effect of $\{Na^+\}$ (data not shown). Coefficients for $\{AlSO_4^+\}$ were not significantly different from zero ($p > 0.10$), suggesting that this species is nontoxic.

Another set of experiments was conducted to provide greater sensitivity to the relative toxicity of $AlSO_4^+$ and to avoid the need for multiple regression analyses. Activities of Al^{3+}, Ca^{2+}, and Na^+ were fixed at the values indicated in Figure 3, while $\{AlSO_4^+\}$ was varied from 0 to 7.5 μ*M*. Relative root growth (RRG) was not significantly ($p > 0.10$) affected by increasing $\{AlSO_4^+\}$ (Figure 3). The dashed line in Figure 3 represents the approximate response to increasing $\{Al^{3+}\}$ based on data from other experiments, and provides a sharp contrast to the negligible effects of $AlSO_4^+$. Similar experiments were conducted with red clover wherein $\{Al^{3+}\}$ was maintained at 0.5 or 1.0 μ*M* and $\{AlSO_4^+\}$ varied from 0 to 5 μ*M*. Again, root growth was unaffected by increases in $\{AlSO_4^+\}$ (data not shown), suggesting that this species is generally nontoxic.

Heretofore, we have assumed that log *K* for the formation of $AlSO_4^+$ is 3.20, the most oft-cited value.[2,17,30,34] However, the primary literature includes reported values of 3.21,[56] 2.04,[57] 3.01,[58] and 3.73,[59] as well as a value that we estimate as *ca.* 2.9[60] when corrected for ionic strength. In addition, the reaction:

$$Al^{3+} + 2SO_4^{2-} = Al(SO_4)_2^-$$

has been proposed with log K's of 5.11[56] and 4.90.[58] We took several steps to test the dependence of the apparent nontoxicity of sulfate-complexed Al on these disparate thermodynamic data. First, we reanalyzed two representative experiments, one for wheat and one for red clover, that were factorial in $\{Al^{3+}\}$ and $\{AlSO_4^+\}$ based on log K = 3.20. Reanalysis consisted of computing new activities based on concentrations of the added salts but using log K's ranging from 2.8 to 3.6, and rerunning the regression analyses. At any assumed log K over this range, coefficients for $\{AlSO_4^+\}$ were quite close to 0 and certainly of lesser magnitude than those for $\{Al^{3+}\}$ (Figure 4). Similarly, reanalysis of these data (and others) with the $Al(SO_4)_2^-$ species included in the computations indicated that only $\{Al^{3+}\}$ adversely affected growth (data not shown). Lastly, the experiment depicted in Figure 3 was repeated assuming log K = 5.11 for the formation of $Al(SO_4)_2^-$. Although this entailed increasing Al_t to 17.2 μM (instead of 13.3) at the highest level of added SO_4^{2-}, the slopes of relative root growth vs both $\{AlSO_4^+\}$ and $\{Al(SO_4)^{2-}$ were not significantly different from zero (data not shown).

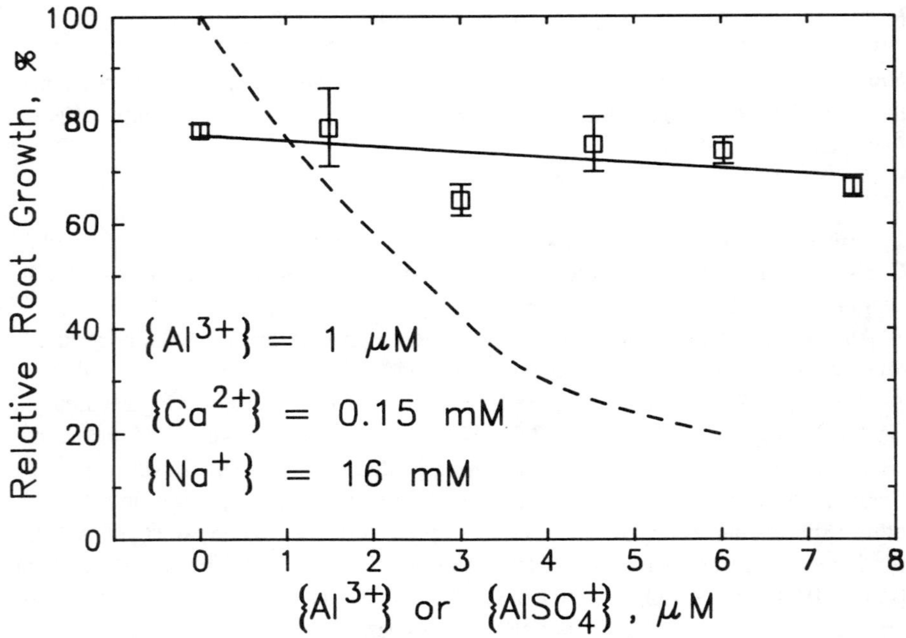

Figure 3. Relative root growth of Tyler wheat vs. $AlSO_4^+$ activity, with activities of Al^{3+}, Na^+, and Ca^{2+} held constant, and solution pH = 4.3 (data points and solid line). Means are for two experiments with two replications each. The slope of the best-fit solid line is not significantly different from zero (p > 0.1). Dashed line is approximate response to $\{Al^{3+}\}$ based on other experiments.

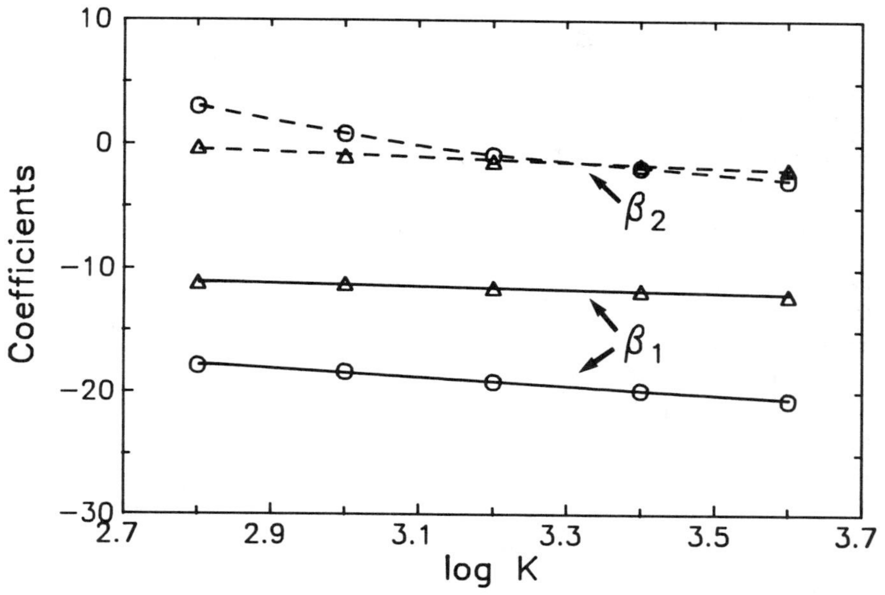

Figure 4. Coefficients from the regression model RRG = $\beta_0 + \beta_1\{Al^{3+}\} + \beta_2\{AlSO_4^+\}$ vs log K for the formation of $AlSO_4^+$. The solution speciation for two representative experiments (one for red clover and one for wheat) was computed for each value of log K, and the regression analysis performed. Triangles indicate wheat and circles indicate red clover.

Computations of species activities were based on the assumption that all Al was mononuclear -- a reasonable assumption given that test solution preparation involved acidification and that the Al stock solution itself contained an equimolar amount of HCl.[13,50] Nonetheless, polynuclear hydroxy-sulfate-Al complexes have been postulated,[61] and we utilized the ferron assay to check solutions typical of those used in growth solutions. Solutions with 20 μM Al_t at pH 4.3, and either 10 mM NaCl or 5 mM Na_2SO_4, exhibited identical reaction kinetics with ferron (data not shown). In contrast, additions of F^- or OH^- markedly slowed the reaction, so that maximal absorbance was not achieved even after 30 minutes. Thus, even though some 60% of the Al was predicted to be sulfate-complexed in the solution described, these complexes were readily degraded by ferron,[62] and there was no evidence of any polynuclear complexes in the sulfate system.

An experiment was also conducted to evaluate the toxicity of fluoride-complexed Al. Solutions contained 1.0 mM $CaCl_2$ at pH 4.3. Total Al ranged from 4 to 27 μM and added NaF from 0 to 34.5 μM. Using the constants from Table 1, $\{Al^{3+}\}$ was computed to be 1.0 μM for all solutions. Root growth was about 70% of control growth, but was unaffected ($p > 0.09$) by increasing Al_t (Figure 5). The dashed line and circles in Figure 5 represent the response to the same levels of Al_t in the absence of F, and clearly indicates the relative nontoxicity of fluoride-complexed Al. Thus, we

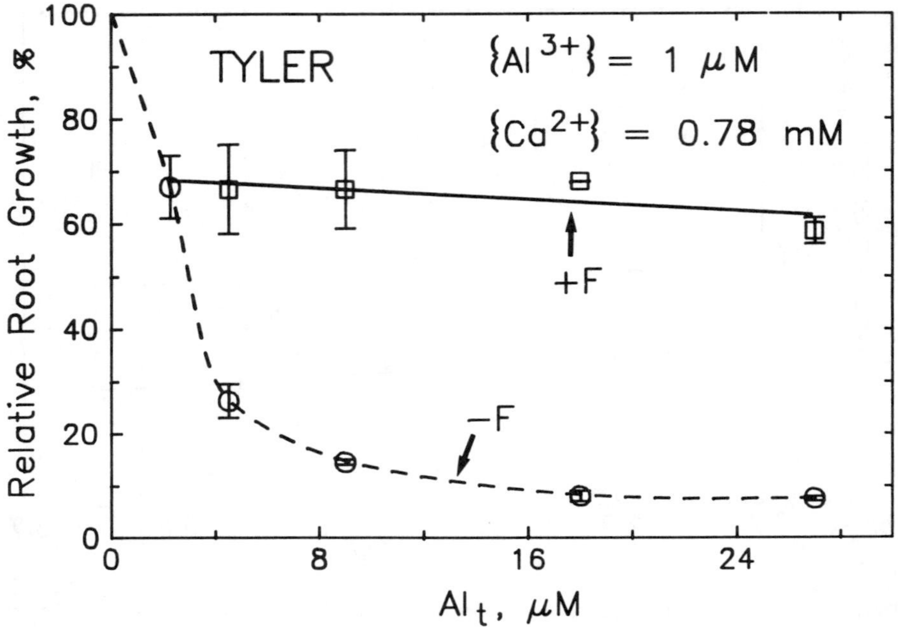

Figure 5. Relative root growth of Tyler wheat vs. Al_t where additions of NaF maintained computed $\{Al^{3+}\}$ at a constant value of 1.0 μ*M* (squares and solid line). $[CaCl_2]$ = 1.0 m*M* and solution pH was 4.3. Means are for a single experiment with two replications. The slope of the best-fit solid line is not significantly different from zero ($p > 0.09$). Dashed line and circles are response to Al_t in the absence of added NaF.

conclude that both sulfate- and fluoride-complexed Al are nontoxic, or at least tenfold less toxic than Al^{3+}.

Hydroxy-Al Solutions

We conducted some initial experiments using Tyler wheat that were factorial in pH (4.2, 4.4, 4.6, 4.8, and 5.0) and Al_t (0, 4, 8, and 12 μ*M*). $[CaCl_2]$ was 1.5 m*M*. There was general improvement in root growth with increasing pH, but evidence of decline at the highest pH where Al_t = 8 or 12 M (data not shown). Root length was closely associated with computed Al^{3+}, but not with Σa_{mono} as reported by others.[30,34] However, these experiments were flawed in that we assumed that no polynuclear Al would form in the absence of added base, and thus did not assay the solutions for these species. Subsequent research[50] revealed that, at low concentrations of Al, significant quantities of polynuclear Al may be present without added base, and even in slightly acidified solutions. Consequently, we designed new experiments wherein polymerization could be accounted for.

For the following experiments, several series of 0.5 m*M* Al stock solutions with basicities (initial molar OH/Al ratio, n_0, where a negative value indicates added H^+) ranging from -4 to 2.9 were prepared. Through trial and error, we arrived at dilutions that would yield test solutions with predictable pH levels and distributions of mono- and polynuclear Al. Total Al content of these stock solutions was checked by inductively-coupled plasma emission spectroscopy (ICPES) using standards prepared from pure Al wire. The stock solutions were always between 14 and 30 d old at the initiation of each experiment to minimize aging effects,[14,43,50] and final dilutions were prepared one day before transfer of seedlings to promote stabilization of the distribution of Al species. The growth solutions were assayed for mononuclear Al with ferron at the beginning and end of each experiment, and the average value is reported here. A 5-min reaction was employed, and the data fit to a simultaneous kinetics model using a pseudo second-order rate law for monomers and a pseudo first-order law for polymers.[50] For this portion of the study, all nonmononuclear Al is included in the fraction of polynuclear Al, f_p; no distinction is made between reactive polymers, unreactive polymers, or precipitated hydroxy-Al.[14,43] In addition, solution pH was more tightly controlled via periodic measurement and adjustment using dilute HCl during the growth period.

First, we used an approach similar to that described above by determining a set of dilutions that yielded a range of pH values (4.20, 4.45, 4.70, and 4.95) but a constant value of Σa_{mono} of 10 μ*M*. The distributions of Al_t, $[Al_a]$, and $\{Al^{3+}\}$ are given in Figure 6a. Inspection of the root growth response to pH (increasing, flat, or decreasing) thus yielded immediate clues as to the species responsible for toxicity, while avoiding the need for statistical analyses (see Discussion). In the presence of Al, root length of Seneca wheat increased with increasing pH up to 4.70, but declined markedly at pH 4.95 (Figure 6b). Only the pH 4.95 solution contained significant polynuclear Al (Figure 6a), suggesting that these complexes were indeed toxic. Inhibition of root growth did not appear to be associated with Σa_{mono}. Virtually identical results were obtained with Tyler wheat at Σa_{mono} = 8 μ*M* and $[CaCl_2]$ = 3 m*M* (data not shown).

Next, we designed experiments wherein sufficient HCl could be added to suppress polymer formation, but a pH range of *ca.* 4.2 to 5.0 along with measurable inhibition of Tyler root growth would be obtained. Total Al was 4 μ*M*, $n_0 \leq -2$, and $[CaCl_2]$ = 1.5 m*M*. Within detection limits, ferron assays indicated that the solutions were 100% mononuclear. Root growth increased with increasing pH, reaching 100% at pH 4.95 (Figure 7). The decline in $\{Al^{3+}\}$ over this pH range suggested a close association between toxicity and this parameter. In contrast, Σa_{mono} increased as root growth improved, again indicating that this quantity is unreliable as a predictor of phytotoxicity (Figure 7).

We then conducted experiments with Al_t = 15 μ*M*, n_0 values of -1, 0, 1.0, 2.0, 2.5, and 2.9, and $[CaCl_2]$ = 3 m*M*. Solution pH ranged from 4.7 and 5.6 and control growth was constant over this range. Relative root growth of Tyler wheat decreased sharply with increasing n_0, reaching a minimum of *ca.* 15% at n_0 = 2.0 (Figure 8). Toxicity was only partially alleviated when the Al was almost completely neutralized (n_0 = 2.9, Figure 8). That at least some of the polynuclear species were quite toxic can be inferred

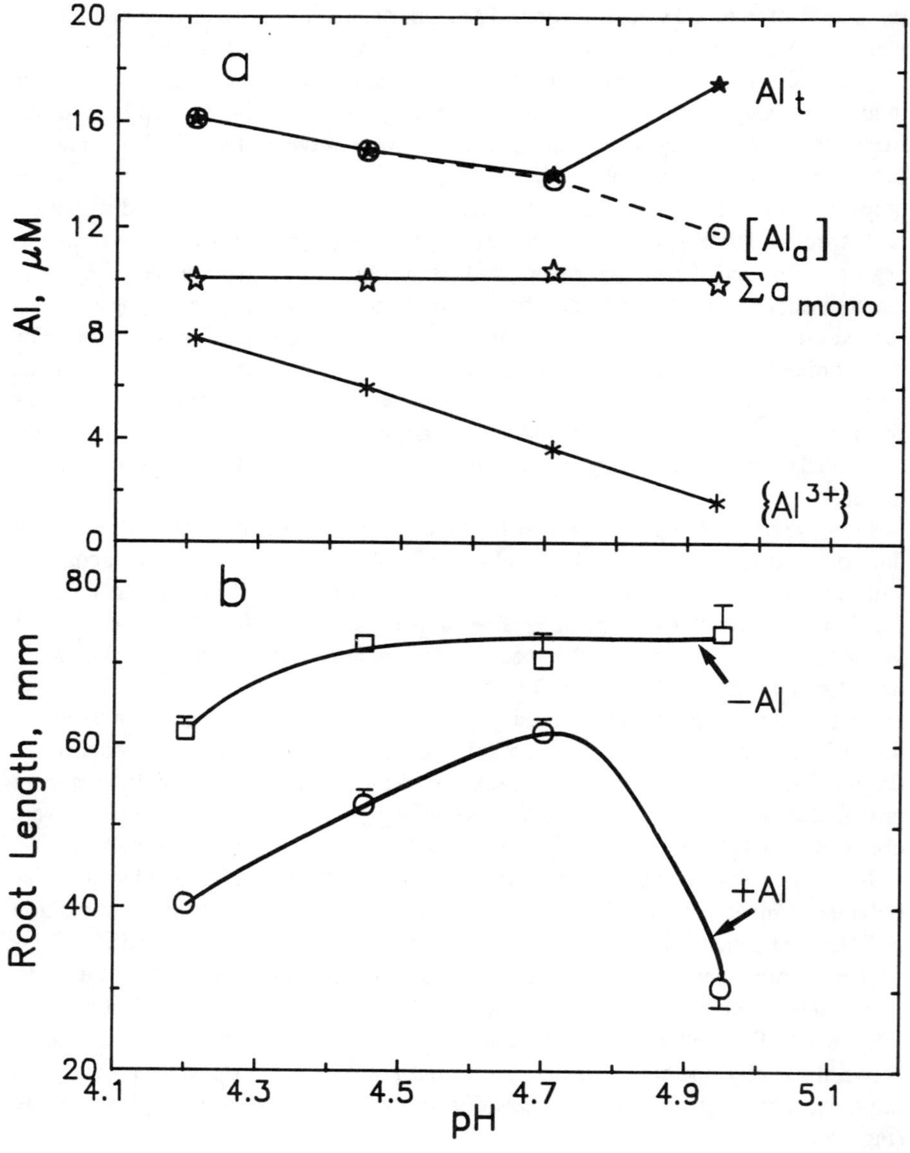

Figure 6. Distribution of Al species (a) and root lengths of Seneca wheat (b) vs. pH for experiments designed to yield solutions with constant Σa_{mono}. [$CaCl_2$] = 1.0 m*M* and points represent means of two identical experiments with two replications each. Lines are hand-fitted.

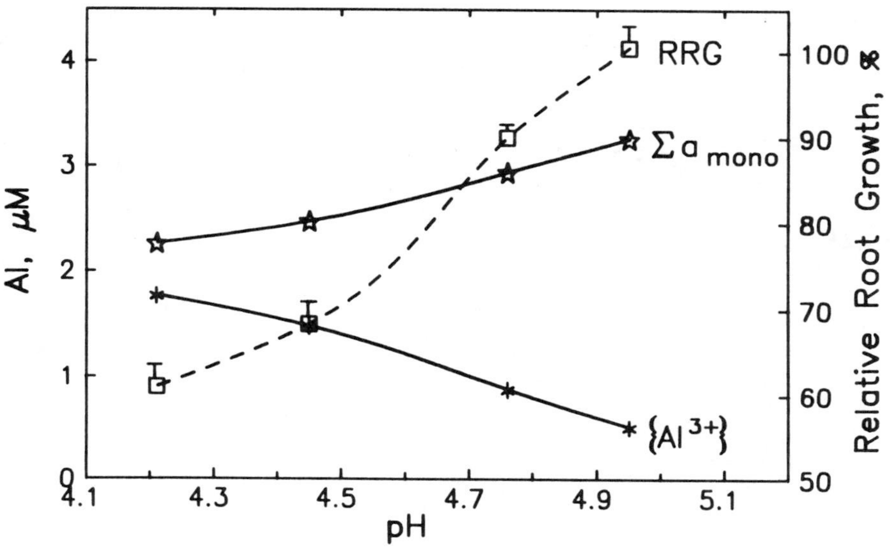

Figure 7. Distribution of Al species and relative root growth (RRG) of Tyler wheat vs. pH at Al_t = 4 µM and $[CaCl_2]$ = 1.5 mM. No polynuclear Al could be detected in any solutions. Controls were included at each pH. Means are for two experiments with two replications each, and lines are hand-fitted.

from two considerations. First, the n_0 = 2.0 solutions in Figure 8 had mononuclear Al concentrations of 2.2 µ*M* and pH levels of 5.2 to 5.3 -- a seemingly nontoxic combination when compared with Figure 7. Second, computed activities of all mononuclear species, as well as Σa_{mono}, declined over the entire range of n_0. We attribute the partial amelioration at n_0 = 2.9 to formation of larger polymers or colloidal $Al(OH)_3$ that was nontoxic. Very similar results were obtained using Seneca wheat at Al_t = 15 µ*M*, $[CaCl_2]$ = 1.0 m*M*, and the same values of n_0 (data not shown).

Further Studies of Polymer Toxicity

For reasons enumerated in the Discussion, we conducted additional experiments to examine any variability in toxicity of polynuclear Al due to experimental conditions. Methodology was similar to that described in the last section with the following differences. Two liters of each test solution were prepared from stocks that were 0.5 m*M* in Al_t at n_0 = 2.0. Three 500-mL aliquots were placed in beakers for measuring root growth, and the remainder retained for analysis. The ferron assay was performed in duplicate on this sample, but not on individual beakers. Here, we carried the ferron scans out to one hour such that three fractions could be distinguished: Al_a, Al_b, and Al_c

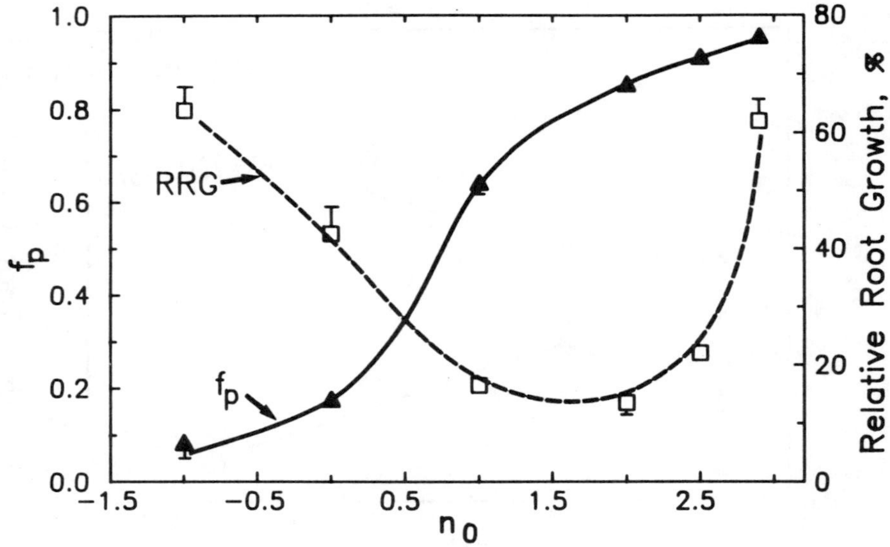

Figure 8. Fraction of polynuclear Al (f_p) and relative root growth (RRG) of Tyler wheat vs. basicity at Al_t = 15 μ*M* and $[CaCl_2]$ = 3 m*M*. Solution pH ranged from 4.7 to 5.6, and control root lengths were constant over this range. Means are for two experiments with two replications each, and lines are hand-fitted.

in the parlance of Smith,[14] corresponding to mononuclear Al, reactive polymers, and unreactive polymers or solid phase $Al(OH)_3$, respectively. The Al reacting within 30 min was taken to be $[Al_a + Al_b]$, and the first 24 min used to fit the binary-species kinetics function to obtain $[Al_a]$ (see 50 for details). Total Al was nominally 15 μ*M* throughout, but was computed exactly for each solution based on the ICPES analysis of the corresponding stock solution. The quantity of Al_c was computed as the difference between Al_t and $[Al_a + Al_b]$. Assays performed on selected solutions at the end of each experiment indicated that, as before, $[Al_a]$ was fairly constant during the 2-d growthperiod. There were significant decreases in $[Al_b]$ and corresponding increases in $[Al_c]$ for most solutions. However, without information on the details of these temporal changes, or on how the plants respond to speciation over time, we opted to use the initial speciation data for correlation with phytotoxicity. Initial pH values (pH_i) are reported here, but final values generally differed by 0.1 unit or less. A single control treatment at pH 5.1 was included because there was no apparent response to pH above 4.5 in the absence of Al (Figure 6). In all experiments, pH exceeded 5.0 and mononuclear Al did not exceed 3 μ*M*, so toxicity is ascribed solely to nonmononuclear Al (see above).

Heretofore, partially-neutralized stocks had been prepared with base addition rates of 0.50 mL min^{-1} . We prepared 500-mL stocks using rates of 0.05, 0.5, 3.0, and 12.0 ml min^{-1}, so that it took *ca.* 196, 19.6, 3.3, or 0.8 min to achieve a basicity of 2.0. For

Table 3. Influence of rate of base addition to Al stock solutions on speciation and root growth of Tyler wheat

Rate of Addition mL min^{-1}	Relative Root Growth %	Ferron Assay $[Al_a]$	$[Al_b]$	$[Al_c]$	k_b min^{-1}	pH_i
		---------μM---------				
0.05	13 ± 1†	2.5	9.1	3.2	0.17	5.21
0.50	16 ± 2	2.5	7.2	5.2	0.17	5.16
3.0	17 ± 1	2.2	6.3	6.2	0.17	5.11
12.0	18 ± 0	2.2	5.5	6.9	0.18	5.10

Note: All growth solutions contained 15 μM Al_t at $n_o = 2.0$, and 3.0 mM $CaCl_2$. Stocks were 6 d old when growth solutions were prepared. Results are from a single experiment with three replications.
† ± standard error of the mean.

the fastest rate, the delivery tip was suspended above the stirred solution but was submerged for the other three rates. Increasing rate of base addition decreased $[Al_b]$ and increased $[Al_c]$, while $[Al_a]$ was relatively constant (Table 3), consistent with other findings.[14] The fitted rate coefficients, k_b, for the Al_b fraction were essentially constant, indicating a similar polymer or suite of polymers in all solutions.[43] However, relative root growth of Tyler wheat was < 20% for all rates (Table 3), suggesting that levels of toxic Al were essentially saturating for all solutions. The 0.5 mL min^{-1} rate was used for all subsequent experiments.

Experiments were then conducted where the Al stock solutions had been aged from 1 to 153 d. Again, $[Al_b]$ decreased and $[Al_c]$ increased (Table 4) in a manner consistent with other reports.[14,43,49] Quantities of $[Al_a]$ were somewhat more variable, but were < 3 μM throughout, while k_b was again quite constant (Table 4). Here root growth improved markedly in the solution prepared from the oldest stock, and some amelioration was evident for the 65-d old stock (Table 4). These results suggest that $[Al_b]$ may reflect the toxic fraction, and that only about 4 to 5 μM is required to fully inhibit root growth of Tyler wheat.

Lastly, we added small quantities of phosphate (as KH_2PO_4) to the growth solutions to achieve molar P/Al ratios of 0.1, 0.2 and 0.3. The Al stock used was 6 d old when growth solutions were prepared, and these were further aged 2 d instead of the usual 1 d before transfer of seedlings. Two control treatments were included -- 0 P and 4.5 μMP at pH 5.1 -- and average root length differed by < 1 mm for these solutions. Increasing concentrations of P decreased $[Al_b]$, increased $[Al_c]$, while $[Al_a]$ was again fairly constant (Table 5). The constancy of k_b indicates that the reactivity of Al_b is largely unaffected by additions of P, suggesting that most of the P is incorporated into

Table 4. Influence of stock age on Al speciation and root growth of Tyler wheat

Stock Age	Relative Root Growth	Ferron Assay				
		$[Al_a]$	$[Al_b]$	$[Al_c]$	k_b	pH_i
days	%	---------μM----------			min^{-1}	
1	13 ± 1†	1.0	9.0	4.9	0.16	5.25
25	15 ± 2	2.4	4.7	7.2	0.17	5.19
65	32 ± 0	2.2	2.2	9.3	0.16	5.12
153	85 ± 6	3.0	0.5	11.2	--‡	5.03

Note: All growth solutions contained 15 m*M* Al_t at $n_o = 2.0$, and 3.0 m*M* $CaCl_2$. Results are from a single experiment with three replications.
† ± standard error of the mean.
‡ Could not fit an accurate value due to low level of $[Al_b]$.

Table 5. Influence of added phosphate on Al speciation and root growth of Tyler wheat

P Added	Relative Root Growth	Ferron Assay				
		$[Al_a]$	$[Al_b]$	$[Al_c]$	k_b	pH_i
μM	%	---------μM----------			min^{-1}	
0	13 ± 2†	1.5	7.8	5.4	0.18	5.19
1.5	14 ± 1	2.1	5.1	7.5	0.17	5.25
3.0	26 ± 3	2.6	2.5	9.6	0.15	5.25
4.5	110 ± 2	2.3	0.8	11.6	--‡	5.20

Note: All growth solutions contained 15 μM Al_t at $n_o = 2.0$, and 3.0 m*M* $CaCl_2$. Stocks were 12 days old when growth solutions were prepared. Results are from a single experiment with three replications.
† ± standard error of the mean.
‡ Could not fit an accurate value due to low level of $[Al_b]$.

the Al_c fraction, probably as a colloidal precipitate.[42] Root growth of Tyler wheat was slightly improved at P/Al = 0.2, and was dramatically increased at P/Al = 0.3 (Table 5). The anomalously high root growth at the highest P/Al ratio was significantly greater than both of the controls, suggesting some stimulation of growth when both P and Al are present, but only when $[Al_b]$ is rather low.

DISCUSSION

There are some implications for the observed effects of cations on Al toxicity. Inasmuch as activity coefficient effects have been accounted for, there is clearly a physiologically-based amelioration for both mono- and divalent cations. The amelioration is nonspecific within a cation valence (with the exception of Ca), suggesting that it does not involve meeting nutritional requirements, which are specific,[63] and thus may involve an external mechanism. Competition for external binding sites is an appealing model, and would account for the large difference in efficacy observed for mono- vs divalent cations. Such binding could be critical as either a primary lesion, for example on membrane surfaces, or as a requisite step in the uptake of Al by cells followed by an intracellular event that inhibits root growth (see Foy[18] for a review of the many proposed mechanisms). Although Ca is particularly ameliorative, and no other cation can substitute for required Ca (which is critical for maintenance of membrane integrity among other things[64]), it seems unlikely that Al toxicity involves simple displacement of Ca from Ca-requiring binding sites.[65] If that were so, then Ca should be the only ameliorative cation, there being no apparent means whereby increasing levels of K, for example, could both displace Al from, and restore Ca to, binding sites.

On a more practical level, it would seem that estimates of toxic Al must take into account both ionic strength effects on activities, and the concentrations (or activities) of mono- and divalent cations. At present, we can offer no explanation for other reports wherein toxicity has been correlatable with Al activities alone.[3,6,30,31] In acid-soil solutions, Ca and Mg are the likely ameliorative cations. Based on our results, something on the order of 10 m*M* or greater K (or Na) would be required to achieve any substantial amelioration. Typical soil solution concentrations would appear to be lower,[2,7,8,66] although monovalent cation concentrations may be this high in nutrient solutions used in the laboratory.

Our studies with sulfate and fluoride were beset with the usual difficulties in Al speciation given the uncertainties in thermodynamic constants (see Ref. 67 for an interesting analysis of this problem). However, we have attempted to protect against a false conclusion via the tests outlined above, and the relative nontoxicity of sulfate-complexed Al seems rather certain. That this was observed for both a monocot (wheat) and a dicot (red clover) suggests that the nontoxicity is general. Although the slopes of the best-fit lines in Figures 3 and 5 are not statistically significant, there is a downward trend for both. We are thus amenable to an alternative hypothesis that complexed Al is much less toxic than Al^{3+}, but not totally innocuous. However, any deviations from the horizontal in Figures 3 and 5 could also be due to errors in speciation and, if formation constants were known exactly, any hints of toxicity for complexed Al might disappear. This is especially true for F, where at least four constants must be known (Table 1) for accurate speciation over the range of Al and F used here, a fact that precluded performing the types of tests done for SO_4^{2-}.

Based on these and other findings,[36] Al complexed with F and SO_4^{2-} should be computed and subtracted from estimates of toxic Al. It remains to be seen whether any of the proposed empirical methods[30,49,55] can differentiate between these complexes and toxic forms. Our results and others[62] using ferron would suggest that this may remain problematic, and that computational approaches may prove superior. Moreover, the ferron method has been shown to suffer from interferences when used with soil solutions.[7] In soil solution, both F and SO_4^{2-} can be found at potentially ameliorative concentrations.[2,7,8,66] Clearly only low concentrations of F are needed to dramatically decrease $\{Al^{3+}\}$. When $\{SO_4^{-}\} > 0.63$ m*M*, the activity of $AlSO_4^+$ will exceed that of Al^{3+} and, if the pH were also low, then $\{AlSO_4^+\}$ could constitute a large fraction of the total mononuclear Al activity. Soil amendments which are largely gypsum ($CaSO_4$) such as fluidized bed combustion residue[68] and byproducts from manufacture of HF or H_3PO_4[69] may provide inexpensive means of promoting plant growth in acid soils.[6,69,70,71] The effectiveness of such amendments should not be ascribed solely to correction of Ca deficiency[2] given the ameliorative action of Ca, the nontoxicity of $AlSO_4^+$, and the presence of F, phosphate, and Ca oxides in these materials.[68,69]

A highly reliable set of published hydrolysis constants also seems to be lacking, especially for the formation of $Al(OH)_2^+$ and $Al(OH)_3^0$.[12,72] Although our data indicate a close relationship between the activity of Al^{3+} and toxicity to wheat, the exact correlations obtained would depend on the choice of constants, and it would be risky to contend that Al^{3+} is necessarily the sole toxic species. But, regardless of which hydrolysis constants are used, at constant $[Al_r]$, $\{Al^{3+}\}$ must decrease and Σa_{mono} increase (due to the effect of reduced valence on activity coefficients) with increasing pH (Figure 7). Thus, our results are at variance with others who have reported a high correlation between toxicity and Σa_{mono}.[30,34] If the primary toxic species were the hydrolysis products $AlOH^{2+}$ [20,34] or $Al(OH)_2^+$ [34], root growth should not exhibit a monotonous increase as pH is increased from 4.2 to 5.0 (Figure 7). This presumes that the hydrolysis constants we have chosen are at least close to correct; if a drastically different set of constants were in fact correct, other interpretations would be possible. For the moment, we would thus concur with other investigators[3,5,31] that Al^{3+} is the most toxic of the mononuclear Al species and that $\{Al^{3+}\}$ is the best single predictor of phytotoxicity in the absence of polynuclear Al.

Based on arguments presented in "Results", there is little doubt that at least a portion of the polynuclear hydroxy-Al can be phytotoxic. While small differences are often observed when empirical methods for Al speciation are compared,[50,73,74] a survey of published reports reveals remarkable consistency in f_p for a given value of n_0.[14,49,50,73] Our values are also in general agreement with other reports, and it seems unlikely that any uncertainties in f_p due to methodology are large enough to alter our conclusions. Thus, we would concur with the findings of others[38,39,40] that polynuclear Al can, under certain conditions, be at least as toxic as Al^{3+}. Research on speciation methodology has often focused on quantification of mononuclear Al as a measure of toxic Al,[55,62,74] a practice that must be reevaluated in light of these findings.

Our findings do contrast with several published reports. Moore[20] ascribed maximal inhibition of wheat root growth at pH 4.5 to a greater toxicity of $AlOH^{2+}$ relative to the other species. However, no attempt was made to account for polynuclear Al, despite evidence that several solutions were supersaturated with respect to $Al(OH)_3$, and we suspect that Moore's [20] results reflect toxicity of polynuclear Al, not differential toxicity of monomeric species. It is more difficult to rationalize the discrepancies between our results and those of Blamey et al.[30] and Alva et al.,[34] who, having estimated mononuclear Al via a modified aluminon procedure, noted a high correlation between root growth of soybeans (*Glycine max* (L.) Merr.) and Σa_{mono}. A number of factors could contribute to this discrepancy including the presence of reactive ligands (phosphate, sulfate, citrate) in their more complete nutrient media that could have introduced additional uncertainties in speciation, and their choice of a different set of thermodynamic constants. Moreover, it is possible that there are differences in the response of monocots and dicots to Al speciation. An additional problem is that of collinearity which was discussed by Blamey et al.[30] when they noted that *R* values exceeded 0.9 for the regression of relative root length on any one of the mononuclear species, as well as on Σa_{mono}. Data presented by Alva et al.[34, Table 1] appear to suffer from similar complications in that the simple correlation coefficients between many pairs of species activities were highly significant. This is a general problem whenever Al_t and pH are both varied, and greatly confounds differentiation of toxic and nontoxic mononuclear species. The approach taken in this study, wherein pH is varied at a single value of Al_t, $[Al_a]$, or Σa_{mono}, is less subject to statistical difficulties. The failure of others[30,32,34] to observe the toxicity of polynuclear Al is more difficult to resolve. There is ample evidence that the products formed upon partial neutralization of Al can depend on Al_t, solution age, rate of base addition, presence of ligands such as phosphate and sulfate, and other factors.[13,14,41,42,43] Comparisons are further confounded by a lack of unambiguous definitions. For example, Al_c has been regarded as either a large polymer[49] or colloidal $Al(OH)_3$.[14] The nonmononuclear fraction has been divided into anywhere from two to four subfractions using a number of diverse techniques.[14,41,43,49,73,75] We have employed the ferron method to differentiate three fractions of Al, and our results are consistent with those where the same method has been used to study "pure" partially neutralized Al solutions.[43,49,73] Blamey et al.[30] and Alva et al.[32,34] utilized a modified aluminon method wherein the absorbance at 30 min was taken as mononuclear Al, although we have shown that this approach may include some polymeric Al as well.[50] The aluminon procedure for "total" Al (wherein the sample is acidified and heated prior to assay) was used to differentiate soluble and solid phase Al, and the difference between this value and the monomeric assay taken to be polymeric Al. It is impossible to say how these results compare with our ferron assays, or even if reactive (and in our view, phytotoxic) polymers were present in their more complex nutrient solutions. These uncertainties in definition and differences in methodology prompted our additional studies to see how toxicity of polynuclear Al might vary with experimental conditions. Increasing rate of base addition did not yield as pronounced effects as did increasing age of the stock, but both decreased [Al_b]. Phosphate was also effective in this regard, an observation that is generally consistent with other reports.[30,42] Taken together, the data from these experiments suggest a close association between [Al_b] and toxicity, although most of the treatments seem to be at saturating levels for Tyler wheat (Figure 9). Thus, more experiments are needed, perhaps employing a more tolerant variety or species, to better define the response to varying levels of Al_b.

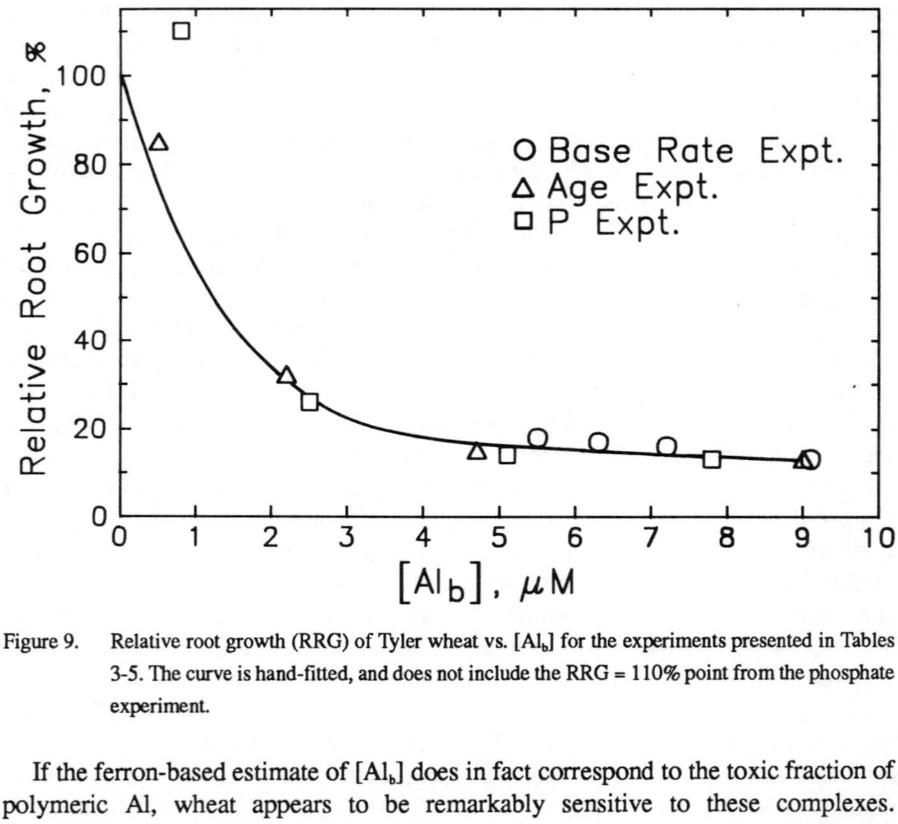

Figure 9. Relative root growth (RRG) of Tyler wheat vs. [Al_b] for the experiments presented in Tables 3-5. The curve is hand-fitted, and does not include the RRG = 110% point from the phosphate experiment.

If the ferron-based estimate of [Al_b] does in fact correspond to the toxic fraction of polymeric Al, wheat appears to be remarkably sensitive to these complexes. Computation of activities for Al_b is precluded by the large size, polyvalent charge, and unknown structure of the polymers, and it is thus difficult to compare relative toxicities of Al_b and Al^{3+}. However, relative root growth approached a minimum at about 3 μM [Al_b] (Figure 9). In contrast, 3 μM [Al^{3+}] (1.2 μM {Al^{3+}}) would result in 85 to 90% RRG at this Ca level (Figure 2, Table 2). Thus, when expressed on a concentration basis, Al_b appears to be considerably more toxic than Al^{3+}. The very low concentrations of Al_b required to substantially inhibit wheat root growth suggest that detection of polymer toxicity may at times be difficult, and may thus explain the infrequency of reports on this effect. Toxicity of polymers has also been suggested in studies of fish[9] and the symbiotic bacterium *Rhizobium trifolii*,[76] suggesting that the toxicity may be general. Future studies on Al toxicity should carefully consider formation of polynuclear species and their possible adverse effect on growth and survival.

It remains to be seen whether the anomalously high root growth observed for the P/Al = 0.3 treatment (Table 5, Figure 9) is a consistent phenomenon, and not just a result of experimental variation. White[44] postulated (but did not conclusively demonstrate) that, at certain combinations of pH (or basicity), and concentrations of P and Al, polynuclear hydroxy-phosphate-Al complexes of low or neutral charge could form that would be readily taken up by plants. If the Al could be internally detoxified, the P could become metabolically available, and thus explain occasional reports of growth stimulation by low levels of Al.[18,44] We are reluctant to support this model based on a single experiment, but it may provide an explanation for our results.

The foregoing discussion has ignored the third hypothesis presented in the introduction -- that changes in sensitivity to Al may account for some of the pH-dependent responses to Al. Unfortunately, there seems to be no direct way of separating speciation and physiological effects of pH. However, we thought that lanthanum, a trivalent metal that does not hydrolyze over the pH range of interest,[12] might offer a means to address this question. Clarkson[77] reported that the trivalent metals Al, La, gallium, and indium induced similar toxicities in onion (*Allium cepa*) roots. Initial results in our studies were encouraging in that La and Al caused similar anatomical symptoms (swelling of the epidermal and outer cortical cells in the zone of elongation), and in that toxicity of La could also be ameliorated by additions of Ca or Na. However, when grown in solutions with equal activities of Al^{3+} or La^{3+}, Tyler and Seneca wheat did not exhibit the differential tolerance to La that they did to Al (data not shown), and it is improbable that the physiological mechanisms for the two metals are exactly the same. Experiments were conducted wherein pH was varied from 4.2 to 5.6 in the presence of 5 or 10 μM La_t, and 1.5 mM $CaCl_2$. There was actually a slight decline in root length of Tyler wheat with increasing pH in the presence of La (data not shown) which at this time is unexplained. However, the data suggest that, to the extent that we can compare Al and La, there is no tendency for low pH to predispose roots to trivalent metal injury.

Our results with hydroxy-Al solutions also have some practical implications for many Al toxicity studies. When solution culture methods are used, consideration of possible polymer formation is critical, especially when discrete levels of toxic Al are desired. In the absence of toxic polymers, $\{Al^{3+}\}$ appears to provide the most reliable estimate of toxic Al (as long as cation amelioration is also considered). At present, it is not known whether significant quantities of toxic polynuclear Al exist in soil solution. In general, soil clays and organic matter are thought to have such an affinity for polymers that it is quite difficult to exchange these species into solution.[78] If so, only an estimate of $\{Al^{3+}\}$ in soil solution may be needed to assess toxic Al status.[6,7] However, Bache and Sharp[79] proposed that Al extracted by dilute salt solutions, and unreactive with the 8-hydroxyquinoline reagent, represented polynuclear hydroxy-Al that could also exist in soil solution. More data are thus needed to assess the ubiquity of polynuclear Al in soil solution, and its potential role in acid soil infertility.

We can now generalize that, regarding mononuclear Al, complexation (with a concomitant reduction in charge) will ameliorate toxicity -- regardless of whether the ligand is SO_4^{2-}, F^-, OH^-, or an organic acid.[16] However, charge reductions of ionic species are usually predicted to enhance diffusional transport across the plasmalemma of cells.[80] Although largely speculative at this point, our current interpretation is that transport of Al into the cell is not necessarily a requisite step in the expression of toxicity, and that primary lesions may be extracellular. The highly charge-dependent amelioration by cations also suggests competition for external, negatively-charged binding sites. While differential tolerance to Al has sometimes been ascribed to exclusion from the cytoplasm or to intracellular detoxification,[18,20] more recent emphasis has been placed on effects of Al binding on membrane structure and function, although a unified model for a molecular basis has yet to emerge.[15,19] Higher plants pose the added complication of the cell wall, which may also represent an extracellular site for Al toxicity.[18]

That polynuclear hydroxy-Al can also be phytotoxic may be complementary with the above hypothesis. While polymerization does result in a reduction in average positive charge per Al atom, the overall polyvalent charge combined with steric factors could lead to a high surface affinity for these species, much as they do with clay minerals.[78] Moreover, the seemingly large size (from perhaps 13 up to 100 Al atoms per unit[14,41]) and high charge of the polymers would argue against ready transport across the plasmalemma. We have conducted experiments with several other polyvalent cations, including La and Sc (as mentioned previously), and two polyamines -- spermine ($H_3N(CH_2)_3H_2N(CH_2)_4H_2N(CH_2)_3NH_3^{4+}$) and spermidine ($H_3N(CH_2)_3H_2N(CH_2)_4NH_3^{3+}$). All were quite inhibitory to wheat and red clover root growth at concentrations in the micromolar range (data not shown), and it would appear that inhibition of root growth by polyvalent cations is a general phenomenon.[77] Surface binding to membranes or cell walls with subsequent disruption of structure or metabolic function might represent one cause for this inhibition.

The question of internal vs. external lesion remains to be answered, but our results may prove useful in that regard. For example, if roots grown in solutions of increasing $\{AlSO_4^+\}$, but low, constant $\{Al^{3+}\}$, could be accurately analyzed for intracellular levels of Al, then increasing accumulation would suggest that intracellular lesions are not of primary importance. Such a conclusion hinges on the weakness of the $AlSO_4^+$ complex, and the presence of numerous cytoplasmic ligands such as phosphate, OH^-, organic acids, nucleotides, and proteins that have high affinities for Al. Thus, if Al enters the cell as $AlSO_4^+$, it is hardly conceivable that its nontoxicity is due to continued stability of this complex.

Relationships between Al speciation and phytotoxicity have largely been investigated using plant species of agronomic importance. We have used wheat for most of our studies because it demonstrates rapid and reproducible root growth that is sensitive to changes in speciation. There seems to be little comparable information for plants, particularly forest species, that might be susceptible to acid deposition-related increases in soluble soil Al. Moreover, it remains to be seen whether speciation-toxicity relationships are generally consistent across plant groups or species that are physiologically diverse. Exploration of this question, along with elucidation of the molecular basis for Al toxicity, represent profitable future endeavors for soil chemists, plant physiologists, and scientists in related disciplines.

REFERENCES

1. Hartwell, B. L., and F. R. Pember. "The presence of aluminum as a reason for the difference in the effect of so-called acid soil on barley and rye", *Soil Sci.* 6(4):259-279 (1918).

2. Adams, F., and P. J. Hathcock. "Aluminum toxicity and calcium deficiency in acid subsoil horizons of two coastal plains soil series", *Soil Sci. Soc. Am. J.* 48(6):1305-1309 (1984).

3. Adams, F., and Z. F. "Effect of chemical activity of soil solution aluminum on cotton root penetration of acid subsoils", *Soil Sci.* 101(3):193-198 (1966).

4. Adams, F., and B. L. Moore. "Chemical factors affecting root growth in subsoil horizons of coastal plain soils", *Soil Sci. Soc. Am. J.* 47(1):99-102 (1983)

5. Brenes, E., and R. W. Pearson. "Root responses of three graminae species to soil acidity in an oxisol and an ultisol", *Soil Sci.* 116(4):295-302 (1973).

6. Pavan, M. A., F. T. Bingham, and P. F. Pratt. "Toxicity of aluminum to coffee in ultisols and oxisols amended with $CaCO_3$, $MgCO_3$, and $CaSO_4 \cdot 2H_2O$". *Soil Sci. Soc. Am. J.* 46(6):1201-1207 (1982).

7. Wright, R. J., V. C. Baligar, and S. F. Wright. "Estimation of phytotoxic aluminum in soil solution using three spectrophotometric methods", *Soil Sci.* 144(2):224-232 (1987).
8. Wright, R. J., and S. F. Wright. "Effects of aluminum and calcium on the growth of subterranean clover in Appalachian soils", *Soil Sci.* 143(5):341-348 (1987).

9. Baker, J. P., and C. L. Schofield. "Aluminum toxicity to fish in acidic waters", *Water Air Soil Pollut.* 18(1-2-3):289-309 (1982).

10. Driscoll, C. T., Jr., J. P. Baker, J. J. Bisogni, Jr., and C. L. Schofield. "Effect of aluminum speciation on fish in dilute acidified waters", *Nature* 284(5752):161-164 (1980).

11. Ulrich, B., R. Mayer, and P. K. Khanna. "Chemical changes due to acid precipitation in a Loess-derived soil in central Europe", *Soil Sci.* 130(4):193-199 (1980).

12. Baes, C. F., Jr., and R. E. Mesmer. *The Hydrolysis of Cations.* (New York: John Wiley and Sons, Inc., 1976).

13. Hsu, P. H. "Aluminum hydroxides and oxyhydroxides", in *Minerals in Soil Environments*, J. B. Dixon and S. B. Weed, Ed. (Madison, WI: Soil Science Society of America, 1977), p. 99-143.

14. Smith, R. W. "Relations among equilibrium and nonequilibrium aqueous species of aluminum hydroxy complex", *Adv. Chem Ser.* 106:250-279 (1971)

15. Haug, A. "Molecular aspects of aluminum toxicity", *CRC Crit. Rev. Plant Sci.* 1:345-373 (1984).

16. Hue, N. V., G. R. Craddock, and F. Adams. "Effect of organic acids on aluminum toxicity in subsoils", *Soil Sci. Soc. Am. J.* 50(1):28-34 (1986).

17. Lindsay, *W. L. Chemical Equilibria in Soils*. (New York: John Wiley and Sons, 1979).

18. Foy, C. D. "Physiological effects of hydrogen, aluminum, and manganese toxicities in acid soil", *Soil Acidity and Liming*, 2nd ed., F. Adams, Ed. (Madison, WI: American Society of Agronomy, 1984) pp. 57-97.

19. Haug, A., and C. R. Caldwell. "Aluminum toxicity in plants: The role of the root plasma membrane and calmodulin", in *Frontiers of Membrane Research in Agriculture*, J. B. St. John et al., Ed. (Totowa, NJ: Rowman and Allanheld, 1985), p. 359-381.

20. Moore, D. P. "Physiological effects of pH on roots", in *The Plant Root and its Environment*, E. W. Carson, Ed. (Charlottesville, VA: Univ. Press of Virginia, 1974), p. 135-151.

21. Kinraide, T. B., R. C. Arnold, and V. C. Baligar. "A rapid assay for aluminum phytotoxicity at submicromolar concentrations", *Physiol. Plant*. 65(3):245-250 (1985).

22. Moore, D. P., W. E. Kronstad, and R. J. Metzger. "Screening wheat for aluminum tolerance", in *Plant Adaptation to Mineral Stress in Problem Soils*, M. J. Wright, Ed. (Ithaca, NY: Cornell Univ. Agric. Exp. Sta., 1976), p. 287-295.

23. Cramer, G. R., and A. Lauchli. "Ion activities in solution in relation to Na-Ca interactions at the plasmalemma", *J. Expt. Bot*. 37(176):321-330 (1986).

24. Khasawneh, F. E. "Solution ion activity and plant growth", *Soil Sci. Soc. Am. Proc.* 35(3):426-436 (1971).

25. Kinraide, T. B., and D. R. Parker. "Cation amelioration of aluminum toxicity in wheat", *Plant Phys*. 83(6):546-551, (1987).

26. Ali, M. E. "Influence of cations on aluminum toxicity in wheat (*Triticum aestivum* Vill. Host.)", Ph.D. Thesis, Oregon State University, (1973).

27. Foy, C. D., A. L. Fleming, and W. H. Armiger. "Aluminum tolerance of soy bean varieties in relation to calcium nutrition", *Agron. J*. 61(4):505-511 (1969).

28. Munns, D. N. "Soil acidity and growth of a legume. II. Reactions of aluminum and phosphate in solution and effects of aluminum, phosphate, calcium, and pH on *Medicago sativa* L. and *Trifolium subterraneum* L. in solution cuture", *Aust. J. Agric. Res.* 16(5):743-755 (1965).

29. Rhue, R. D., and C. O. Grogan. "Screening corn for Al tolerance using different Ca and Mg concentrations", *Agron. J*. 69(5):755-760 (1977).

30. Blamey, F. P. C., D. G. Edwards, and C. J. Asher. "Effects of aluminum, OH:Al and P:Al molar ratios, and ionic strength on soybean root elongation in solution culture", *Soil Sci.* 136:(4)197-207 (1983).

31. Pavan, M. A., and F. T. Bingham. "Toxicity of aluminum to coffee seedlings grown in nutrient solutions", *Soil Sci. Soc. Am. J*. 46(5):993-997 (1982).

32. Alva, A. K., D. G. Edwards, C. J. Asher, and F. P. C. Blamey. "Effects of phosphorus/aluminum molar ratio and calcium concentration on plant response to aluminum toxicity", *Soil Sci. Soc. Am. J*.240(1):133 (1986).

33. Franco, A. A., and D. N. Munns. "Acidity and aluminum restraints on nodulation, nitrogen fixation, and growth of *Phaseolus vulgaris* in solution culture", *Soil Sci. Soc. Am. J.* 46(2):296-301 (1982).

34. Alva, A. K., D. G. Edwards, C. J. Asher, and F. P. C. Blamey. "Relationships between root length of soybean and calculated activities of aluminum monomers in solution", *Soil Sci. Soc. Am. J.* 50(4)959-962 (1986).

35. Kinraide, T. B., and D. R. Parker. "Nonphytotoxicty of the aluminum sulfate ion, $AlSO_4^+$", *Physiol. Plant.* 71(2):207-212 (1987).

36. Cameron, R. S., G. S. P. Ritchie, and A. D. Robson. "Relative toxicities of inorganic aluminum complexes to barley", *Soil Sci. Soc. Am. J.* 50(5):1231- 1236 (1986).

37. Konishi, S., and S. Miyamato. "Alleviation of aluminum stress and stimulation of pollen tube growth by fluorine", *Plant Cell Physiol.* 24:857- 862 (1983).

38. Bartlett, R. J., and D. C. Riego. "Toxicity of aluminum in relation to pH and phosphorus", *Soil Sci.* 114(3):194-201 (1972).

39. Wagatsuma, T. and Y. Ezoe. "Effect of pH on ionic species of aluminum in medium and on aluminum toxicity under solution culture", *Soil Sci. Plant Nutr.* 31(4):547-561 (1985).

40. Wagatsuma, T. and M. Kaneko. "High toxicity of hydroxy-aluminum polymer ions to plant roots", *Soil Sci. Plant Nutr.* 33(1):57-67 (1987).

41. Bertsch, P. M. "Conditions for Al_{13} polymer formation in partially neutralized aluminum solutions", *Soil Sci. Soc. Am. J.* 51(3):825-828 (1987).

42. Hsu, P. H. "Interactions between Aluminum and phosphate in aqueous solution", *Adv. Chem. Ser.* 243:115-127 (1968).

43. Turner, R. C. "Effect of aging on properties of polynuclear hydroxyaluminum cations", *Can. J. Chem.* 54(10):1528-1534 (1976).

44. White, R. E. "Studies on mineral adsorption by plants. III. The interaction of aluminum, phosphate, and pH on the growth of *Medicago sativa*", *Plant Soil* 46(1):195-208 (1976).

45. Bartlett, R. J., and D. C. Riego. "Effect of chelation on toxicity of aluminum", *Plant Soil.* 247(2):419-423 (1972).

46. DeKock, P. C., and R. L. Mitchell. "Uptake of chelated metals by plants", *Soil Sci.* 84(1):55-62 (1957).

47. Foy, C. D., and J. C. Brown. "Toxic factors in acid soils: II. Differential tolerance of plant species", *Soil Sci. Soc. Am. Proc.* 28(1):27-32 (1964).

48. McLean, F. T., and B. E. Gilbert. "The relative aluminum tolerance of crop plants", *Soil Sci.* 24(3):163-171 (1927).

49. Jardine, P. M., and L. W. Zelazny. "Mononuclear and polynuclear aluminum speciation through differential kinetic reactions with ferron". *Soil Sci. Soc. Am. J.* 50(4):895-900 (1986).

50. Parker, D. R., L. W. Zelazny, and T. B. Kinraide. "Comparison of three spectrophotometric methods for differentiating mono- and polynuclear hydroxy-aluminum complexes", *Soil Sci. Soc. Am. J.* 52(1):67-75 (1988).

51. Parker, D. R., T. B. Kinraide, and L. W. Zelazny. "Aluminum speciation and phytotoxicity in dilute hydroxy-aluminum solutions", *Soil Sci. Soc. Am. J.* 52(2):438-445 (1988).

52. Parker, D. R., L. W. Zelazny, and T. B. Kinraide. "Improvements to the program GEOCHEM", *Soil Sci. Soc. Am. J.* 51(2):488-491 (1987).

53. Sposito, G., and S. V. Mattigod. *GEOCHEM: A Computer Program for the Calculation of Chemical Equilibria in Soil Solution and Other Natural Water Systems* (Riverside, CA: Kearney Foundation of Soil Science, Univ. of California, 1980).

54. Baumann, E. "Determination of stability constants of hydrogen and aluminum fluorides with a fluoride-selective electrode", *J. Inorg. Nucl. Chem.* 31(10):3155-3162 (1969).

55. James, B. R., C. J. Clark, and S. J. Riha. "An 8-hydroxyquinoline method for labile and total aluminum", *Soil Sci. Soc. Am. J.* 47(5):893-897 (1983).

56. Behr, B., and H. Wendt. "Fast ion reactions in solutions. I. Formation of aluminum sulfate complexes", *Z. Elektrochem.* 66:223-228 (1962).

57. Nanda, R. K., and S. Adity. "Studies on ion association equilibria. spectrophotometric determination of the thermodynamic instability constants of $AlSO_4^+$, $GaSO_{4+}$, and $InSO^{4+}$". *Z. Phys. Chem.* 35:139-145 (1962).

58. Izatt, R. M., D. Eatough, J. J. Christensen, and C. H. Bartholomew. "Calorimetrically determined log K, ΔH^0, and ΔS^0 values for the interaction of sulphate ion with several bi- and ter-valent ions", *J. Chem. Soc. A.* 47-53 (1969).

59. Nishide, T., and R. Tsuchiya. "The formation of the Al^{3+}-SO_4^+ ion pair in an aqueous solution of potassium aluminum alum", *Bull. Chem. Soc. Japan* 38(8):1398-1400 (1965).

60. Stryker, L. J., and E. Matijevic. "Counterion complexing and sol stability. II. coagulation effects of aluminum sulfate in acidic solutons", *J. Phys. Chem.* 73(5):1484-1487 (1969).

61. Matijevic, E., and L. J. Stryker. "Coagulaton and reversal of charge of lyophobic colloids by hydrolized metal ions. III. Aluminum sulfate", *J. Colloid Int. Sci.* 22:68-77 (1966).

62. Jardine, P. M., and L. W. Zalazny. "Influence of inorganic anions on the speciation of mononuclear and polynuclear aluminum by ferron", *Soil Sci. Soc. Am. J.* 51(4):889-892 (1987).

63. Luttge, U., and N. Higinbotham. *Transport in Plants* (New York: Springer- Verlag, Inc., 1979).

64. Hanson, J. B. "The functions of calcium in plant nutrition", in *Advances in Plant Nutrition*, Vol. I. P. B. Tinker and A. Lauchli, Ed. (New York: Praeger Publishers, 1984), pp. 149-208.

65. Deleers, M. "Cationic atmosphere and cation competition binding at negatively charged membranes: Pathological implications for aluminum", *Res. Commun. Chem. Pathol. Pharmacol.* 49(2):277-294 (1985).

66. Adams, F. "Soil solution", in *The Plant Root and its Environment*, E. W. Carson, Ed. (Charlottesville, VA: Univ. Press of Virginia, 1974), p. 441-481.

67. Schecher, W. D., and C. T. Driscoll. "An evaluation of uncertainty associated with aluminum equilibrium calculations", *Water Resources Res.* 23(4):525-534 (1987).

68. Hern, J. L., W. L. Stout, R. C. Sidle, and O. L. Bennett. "Characterization of fluidized combustion wastes: Composition and variability as they relate to disposal on agricultural land", in *Proceedings of the 5th International Conference on Fluidized Bed Combustion*, Volume II (McLean, VA: Mitre Corp., 1978), pp. 833-839.

69. Oates, K. M., and A. G. Caldwell. "Use of by-product gypsum to alleviate soil acidity", *Soil Sci. Soc. Am. J.* 49(4):915-918.

70. Ritchey, K. D., D. M. G. Souza, E. Lobato, and O. Correa. "Calcium leaching to increase rooting depth in a Brazilian savannah oxisol", *Agron. J.* 72(1):40-44 (1980).

71. Sumner, M. E., H. Shahandeh, J. Bouton, and J. Hammel. "Amelioration of an acid soil profile through deep liming and surface application of gypsum", *Soil Sci. Soc. Am. J.* 50(5):1254-1258 (1986).

72. Sadiq, M., and W. L. Lindsay. "Selection of standard free energies of formation for use in soil chemistry", *Colorado Exp. Sta. Tech. Bull.* 134. (1979) pp. 79-128.

73. Bertsch, P. M., W. J. Layton, and R. I. Barnhisel. "Speciation of hydroxy- aluminum solutions by wet chemical and aluminum-27 NMR methods", *Soil Sci. Soc. Am. J.* 50(6):1449-1454 (1986).

74. Hodges, S. C. "Aluminum speciation: A comparison of five methods", *Soil Sci. Soc. Am. J.* 51(1):57-64 (1987).

75. Bersillon, J. L., P. H. Hsu, and F. Fiessinger. "Characterization of hydroxy-aluminum solutions", *Soil Sci. Soc. Am. J.* 44(3):630-634 (1980).

76. Wood, M., and J. E. Cooper. "Aluminum toxicity and multiplication of *Rhizobium trifolii* in a defined growth medium", *Soil Biol. Biochem.* 16(6):571-576 (1984).

77. Clarkson, D. T. "The effect of aluminum and some other trivalent metal cations on cell division in the root apices of *Allium cepa*", *Ann. Bot.* 29(114):309-315 (1965).

78. Zelazny, L. W., and P. M. Jardine. "Surface reactions of aqueous aluminum", in *Environmental Chemistry of Aluminum*, G. Sposito, Ed. (Boca Raton, FL: CRC Press, in press).

79. Bache, B. W., and G. S. Sharp. "Soluble polymeric hydroxy-aluminum ions in acid soils", *J. Soil Sci.* 27(2):167-171 (1976).

80. Marschner, H. *Mineral Nutrition of Higher Plants* (London: Academic Press, 1986).

CHAPTER 8

DYNAMIC PROPERTIES OF CALMODULIN IN RESPONSE TO ALUMINUM

Shixing Yuan, Chris Weis, Alfred Haug

INTRODUCTION

Aluminum, the most abundant metal in the earth's crust, has been implicated as early as 1918 as a cause of root-growth retardation in crop plants grown on acidic soil.[1] Today, aluminum toxicity is recognized as a serious global problem for crop productivity because vast areas of the world suffer from soil acidity which mobilizes soil aluminum available for plants.[2,3] Besides natural soil acidity, anthropogenic perturbation of hydrogeochemical processes has further contributed to the acidification of terrestrial and aquatic regions previously not suffering from aluminum toxicity.[4] For example, so-called acid rain has substantially increased over wide parts of the eastern United States, Canada, and western Europe.[5] As to human health, clinical and experimental findings suggest a link between aluminum uptake and a variety of disorders in man, e.g., osteomalacia-type diseases and dialysis encephalopathy.[6]

Little is known about molecular mechanisms regarding entry of aluminum species into cells via the plasma membrane, about primary targets for these species, or about aluminum-induced cellular lesions. Deeper insight into cellular and molecular processes involving aluminum ions is crucial for the biotechnological development of plant resistance to aluminum toxicity. Despite this dearth of information, cellular membranes and calcium-regulated biochemical processes appear to emerge as potential targets for toxic aluminum ions.

Concerning aluminum interactions with membranes, a body of evidence suggests pronounced aluminum-induced phase changes in the plasma membrane of thermophilic Archaebacteria[7] and in plasma membrane-enriched microsomal fractions isolated from maize roots.[8] Furthermore, micromolar aluminum concentrations induced membrane rigidification and membrane fusion in unilamellar lipid vesicles containing phosphatidylserine.[9]

Besides interfering with the plasma membrane, physiological studies indicated that aluminum ions effect changes in calcium regulation.[2,6] Therefore, a potential target for aluminum interaction might be calmodulin, a pivotal, calcium-dependent regulatory protein (MW≈17,000). This acidic protein has been highly conserved during evolution of eukaryotic cells, where it participates in a multitude of cellular processes.[10] Given the significance of calmodulin in integrating calcium-dependent processes, aluminum-induced changes of the protein's structure would therefore have severe repercussions on cellular functions. Indeed, recent investigations indicate that binding of aluminum ions to bovine brain calmodulin results in pronounced structural and dynamic changes which, in turn, lead to the inhibition of crucial calcium- and calmodulin-dependent enzymes.[3,11,12]

As to aluminum-induced structural changes, the metal binds stoichiometrically and cooperatively to the protein, as determined by fluorescence and circular dichroism studies. Binding of aluminum occurs irrespective of the presence or absence of calcium ions.[11,13] When bound to the protein, aluminum ions trigger a helix-coil transition whereby the helix content decreases by about 30%. It is noteworthy, that addition of calcium ions promotes helix formation,[10] with aluminum absent. Atomic absorption experiments indicated that the application of excess citrate to an [aluminum]/[calmodulin]=(3:1) complex resulted in a time-dependent, quantitative removal of the metal from the protein.[14] As judged by spectroscopic techniques, this citrate-mediated protein restoration is somewhat incomplete in terms of helical content. The incomplete rearrangement of calmodulin structure is probably related to aluminum-induced changes in solvation. The rearranged water molecules may compete for hydrogen bonding sites on calmodulin when the protein refolds as a result of citrate-mediated aluminum removal.

Concomitant with aluminum-induced changes in helix content, the hydrophobic surface exposure of calmodulin is enhanced,[3] relative to the area observed in the presence of stoichiometric quantities of calcium ions.[10] This hydrophobic region or its immediate vicinity probably plays a key role in the interaction of calmodulin with specific partner proteins, e.g., an ATPase.[10] Aluminum-related changes at the interface between calmodulin and a partner protein were illustrated in experiments with melittin,[15] a small protein which has a high affinity with calcium-calmodulin.[16] As monitored at the single tryptophanyl residue of melittin, the fluorophore's average microenvironment is modified such that its apparent lifetime is shortened when aluminum is present. Moreover, the interface between the two proteins seems to become more polar when aluminum is bound to calmodulin. In the presence of aluminum, calmodulin's solvation structure is possibly altered, which may be unfavorable for a proper fit between calmodulin and the target protein.

We present in this article further results on aluminum-related structural and dynamic changes monitored by a fluorescent dye attached to the single cysteinyl residue of spinach calmodulin. By measuring molecular parameters like fluorescence anisotropy, motional parameters were derived which indicate that aluminum altered calmodulin differs from native calcium-calmodulin.

MATERIAL AND METHODS

Mastoparan-X and mastoparan were purchased from Peninsula Laboratories (Belmont, CA). N-iodoacetyl-N'-(5-sulfo-1-naphthyl) ethylene-diamine (1,5-I-AEDANS) was obtained from Sigma Chemical Company (St. Louis, MO). Affinity chromatography material, phenothiazine Affi-gel, and a protein assay kit were purchased from Bio-Rad Laboratories (Richmond, CA). All other chemicals used were of the highest quality available.

Calmodulin was isolated from fresh spinach using methods previously described.[14,17,18] Spinach calmodulin obtained in this manner was found to be free of tryptophan as judged by UV absorbance and lack of fluorescence emission at 340 nm following excitation with 295 nm light. In addition, the protein's UV absorption spectrum showed a partially resolved vibrational structure typical of that known for bovine brain calmodulin.[10] Furthermore, analysis of the spinach calmodulin preparation on 15% polyacrylamide gels followed by Coomassie blue staining indicated a single band migrating at an apparent molecular weight of 17,000, this result being consistent when up to 45 μg protein were loaded per gel lane.

To attach the fluorescence probe, 1,5-I-AEDANS, to spinach calmodulin, the protein was first dissolved in Tris buffer (10 m*M*, pH 7.0) at a concentration of about 0.2 mg/mL buffer, then a thiol-reducing agent, dithiothreitol, was added to the solution to a final concentration of 0.2 m*M*. The reaction was allowed to proceed at room temperature for 18 hours. The reacted sample was then extensively dialysed against the same Tris buffer to remove excess thiol-reducing agent. The fluorescence label, 1,5-I-AEDANS, was added and reacted for about 2 hours with the sample as described.[19] Finally, the reaction mixture was extensively dialyzed against Mops buffer (10 m*M* at pH 6.5 and 0.1 M KCl), which was used in all experiments. The molar ratio of AEDANS bound per protein was found to be 0.95 to 1, as determined by dye absorbance (= 6100 $cm^2 \cdot mM^{-1}$ at 337 nm)[20] and Bradford assay[21], respectively. Unless indicated otherwise, the protein concentrations employed were 10 μ*M*. To occupy all four Ca^{2+} binding domains of calmodulin, Ca^{2+} ions were added to the protein at a molar ratio of 5:1, to achieve saturating conditions.

As described previously[15,18], fluorescence experiments (anisotropy, lifetime) were performed on an SLM spectrofluorimeter, model 4800, manufactured by SLM Instruments, Inc. (Urbana, IL). For each assay a Beckman cuvette with 1 cm optical pathlength and a sample volume of 2.5 mL was used. Furthermore, at the low protein concentrations used, the anisotropy and the lifetime were independent of protein concentration.[22] Each data point represents the average value for 100 counts per sample which, in turn, was four times replicated. Briefly, the steady-state fluorescence polarization anisotropy, r, was calculated from

$$r = (I_{\parallel} - I_{\perp})/(I_{\parallel} + 2I_{\perp}) \qquad (1)$$

where $I_{\parallel}$, and $I_{\perp}$ are the fluorescence intensity values of the polarized emission measured parallel and perpendicular with respect to excitation polarization. The fluorescence anisotropy is related to the lifetime of the fluorescence, ζ, and rotational correlation time, ϕ, according to the Perrin equation[22]:

$$r = r_o(1 + \zeta/\phi)^{-1} \tag{2}$$

where r_o refers to the anisotropy in the absence of whole molecular motion. Applying phase shift and demodulation techniques[22], fluorescence lifetimes could be determined. Using eqn. 2, the rotational correlation time could thus be evaluated. The lifetime data obtained could not be described by a monoexponential decay process.[22] Therefore, a biphasic analysis was applied which approximated the data well.

Fluorescence energy transfer was evaluated between a donor, D, viz., the single tryptophanyl residue on mastoparan-X (MW = 1,557) and an acceptor, A, viz., the fluorescence label, AEDANS, residing at the single cysteinyl residue, Cys-26[23], of spinach calmodulin. As opposed to mastoparan-X, mastoparan (MW = 1,479) does not harbor any tryptophanyl residue. The energy transfer efficiency was measured by the donor quenching method[24], i.e.,

$$E = 1 - Q_{DA}/Q_D \tag{3}$$

where E stands for the energy transfer efficiency. Q_{DA} and Q_D refer to the fluorescence quantum efficiency of the donor, in the presence and in the absence of the acceptor, respectively. The ratio Q_{DA}/Q_D can be evaluated from corresponding fluorescence intensities, f_{DA} and f_D, according to

$$Q_{DA}/Q_D = [f_{DA}/f_D]_{av} \times [D]_D/[D]_{DA} \tag{4}$$

Here, both fluorescence intensities, f_{DA} and f_D[22], are measured upon excitation of the donor at 299 nm. The quantities, f_D and f_{DA}, are the fluorescence intensities of the donor, measured in the absence of the acceptor, and in the presence of the acceptor within a wavelength range free from acceptor emission, respectively. Subsequently, the intensity ratios were averaged, thus yielding $[f_{DA}/f_D]_{av}$. $[D]_D$ is the concentration of donor molecules in the absence of any acceptor, whereas $[D]_{DA}$ is that in the presence of both acceptor and donor molecules.

The distance, R, from the donor, D, to the acceptor, A, was calculated from the following equation,

$$R = R_o(1/E - 1)^{1/6} \tag{5}$$

where R_o is the Förster critical distance, evaluated from

$$R_o^6 = (8.79 \times 10^{-5}) \times k^2 \times n^{-4} \times Q_D \times J_{DA} \tag{6}$$

Here, k^2 is the orientation factor and a value of 2/3 was used in these experiments. Usage of this value is justified if donor and acceptor molecules randomize prior to energy transfer.[25] Since the aromatic ring systems of the donor (tryptophan)[22] and the acceptor (AEDANS) have two linear transition moments each, the assumption of random orientation appears to be reasonable. Moreover, n is the refractive index and J_{DA} is the spectral overlap integral defined as

$$J_{DA} = \{\Sigma_\lambda f_D(\lambda)\cdot\varepsilon_A(\lambda)\cdot\varepsilon^4\cdot\Delta\lambda\}/\{\Sigma_\lambda f_D(\lambda)\cdot\Delta\lambda\} \quad (7)$$

where ε_A is the extinction coefficient of the acceptor in $M^{-1}cm^{-1}$; λ is the wavelength in nm.

Four kinds of samples have to be prepared for these energy transfer studies: (a) calmodulin plus mastoparan (neither donor nor acceptor present), (b) calmodulin plus mastoparan-X (donor present, D); (c) AEDANS-labelled calmodulin plus mastoparan (acceptor present, A); and (d) AEDANS-labelled calmodulin plus mastoparan-X (both donor and acceptor present, DA).

RESULTS

Our experiments show that the emission of the AEDANS-spinach calmodulin conjugate reaches a maximum at 490 nm and is considerably blue-shifted from the emission maximum of 520 nm reported for the aqueous free probe.[20] An emission maximum observed at 490 nm corresponds to that reported for AEDANS dissolved in 60% ethanol/water.[20] When covalently bound to spinach calmodulin, the probe apparently resides in an environment less accessible to water. Increased rigidity or further water removal from the probe's microenvironment may be responsible for the blue shift. The fluorescence spectra of dye-labelled spinach calmodulin, measured in the presence or absence of micromolar aluminum or calcium concentrations, virtually coincide.

Table 1 lists the values of various spectroscopic parameters of the dye-labelled calmodulin in the presence and absence of calcium and aluminum. The average fluorescence lifetime of the probe remains relatively constant through serial calcium titration and upon addition of aluminum up to a molar ratio of 3 aluminum ions per protein. However, the rotational correlation time decreases with increasing calcium concentration, consistent with reports that the protein adopts a more compact conformation upon calcium binding.[26] At a molar ratio of [Al]/[calcium saturated calmodulin] = 3:1, the rotational correlation time significantly increases relative to the time determined in the absence of aluminum. The rotational correlation times were calculated from eqn. 2 using an r_o value of 0.0910 ± 0.0035. This value was obtained from a Perrin plot by extrapolation with a correlation coefficient of 0.988.

Our fluorescence lifetime data can be interpreted in terms of a biphasic fluorescence decay. The lifetime of the long-lived component roughly coincides with that measured for free AEDANS in water.[20] Since our instrument does not permit measurements of time-resolved anisotropy decays, an average rotational correlation time had to be calculated (Table 1).

To assess the impact of aluminum on the interaction between calmodulin and its target proteins, radiationless energy transfer experiments were performed between the single tryptophanyl residue on the partner peptide, mastoparan-X, and the AEDANS label on Cys-26 on spinach calmodulin. This kind of peptide is known to bind calmodulin with high affinity in a calcium-dependent manner.[16] The wavelength region of the fluorescence emission band of tryptophan shows excellent overlap with that of the acceptor's absorption spectrum, AEDANS- conjugated cysteine (Figure 1), and therefore, energy transfer from tryptophan to the acceptor can be expected. By integrating in steps of 2 nm, the value of the overlap integral, J_{DA}, was calculated according to eqn. 7. From the reported values of the donor quantum yield, Q_D, of 0.155,[27] a refractive index of n = 1.334, corresponding to that of the 0.1 M KCl solution at room temperature,[28] and an orientational factor of k^2 = 2/3, the Förster critical distance, R_o, can be determined (Table 2). The Förster critical distance, R_o, is in accord with that reported for the radiationless energy transfer from Trp-19 of melittin to AEDANS-labelled troponin C at Cys-98, viz., R_o = 23.6 Å.[29] Since the spectral characteristics of donor emission and acceptor absorption did not change when aluminum ions were added to the calcium-containing sample, the overlap integral and therefore R_o remained the same (Table 2). The efficiency of energy transfer between donor and acceptor was higher when aluminum was present in the sample, resulting in a distance of 29.4 Å. In the absence of aluminum, this distance was 30.3 Å (Table 2). Mastoparan-X probably binds to calmodulin's region III, when calcium is present.[30] Therefore, a distance of about 30 Å between donor and acceptor appears to be reasonable, considering the location of AEDANS at Cys-26 in region I.

Table 1. Fluorescence properties of AEDANS-labelled spinach calmodulin at various cation concentrations

Sample	r	a_1^* / a_2	ζ_1^* / ζ_2 (ns)	ζ_{av} (ns)	ϕ_{av} (ns)
AEDANS-calmodulin	0.0579	0.14 0.86	0.88 9.86	9.7	15.0
AEDANS-calmodulin + $2Ca^{2+}$	0.0524	0.12 0.88	0.57 10.44	10.4	12.6
AEDANS-calmodulin + $5Ca^{2+}$	0.0528	0.14 0.86	0.55 9.71	9.6	11.7
AEDANS-calmodulin + $5Ca^{2+}$ + 3Al	0.0575	0.13 0.87	0.08 9.94	9.9	14.9

* The preexponential factors, a_1 and a_2, and the corresponding lifetimes, ζ_1 and ζ_2, were analyzed on the basis of a biphasic decay. The errors in lifetimes are less than ± 0.07 ns.

r is the fluorescence anisotropy (errors less than 0.0004) measured by T-format method as indicated in legend to Figure 2, except at room temperature.

ζ_{av} are average lifetimes calculated from the relations: $\zeta_{av} = (a_1 \zeta_1^2 + a_2 \zeta_2^2)/(a_1 \zeta_1 + a_2 \zeta_2)$

ϕ_{av} values are the rotational correlation times calculated from eqn. 2.

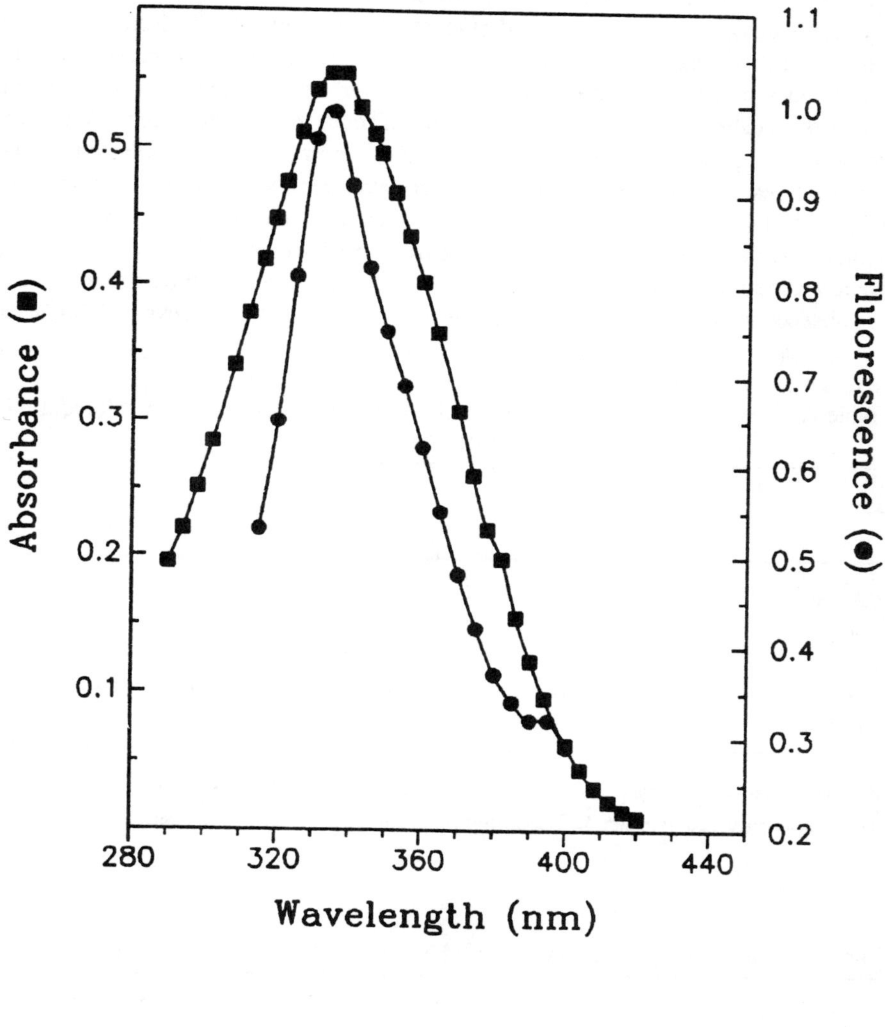

Figure 1. The spectral overlap of the fluorescence emission ("solid circle") of the single tryptophanyl residue in mastoparan-X (donor), associated with spinach calmodulin, and the absorption ("solid square") of AEDANS-conjugated cysteine (acceptor). The samples were dissolved in 4-morpholinopropane sulfonic acid (Mops) buffer (10 m*M*, pH 6.5, 0.1 M KCl and 0.15 m*M* $CaCl_2$). The concentration of mastoparan-X was 20 μ*M*, while that of the spinach calmodulin was 30 μ*M*. The concentration of AEDANS-conjugated cysteine was 9.3 μ*M*. Tryptophan was excited at 299 nm, and both experiments were conducted at room temperature.

To further evaluate aluminum-related dynamic changes of spinach calmodulin, the influence of temperature on the fluorescence anisotropy of AEDANS-conjugated spinach calmodulin was determined (Figure 2). The anisotropy decreased as the temperature increased; however, in the presence of aluminum ions, the anisotropy value was higher at all temperatures tested. Moreover, temperature-induced structural changes of calmodulin are also reflected in molecular parameters representative for those of the protein (Table 3). The molecular parameters were evaluated at two temperatures. At 18.2°C, the rotational correlation time, 29.1 ns, measured for AEDANS-labelled spinach calmodulin in the presence of calcium, is in general accord with values measured at different wavelengths for bovine calmodulin.[31] At 18.2°C, the rotational correlation time is faster, when aluminum is present, compared with that in the absence of aluminum. At 28.5°C, however, this time pattern is reversed (Table 3).

Table 2. Energy transfer characteristics of AEDANS-labelled spinach calmodulin in the absence and in the presence of aluminum ions

sample	energy transfer efficiency (E)	Overlap Integral (J_{DA})	R_o (Å)	R (Å)
Calmodulin + $5Ca^{2+}$	0.194 ± 0.014	6.50×10^{13}	23.9	30.3
Calmodulin + $5Ca^{2+}$ +3Al	0.219 ± 0.009	6.50×10^{13}	23.9	29.4

R_o is the Förster critical distance, as calculated from eqn. 6. R is the distance between donor and acceptor molecule, as calculated from eqn. 5. The experiments were performed at room temperature.

Table 3. Molecular parameters of AEDANS-labelled spinach calmodulin

sample	temperature (°C)	r	ζ_{av} (ns)	ϕ_{av} (ns)
Calmodulin + $5Ca^{2+}$	18.2	0.0571	17.3	29.1
	28.5	0.0486	8.5	9.7
Calmodulin + $5Ca^{2+}$ + 3Al	18.2	0.0616	12.0	25.1
	28.5	0.0525	9.2	12.5

Symbols equivalent to those in Table 1.

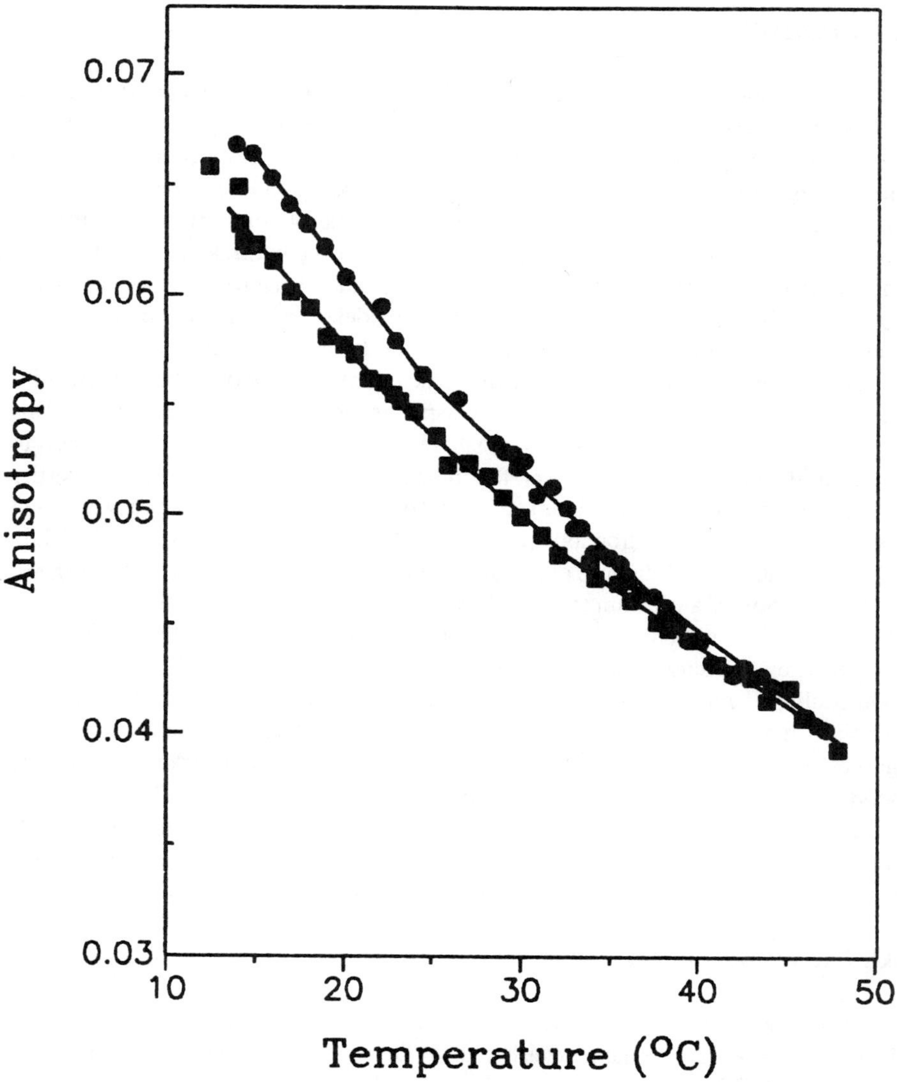

Figure 2. Fluorescence anisotropy values of AEDANS-labelled spinach calmodulin versus temperature, in the presence and in the absence of aluminum ions. The experiments were performed according to the T-format method on a spectrofluorimeter, model SLM 4800. Two identical Corning cut-off filters, CS-72, were placed in the emission beams. The SLM instrument was interfaced with a microcomputer, HP-85, for data analysis. Both samples were excited at 344 nm. Calmodulin concentration was 10 μM.

("solid square"), [calmodulin]:[Ca^{2+}] =1:5;

("solid circle"), [calmodulin]:[Ca^{2+}]:[aluminum] = 1:5:3.

DISCUSSION

X-ray diffraction methods revealed that the equilibrium average structure of calmodulin consists of a dumbbell-shaped protein in which the two lobes are interconnected by an eight-turn α-helix.[32] The C-terminal lobe of calmodulin harbors the two high-affinity calcium binding sites.[10] Like other globular proteins, calmodulin is also a dynamic system which comprises a multitude of conformational substates, easily accessible by rapid structural fluctuations.[33,34] These conformational substates are, in part, dependent on the viscosity of the solvent.[35] Our understanding of the molecular dynamics of proteins is in its infancy; however, evidence is accumulating that at least some of these conformational substates are crucial for protein function.[34,35]

Indeed, we have demonstrated in this article that application of aluminum ions perturbs the conformation of calmodulin existing in the presence of calcium ions. This is illustrated by our findings that application of aluminum ions to [calcium]/[calmodulin]=(5:1) alters the fluorescence probe's motional properties different from those with aluminum absent. Moreover, the probe's microenvironment seems to become more hydrophobic, resembling that observed in the absence of calcium ions. Our data on the influence of temperature on the fluorescence anisotropy suggest that application of aluminum to spinach calmodulin produces a calmodulin structure distinctly different from that in the absence of aluminum, especially at the lower temperatures measured. Since aluminum application breaks hydrogen bonds in calmodulin,[11,15] transitions between certain energy minima of the protein are facilitated which -- in the absence of aluminum -- would be less accessible. In the plot of the fluorescence anisotropy versus temperature, changes may reflect rearrangements in the solid- and liquid-like microdomain structure of the protein.[36] This kind of intramolecular reorganization in aluminum-altered calmodulin is probably responsible for the protein's inability of stimulating enzymes such as cyclic nucleotide phosphodiesterase.[3]

ACKNOWLEDGMENT

This research was supported by a grant from the National Institutes of Health, No. 1 RO1 ES04468-01. The authors wish to thank Dr. K. Poff, Plant Research Laboratory, for use of the SLM instrument.

REFERENCES

1. Hartwell, B. L. and F. R. Pember. "The presence of aluminum as a reason for the difference in the effect of so-called acid soil on barley and rye," *Soil Sci.* 6:259-279 (1918).

2. Foy, C. D., R. L. Chaney, and M. C. White. "The physiology and metal toxicity in plants," *Ann. Rev. Plant Physiol.* 29:511-566 (1978).

3. Haug, A. " Molecular aspects of aluminum toxicity," *CRC Crit. Rev. Plant Sci.* 1:345-373 (1984).

4. Stumm, M., J. J. Morgan and J. L. Schnoor. "Saurer Regen, eine Folge der Störung hydrogeochemischer Kreisläufe," *Naturwiss* 70:216-223 (1983).

5. Evans, L. S. "Biological effects of acidity in precipitation on vegetation: A review," *Environm. Exper. Bot.* 22:155-169 (1982).

6. Ganrot, P. 0. "Metabolism and possible health effects of aluminum," *Environm. Health Perspect.* 65:363-441 (1986).

7. Vierstra, R. and A. Haug. "The effects of Al^{3+} on the physical properties of membrane lipids in *Thermoplasma acidophilum,*" *Biochem. Biophys. Res. Comm.* 84:138-144 (1978).

8. Suhayda, C. G. and A. Haug. "Organic acids reduce aluminum toxicity in maize root membranes," *Physiol. Plant.*24 68:189-195 (1986).

9. Deleers, M., J. P. Servais and E. Wülfert. "Neurotoxic cations induce membrane rigidification and membrane fusion at micromolar concentrations," *Biochim. Biophys. Acta* 855: 271-276 (1986).

10. Klee, C. B. and T. G. Vanaman. "Calmodulin, "*Adv. Protein Chem.* 35:213-321 (1982).

11. Seigel, N. and A. Haug. "Aluminum interaction with calmodulin: Evidence for altered structure and function from optical and enzymatic studies," *Biochim. Biophys. Acta* 744:35-45 (1983).

12. Siegel, N. and A. Haug. "Calmodulin-dependent formation of membrane potential in barley root plasma membrane vesicles: A biochemical model of aluminum toxicity in plants," *Physiol. Plant.* 59:285-291 (1983).

13. Siegel, N. and A. Haug. "Aluminum-induced inhibition of calmodulin-regulated phosphodiesterase activity: Enzymatic and optical studies," *Inorgan. Chim. Acta* 79:230-231 (1983).

14. Suhayda, C. G. and A. Haug. "Citrate chelation as a potential mechanism against aluminum toxicity in cells: The role of calmodulin," *Canad. J. Biochem. Cell Biol.* 63:1167-1175 (1985).

15. Weis, C. and A. Haug. "Aluminum-induced conformational changes in calmodulin alter the dynamics of interaction with melittin," *Arch. Biochem. Biophys.* 254:304-312 (1987).

16. Cox, J. A., M. Comte, J. E. Fitton, and W. F. DeGrado. "The interaction of calmodulin with amphiphilic peptides," *J. Biol. Chem.* 260:2527-2534 (1985).

17. Caldwell, C. R. and A. Haug. "Affinity chromatographic isolation of calmodulin from bovine brain acetone powder," *Anal. Biochem.* 116:325-330 (1981).

18. Yuan, S. and A. Haug. "Ligand-triggered conformational perturbations elicit changes at the single cysteinyl residue of spinach calmodulin," *Eur. J. Biochem.* 175:119-124 (1988).

19. Wu, C. W. and L. Stryer. "Proximity relationships in rhodopsin," *Proc. Nat. Acad. Sci. USA* 69:1104-1108 (1972).

20. Hudson, E. N. and G. Weber. "Synthesis and characterization of two fluorescent sulfhydryl reagents," *Biochem.* 12:4154-4161 (1973).

21. Bradford, M. M. "A rapid and sensitive method for the quantitation of microgram quantities of protein utilizing the principle of protein-dye binding," *Anal. Biochem.* 72:248-254 (1976).

22. Lakowicz, J. R. *Principles of Fluorescence Spectroscopy* (New York: Plenum Press, 1984).

23. Vanaman, T. C. "Chemical approaches to the calmodulin system," *Meth. Enzymol.* 102:296-310 (1983).

24. Fairclough, R. F. and C. R. Cantor. "The use of singlet-singlet energy transfer to study macromolecular assemblies," *Meth. Enzymol.* 48:347-379 (1977).

25. Eisinger, J. "Intramolecular energy transfer in adrenocorticotropin," *Biochem.* 8:3902-3915 (1969).

26. Seamon, K. B. "Calcium- and magnesium-dependent conformational states of calmodulin as determined by NMR," *Biochem.* 19:207-215 (1980).

27. Steiner, R. F., L. Marshall, and D. Needleman. "The interaction of melittin with calmodulin and its tryptic fragments," *Arch. Biochem. Biophys.* 246: 286-300 (1986).

28. Weast, R. C. *Handbook of Chemistry and Physics* (Boca Raton, FL: CRC Press, 1983), p. D-247.

29. Steiner, R. F. and L. Norris, "The interaction of melittin with troponin C," *Arch. Biochem. Biophys.* 254:342-352 (1987).

30. Muchmore, D. C., D. A. Malencik and S. R. Anderson. "^{1}H-NMR studies of mastoparan binding by calmodulin," *Biochem. Biophys. Res. Comm.* 137:1069-1078 (1986).

31. LaPorte, D. C., C. H. Keller, B. B. Olwin, and D. R. Storm. "Preparation of a fluorescent-labeled derivative of calmodulin which retains its affinity for calmodulin binding proteins," *Biochem.* 20:3965-3972 (1981).

32. Babu, Y. S., J. S. Sack, T. J. Greenhough, C. E. Bugg, A. R. Means and W. J. Cook. "Three-dimensional structure of calmodulin," *Nature (London)* 315:37-40 (1985).

33. Careri, G. *Order and Disorder in Matter* (Menlo Park, CA: The Benjamin/ Cummings Publ., 1984).

34. Karplus, M. and J. A. McCammon. "The internal dynamics of globular proteins," *CRC Crit. Rev. Biochem.* 9:293-349 (1981).

35. Beece, D., L. Eisenstein, H. Frauenfelder, D. Good, M. C. Marden, L. Reinisch, A. H. Reynolds, L. B. Sorensen, and K. T. Yue. "Solvent viscosity and protein dynamics," *Biochem.* 19:5147-5157 (1980).

36. Elber, R. and M. Karplus. "Multiple conformational states of proteins: A molecular dynamics analysis of myglobin," *Science* 235:318-321 (1987).

CHAPTER 9

EPISODIC VARIATIONS IN STREAMWATER ALUMINUM CHEMISTRY AT BIRKENES, SOUTHERNMOST NORWAY

Hans M. Seip, Nils Christophersen, Timothy J. Sullivan

INTRODUCTION

The increase in concentrations of aluminum in surface water when soils and water become more acidic is well known.[1,2] However, the detailed mechanisms controlling aluminum concentrations are far from understood.[3,4,5,6,7] Models for water acidification usually assume simple equilibrium with solid aluminum hydroxide (gibbsite), which implies the following relationship:

$$[Al^{3+}]\ [H^+]^{-3} = K \qquad (1)$$

This equation may also be used to describe cation exchange[8], in which case the constant is not restricted to the values corresponding to gibbsite equilibrium. If the silica concentration is nearly constant, as is the case at Birkenes, equilibrium with kaolinite, and other minerals with the same chemical formula, will also give concentrations approximately satisfying eqn. (1).[9] Improved knowledge of aluminum chemistry is of great importance for predicting surface water chemistry and related biological effects under various scenarios of acid deposition. We have, therefore, carried out detailed studies of stream and soil water chemistry at Birkenes, southernmost Norway[4,5,10,11,12].

SITE DESCRIPTION AND METHODS

Birkenes is a small (0.41 km^2) catchment with shallow podzolic soils on granitic bedrock. It lies in the area of maximum deposition of anthropogenic atmospheric compounds in Norway with a loading of about 2 g S m^{-2} yr^{-1}. The vegetation is dominated by Norway spruce. The catchment is drained by three small second-order streams. They combine to form a third-order stream which has been studied since

1972. Streamwater is chronically acid with a volume-weighted pH of 4.5. The data have been used to develop a mathematical model for streamwater chemistry. Further details are given by Christophersen et al.[1]

Concentrations of H^+, aluminum species, total fluoride, total organic carbon (TOC), and major cations and anions have been measured in stream and soil water during autumn rainstorms (1984, 1985, and 1986) and spring snowmelt (1985 and 1986). Soil water was collected on 2-5 occasions per field season using 10-15 porous-tension lysimeters with a 0.2-μm pore size teflon membrane. The lysimeters sample water only under saturated conditions.

Analytical methods were described by Sullivan et al.[13,10] Major anion and cation concentrations were determined by ion chromatography and atomic absorption spectroscopy, respectively. Labile monomeric, mainly inorganic, aluminum (Al_l) and non-labile monomeric, mainly organically complexed, aluminum (Al_O) were obtained by ion-exchange fractionation, complexation with 8-hydroxyquinoline, and subsequent extraction into methyl isobutyl ketone.[14,15] The reproducibility of the method is good, with a standard deviation of replicate analyses of approximately 1%. Comparison with dialysis results indicates, however, that [Al_l] may be slightly (< 10%) overestimated by assuming that it is equal to the labile fraction.[16] Speciation of Al_l was performed by the computer program ALCHEMI.[17] This program includes sulfate-, hydroxy-, and fluoride complexes, but not species with two different ligands. This is not strictly correct.[18] The most important mixed complex at the pH values in streamwater at Birkenes is $Al(OH)F^+$. A few calculations with the program MINEQL[19] were performed comparing results obtained with and without this complex. The differences in the results for [Al^{3+}] were small (<0.1 μM). At low pH values the relative shift in [Al^{3+}] is <1%; at the highest pH value (5.2), for which the concentrations of Al^{3+} are <1 μM, the relative shift may be up to 20%. For the purpose of this chapter these differences are not important. The main change when the mixed complex is introduced is that the concentration of AlF^{2+} is reduced and replaced by $Al(OH)F^+$.

RESULTS

During both autumn and spring episodes the patterns of stream response were similar for most chemical components for all the five field seasons studied. Hydrogen ion, Al_O, and TOC increased with episodic discharge while basic cations and total fluoride decreased. Cation and anion sums decreased during most of the snowmelt. Patterns for labile monomeric aluminum (Al_l), however, were more variable. The first episodes following extended baseflow periods resulted in large pulses of Al_l during both autumn and spring events. As autumn rain or spring snowmelt progressed, [Al_l], though remained high, followed a generally decreasing trend, sometimes with drops in concentration close to discharge peaks (Figure 1). The three major tributaries to the main brook followed similar patterns, although absolute concentrations varied somewhat among the four sites. A tributary which flows through a small bog area had generally lowest Al_l concentrations.[10,11]

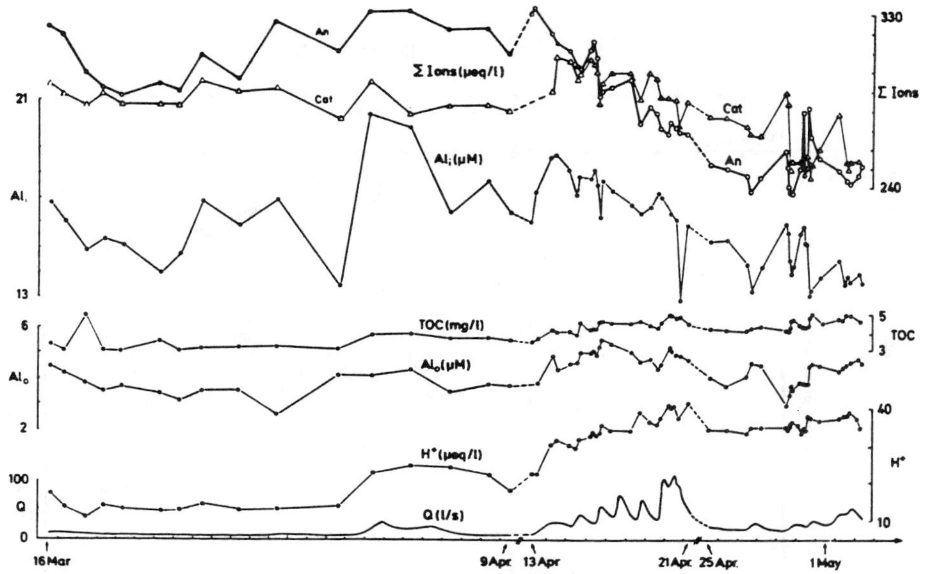

Figure 1. Streamwater discharge (Q) and concentrations of selected aqueous species for the main stream at Birkenes during the spring 1985 snowmelt.[4] Ammonium and organic anions are not included in the ion sums. Ammonium was not measured in 1985, but observations in 1986 showed quite significant concentrations during early snowmelt.

Streamwater was saturated with respect to synthetic gibbsite at baseflow, but consistently undersaturated during wet periods. With respect to microcrystalline gibbsite streamwater is always undersaturated. At high flow Al^{3+} was the dominant inorganic aluminum species in the streams, but at low flow the concentration of the sum of aluminum fluorides (AlF^{2+}, AlF_2^+) may be higher. The concentrations of aluminum hydroxides remained low during high flow periods (Figure 2). The stability constants for the hydroxide complexes decrease with decreasing temperature (e.g., $\log\{([Al(OH)^{2+}][H^+])/[Al^{3+}]\}$ is -5.1 at 20°C and -5.6 at 5°C), while the stability constants for the corresponding fluoride complexes are nearly independent of temperature. The relative concentrations of the Al-hydroxide complexes therefore decrease when streamwater temperature decreases.

Soil water from different tension lysimeters showed widely variable concentrations. Leachate from upper, organic soils tends to have lower concentrations of inorganic aluminum than deeper layers.[6] In general, the concentrations of H^+ and Al_i in soil water showed small temporal variations compared with both soil leachate spatial variations and temporal variations in streamwater (Figure 3). The concentrations of other major ions showed variations between seasons similar to those in streamwater. Sulfate was highest during spring 1985 both in stream and soil water. For the lysimeters included in Figure 3 the average sulphate concentrations for the five seasons were 184, 225, 161, 157, and 156 µeq/L. Lysimeter 18 showed exceptionally high concentrations of Al_i,

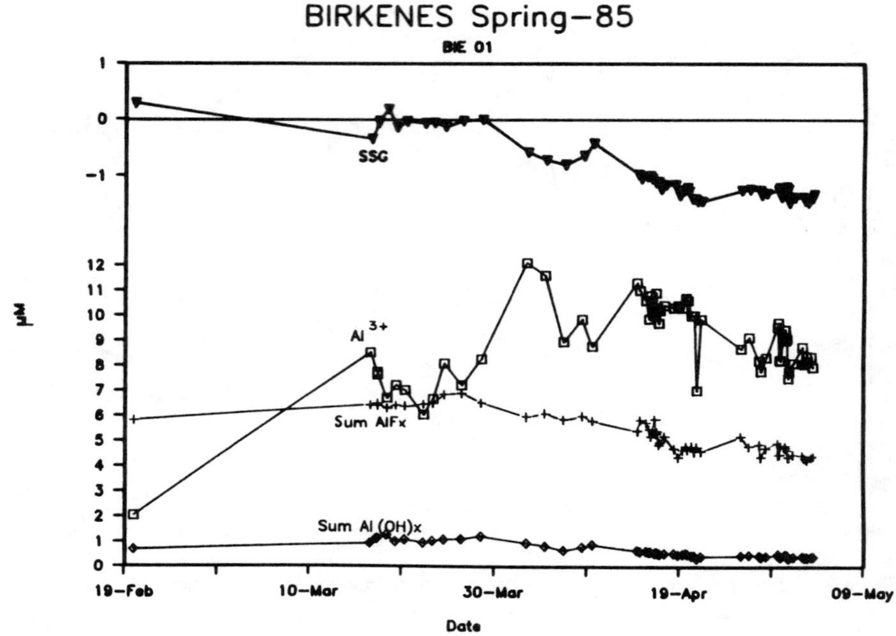

Figure 2. Speciation of Al_i for the main stream at Birkenes during the 1985 snowmelt. Calculations were performed for 2°C. The saturation index for synthetic gibbsite (SSG) is also given. SSG = log{$[Al^{3+}]$ $[H^+]^{-3}$/Solubility Product}

during spring 1985 and for the last sample in spring 1986. These samples had also much higher concentrations of NO_3^- (90 - 135 µeq/L) than found at other times. We have not been able to identify a mineral that controls aluminum concentrations in soil water. The saturation index for gibbsite varies considerably with flow. The index for jurbanite ($Al(OH)SO_4$) shows less variations, but indicates substantial undersaturation. Furthermore, the ratio $[Al^{3+}]/[H^+]$ increased with $[SO_4^{2-}]$ rather than decreased as it should if jurbanite equilibrium prevailed.[12]

Additional measurements were carried out to facilitate interpretation of the results mentioned above. $^{18}O/^{16}O$ ratios were measured in snow, meltwater, and streamwater during spring snowmelt. The results showed that runoff is dominated by "old", pre-event water, particularly during the first phase of snowmelt. Pre-event water is apparently flushed out through a piston effect, implying that initial meltwater composition *per se* probably is relatively unimportant in determining early snowmelt streamwater chemistry at Birkenes.[10] Water table measurements in piezometers installed in a subcatchment in October 1986 generally suggested interception of higher soil layers as the autumn rain season progressed.[5] In some piezometers, later storms (when streamwater $[Al_i]$ decreased) resulted in interception of organic horizons which were not intercepted during earlier storms (when $[Al_i]$ increased).

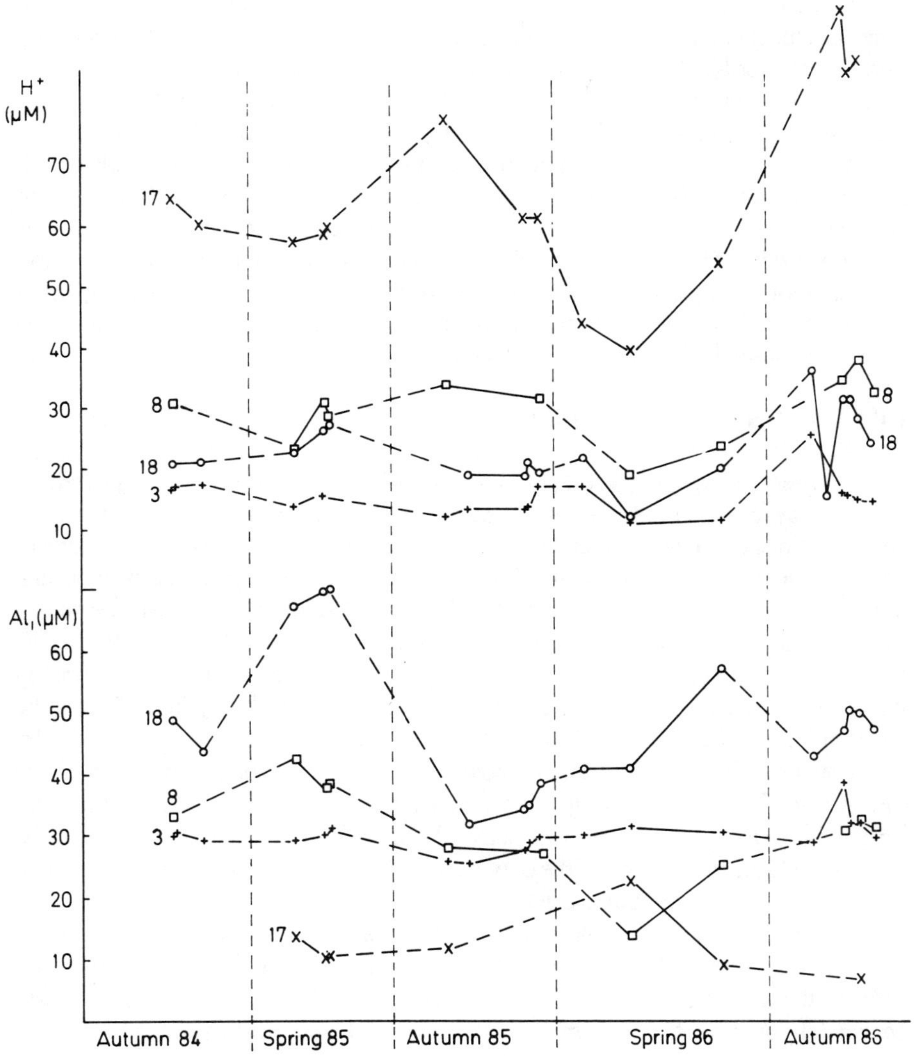

Figure 3. Concentrations of Al_1 and H^+ for selected soil lysimeters. The depths are: 70cm (no 3), 30cm (no 8), 17cm (no 17), and 90cm (no 18).

MODELING

The model previously developed for the Birkenes catchment and applied also to other catchments (see Figure 4) predicts important trends in streamwater chemistry fairly satisfactorily.[1,20] However, an attempt to model episodic changes in streamwater chemistry was only partly successful. The observed concentrations of H^+ and Al^{3+} are compared to two sets of simulated values in Figure 5. The first calculation was carried out using log K = 8.0 (see eqn. 1) in both soil reservoirs (Figure 4). The results for H^+ show the correct trends, although there are discrepancies in peak heights. The simulated concentrations for Al^{3+} are also fairly satisfactory for the autumn period, but not for the spring period. The model particularly underestimates the first aluminum peak concentration during snowmelt. We tried to overcome this difficulty simply by using different values for K in upper and lower soil reservoirs. However, this change gave only minor shifts in the predicted values for $[Al^{3+}]$ and $[H^+]$.

DISCUSSION AND CONCLUSIONS

The observed large *increase* in $[Al_1]$ when discharge increases after long baseflow periods may cause greater biological stress than during later episodes when peak concentrations of Al_1 are similar, but the fluctuations less. The high concentrations of Al_1, in the beginning of the snowmelt (Figures 1 and 2) may have particularly important biological implications. Streamwater at Birkenes is generally saturated with respect to synthetic gibbsite at low flow when the concentrations of both H^+ and Al^{3+} are low. Peak values for $[Al^{3+}]$ were reached in the beginning of the snowmelt. The saturation index decreased somewhat at that time (beginning of April in 1985, Figure 2), but the solution was still fairly close to saturation with respect to synthetic gibbsite. A similar increase in Al^{3+} occurred during early snowmelt 1986, and the solution was even closer to saturation.[5] It has been suggested that saturated or oversaturated solutions are especially toxic to fish.[21] The first stage of the snowmelt is therefore likely to be particularly damaging since it is the only period when high concentrations of inorganic aluminum occur in a solution close to saturation.

The variations in Al-concentrations in the Birkenes stream showed similarities to earlier observations in other catchments. Hendershot et al.[22] found the highest Al-values early in the snowmelt followed by a dilution. They observed, however, increased concentrations after the snowmelt. Neal et al.[23] found that streamwater in two catchments in Wales were highly undersaturated with respect to synthetic gibbsite at low pH. Simulation models were not utilized in these studies.

The relationship given by eqn. (1) is commonly assumed in acidification models.[1,24] We feel that model validation using detailed short-term observations is necessary as in our simulations for the episodic periods in 1985. It is doubtful that any of the existing models will pass such stringent tests very well.

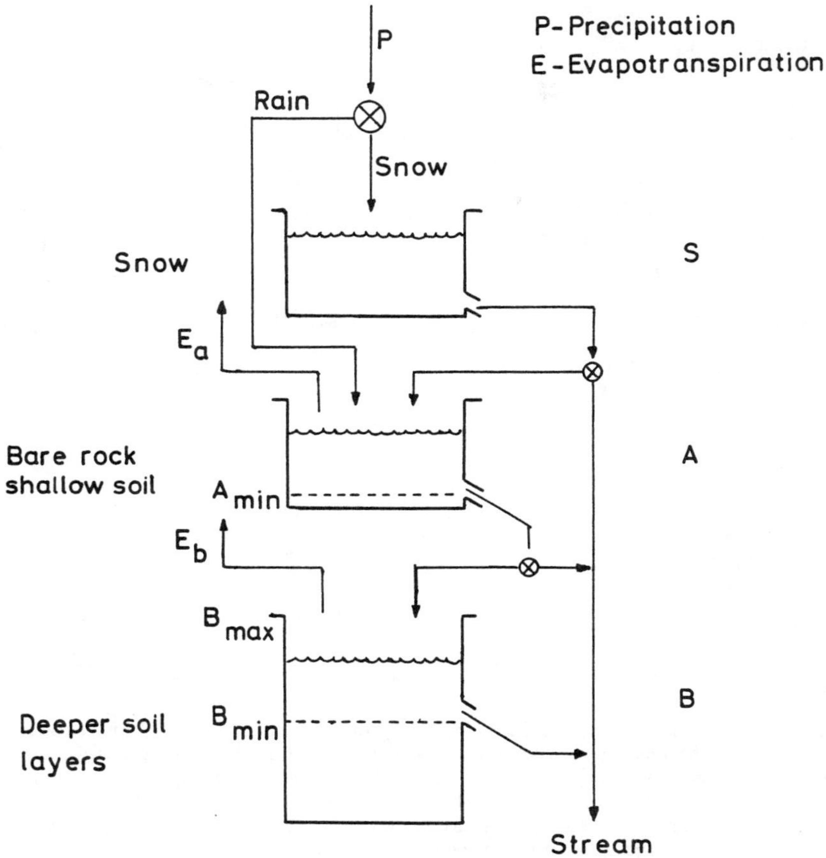

Figure 4. Schematic description of the Birkenes model used to simulate streamwater chemistry.[20]

	Process operating	
	Shallow soil reservoir (A)	Deeper soil reservoir (B)
H_2O	Precipitation, evapotranspiration, infiltration to lower reservoir, discharge to stream	Infiltration, evapotranspiration, discharge to stream, piston flow
SO_4^{2-}	Wet + dry deposition, adsorption/ desorption, mineralization	Adsorption/desorption, reduction during wet periods
$Ca^{2+} + Mg^{2+}$	Ion exchange	Release by weathering, adsorption/ desorption
H^+	Ion exchange, equilibrium with gibbsite	Consumption by weathering, adsorption/desorption, equilibrium with gibbsite
Al^{3+}	Equilibrium with gibbsite or ion exchange	Equilibrium with gibbsite, adsorption/desorption
HCO_3	Equilibrium with a fixed CO_2 pressure	Equilibrium with a CO_2 pressure varying with temperature

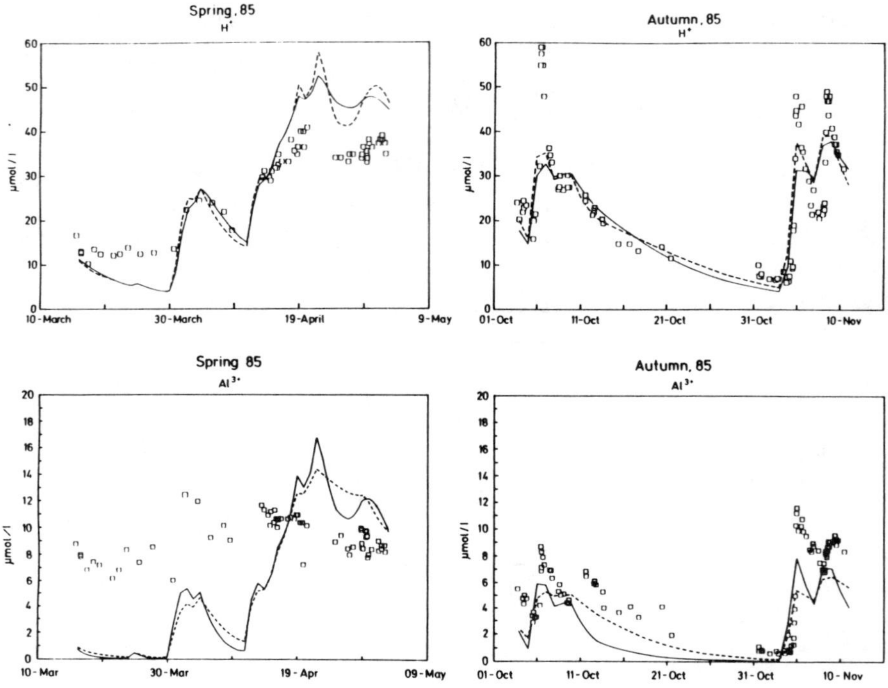

Figure 5. Simulated (lines) and observed (squares) concentrations of H^+ and Al^{3+} for intensively studied periods during spring and autumn 1985. Two sets of simulations were carried out, in one case (full line) K (in eqn 1) was the same in both soil reservoirs (log K = 8), in the second case (dashed line) log K = 7.5 in the upper soil reservoir and log K = 8.5 in the lower reservoir. The discharge during the spring period is given in Figure 1. During the autumn period there were three discharge peaks coinciding with the H^+ peaks.

An improved model for stream and soil water chemistry must take the following observations into account: (i) most of the discharge is pre-event water, (ii) the concentrations of H^+ and Al_1, are generally more stable in soil water than in streamwater, and (iii) during peak discharge the water table intercepts organic horizons in some places, particularly when the episode is preceded by wet conditions. We suggest that changes in flow paths and different aluminum chemistry in various soil layers are the main factors controlling aluminum concentrations in streamwater. In organic soils cation exchange probably dominates, while in deeper soils aluminum in soil solution may be in equilibrium with an aluminum bearing mineral though we have not been able to identify it.[6] The especially high Al_1 concentration in streamwater when an event occurred after a dry period, and in particular in the beginning of the snowmelt, could be due to redissolution of easily dissolved aluminum accumulated in the soil[3] or redissolution of aluminum from the streambed.[25] These mechanisms are not included in the model discussed above. Although the aluminum sub-model needs to be modified it seems more crucial to develop a better hydrological model. Neal et al.[26] have recently suggested a model emphasizing dynamic interchange of water between upper acidic soils and the mineral soils during stormflow. This concept may lead to a more realistic description of water pathways.

ACKNOWLEDGMENTS

This work was supported by the British-Scandinavian Surface Water Acidification Programme (SWAP). A collaboratory research grant from NATO/CCMS is also gratefully acknowledged. R. P. Hooper, I. P. Muniz, and R. D. Vogt have contributed through field work and helpful discussions on aluminum controls. Valuable comments have been given by Colin Neal.

REFERENCES

1. Christophersen, N., H. M. Seip, and R. F. Wright. "A model for streamwater chemistry at Birkenes, a small forested catchment in Southernmost Norway," *Water Resour. Res.* 18:977-996 (1982).

2. Cronan, C. S. and C. L. Schofield. "Aluminum leaching response to acid precipitation: Effects on high elevation watersheds in the northeast US," *Science* 204:304-306 (1979).

3. Hooper, R. P. and C. L. Shoemaker. "Aluminum mobilization in an acidic headwater stream: Temporal variations and mineral dissolution equilibria,"*Science* 229:463-465 (1985)

4. Sullivan, T. J., N. Christophersen, I. P. Muniz, H. M. Seip and P. D. Sullivan. "Aqueous aluminum chemistry response to episodic increases in discharge," *Nature* 323:324-327 (1986)

5. Sullivan, T. J., N. Christophersen, R. P. Hooper, H. M. Seip, I. P. Muniz, P. D. Sullivan and R. D. Vogt. "Episodic variation in streamwater chemistry at Birkenes, Southernmost Norway: Evidence for importance of flow pathways," Proc. UNESCO IHP Symposium "Acidification and water pathways" Vol.I:269-279 (1987)

6. Cronan, C. S., W. J. Walker and P. R. Bloom. "Predicting aqueous aluminum concentrations in natural waters," *Nature* 324:140-143 (1986)

7. Neal, C., R. A. Skeffington, R. Williams and D. J. Roberts. "Aluminum solubility controls in acid waters: The need for a reappraisal," *Earth Planet. Sci. Letter* 86:105-112 (1987)

8. Reuss, J. O., N. Christophersen and H. M. Seip. "A critique of models for freshwater and soil acidification," *Water Air Soil Pollut.* 30:909-930 (1986)

9. Paces, T. "Reversible control of aqueous aluminum and silica during the irreversible evolution of natural waters," *Geochim. Cosmochim. Acta* 42:1487-1493 (1978).

10. Sullivan, T. J., N. Christophersen, and I.P. Muniz. "Soil and streamwater chemistry during snowmelt at the Birkenes Catchment," Report 850701-1, Center for Industrial Research, Oslo (1987).

11. Sullivan, T. J., H. M. Seip, and R. D. Vogt. "Episodic changes in aluminum concentrations and other important parameters in stream and soil water in the Birkenes Catchment, Southernmost Norway" Report 850701-2, Center for Industrial Research, Oslo (1987).

12. Seip, H. M., D. 0. Andersen, N. Christopherson, T. J. Sullivan, and R. D. Vogt. "Variations in concentrations of aqueous aluminium and other chemical species during hydrological episodes at Birkenes, Southernmost Norway," *J. Hydrol.* (in press).

13. Sullivan, T. J., H. M. Seip, and I. P. Muniz. "A comparison of frequently used methods for the determination of aqueous aluminium," *Intern. J. Environ. Anal. Chem.* 26:61-75 (1986).

14. Barnes, R. B. "The determination of specific forms of aluminum in natural water," *Chem. Geol.* 15:177-191 (1975).

15. Driscoll, C. T. "A procedure for the fractionation of aqueous aluminum in dilute acidic waters," *Inter. J. Environ. Anal. Chem.* 16:267-283 (1984).

16. Backes, C. A. and E. Tipping. "An evaluation of the use of cation-exchange resin for determination of organically complexed Al in natural waters," *Int. J. Environ. Anal. Chem.* 30:135-144 (1987).

17. Schecher, W. D. and C. T. Driscoll. "An evaluation of uncertainty associated with aluminum equilibrium calculations," *Water Resour. Res.* 23:525-535 (1987).

18. Couturier, Y. "Contribution a l'etyde des complexes mixtes de l'aluminium (III) avec les ions fluorure et hydroxyde," *Bull Soc. Chim France* 171-177 (1986).

19. Westall, J. C., J. L. Zachary, and F.M.M. Morell. "MINEQL, a computer program for the calculation of chemical equilibrium composition of aqueous systems," Technical note no. 18, Department of Civil Engineering, Massachusetts Inst. of Technology (1976).

20. Rustad, S., Christophersen, N., Seip, H. M., and Dillon, P. J. "Model for streamwater chemistry of a tributary to Harp Lake, Ontario," *Can. J. Fish. Aquat. Sci.* 43:625-633 (1986).

21. Baker, J. P. "Effects on fish of metals associated with acidification" In R. E. Johnson [ed.] Acid rain/fisheries, *Amer. Fish. Soc.* p. 165-176 (1982).

22. Hendershot, W. H., A. Dufresne, H. Lalande, and F. Courchesne. "Temporal variation in aluminum speciation and concentration during snowmelt," *Water Air Soil Pollut.* 31:231-237 (1986).

23. Neal, C., C. J. Smith, J. Walls, and C. S. Dunn. "Major, minor and trace element mobility in the acidic upland forested catchment of Upper River Severn, Mid Wales," *J. Geol. Soc. Lond.* 143:635-648 (1986).

24. Cosby, B. J., G. M. Hornberger, J. N. Galloway, and R. F. Wright. "Modeling the effects of acid deposition: Assessment of a lumped-parameter model of soil water and streamwater chemistry," *Water Resour. Res.* 21:51-63 (1985).

25. Norton, S. A., A. Henriksen, B. M. Wathne, and A. Weidal. "Aluminum dynamics in response to experimental additions of acid to a small Norweigan stream," in *Proceedings of UNESCO-IHP Symposium Acidification and Water Pathways* Vol I, Bolkesjo, Norway (1987) p. 249-258.

26. Neal, C., N. Christophersen, R. Neale, C. J. Smith, P. G. Whitehead, and B. Reynolds. "Chloride in precipitation and streamwater for the upland catchment of River Severn: Some consequences for hydrochemical models," *Hydrol. Processes* 2:155-165 (1988).

CHAPTER 10

ALUMINUM CHEMISTRY OF ACIDIC SANDY SOILS WITH VARIOUS INPUTS OF ACIDIC DEPOSITION IN THE NETHERLANDS AND IN DENMARK

J. Mulder, N. Van Breemen, L. Rasmussen, C. T. Driscoll

INTRODUCTION

In acidic, sandy soil in areas with a precipitation surplus, dissolution, downward transport and subsequent precipitation of Al is a natural process. This process of podzolization is generally believed to be the result of mineral dissolution and complexation of Al by natural organic acids, originating from forest floor and canopy throughfall.[1] At greater depth the organic acids precipitate, probably because of saturation with Al and Fe, resulting in the formation of a so-called spodic B horizon. This podzolization hypothesis was challenged recently by a number of researchers, who suggested that Al and Fe may be mobilized by short-lived organic acids and subsequently transported as Al-Fe-SiO_2 sols.[2,3] Soil solution data from North American Spodosols suggest that both organically complexed Al, and polymeric forms (which may include Al-Fe-SiO_2 sols), contribute to total dissolved Al,[4,5] and that in background situations, with low atmospheric acid inputs, the solubility of Al is low.[6,7]

Recently, increased inorganic monomeric Al concentrations have been observed in drainage water from acidic, sandy soils, in parts of North America and Northwestern Europe, indicating that the Al chemistry of these soils has changed. These changes have been attributed to both acidic deposition and changing land-use resulting in buildup of organic acids.[8,9] By using the proton budget approach, however, Van Breemen et al.[10] (1) demonstrated that atmospheric acidic deposition rather than internal sources of soil acidity is the dominant constituent of the total acid load of many North American and Northwestern European soils, and (2) found evidence for the hypothesis that acidic deposition is the main cause for the observed increase in Al mobilization. Monitoring studies[11] and manipulation experiments on soil columns[12,13] strongly support this hypothesis. However, to date the hypothesis was never tested *in situ* on similar soil vegetation systems, with prolonged inputs of various amounts of acid deposition.

In order to test the hypothesis *in situ* we selected six sandy, acidic soils, which were chemically nearly identical (i.e., similar low contents of basic cations and high concentrations of Si), but which received progressive amounts of atmospherically derived acidity. Data were used from 2 to 4 year hydrochemical monitoring at sites in the Netherlands and in Denmark.

STUDY APPROACH AND PROCEDURES

Experimental Setting

Monitoring was conducted at four sites in the Netherlands and two in Denmark. All sites had similar acidic, sandy soils, but the two members of each pair differed in vegetation (one with and one without forest cover). We anticipated significant variations in acid input (i) between the paired sites, due to the generally higher acid loads in the Netherlands than in Scandinavia,[10] and (ii) between the two members of one pair, due to a more effective entrapment of acidic dry deposition in forest canopies, relative to non-forest vegetation.[14]

In the Netherlands forests were shown to increase the dry deposition of $(NH_4)_2SO_4$, which causes acidification of the surface soil after assimilation of NH_4 by the vegetation or nitrification of NH_4 to a mixture of nitric and sulfuric acid.[14] In non-forest vegetation (e.g., heathland) dry deposition of $(NH_4)_2SO_4$ and therefore the acid load was expected to be lower.

At the Jutland westcoast in Denmark, acidic deposition is considerably lower than in the Netherlands,[15] indeed. So, the two Danish sites together with the four in the Netherlands constitute a sequence of progressive atmospheric acid loads on similar soils. Characteristics concerning parent material, soil, vegetation, and annual precipitation at the various sites are given in Table 1.

Soil Analysis

Total contents of the major oxide components in the parent material are given in Table 2. Elemental analysis, except for MgO and Na_2O, was conducted by X-ray fluorescence spectroscopy of a Li-tetraborate melt of the soil material. MgO and Na_2O were analyzed by atomic absorption after a 10 s HF-H_2SO_4 destruction at 60°C.[16] The chemical composition of all parent materials is characterized by high concentrations of Si and low concentrations of Al and basic cations (Ca, Mg, Na, K). Concentrations of Al and basic cations in the Dutch soils are slightly lower than in the Danish soils.

Soil chemical characteristics of the six sites are given in Table 3. Soil pH was measured in a 1:2.5 (mass to volume) extract in 0.01M $CaCl_2$, organic carbon was determined potentiometrically, after wet combustion in a phosphoric-chromic acid mixture,[16] free Al_2O_3 refers to the dithionite-oxalate extractable fraction, essayed colorimetrically with pyrocatechol-violet.[16] CEC and base saturation (BS) were measured using Bascomb's $BaCl_2$-$MgSO_4$ method[17] for the Dutch soil samples and Li-EDTA[16] for the Danish KH and KF soils.

Table 1. Selected characteristics of the six research sites

Site	Code	Country	Vegetation	Parent Material	Soil Class	Precip.
Klosterhede	KH	DK	*Calluna vulgaris*	poorly sorted fluvioglacial sands	Typic Haplorthod	992
Klosterhede	KF	DK	*Picea abies*	poorly sorted fluvioglacial sands	Typic Haplorthod	992
Gerritsfles	GB	NL	bare sand	well sorted driftsands	Typic Udipsamment	839
Gerritsfles	GF	NL	*Pinus sylvestris*	well sorted driftsands	Typic Udipsamment	839
Hasselsven	HH	NL	*Calluna vulgaris*	well sorted coversands	Typic Haplorthod	800
Tongbersven	TF	NL	*Pinus sylvestris*	well sorted coversands	Typic Haplorthod	785

Note: Soil classification according to Soil Survey Staff,[33] mean annual precipitation (in mm) is from 1 April 1983 to 31 March 1987 (the Netherlands, NL) and 1 June 1985 to 31 May 1986 (Denmark, DK).

Table 2. Total contents expressed in mass fractions -%- of oxides of the parent material (fraction <2 mm) at the six sites

Site	SiO_2	Al_2O_3	MgO	CaO	Na_2O	K_2O
KH	89.2	5.0	0.2	0.4	0.8	1.6
KF	92.6	4.3	0.2	0.3	0.8	1.1
GB	97.2	2.3	0.0	0.1	0.4	0.9
GF	96.8	2.4	0.1	0.1	0.5	0.9
HH	95.2	2.5	0.1	0.1	0.4	0.9
TF	98.3	1.8	0.1	0.1	0.3	0.7

Table 3. Selected soil chemical characteristics of the six research sites

Site	Depth (cm)	Horizon	pH($CaCl_2$)	C-free % Mass	Al_2O_3 Fraction	CEC $mmol_c/kg$	BS %
KH	0-8	Al	3.1	4.22	0.11	46	37
	8-19	E	3.4	0.66	0.07	19	26
	19-43	Bhs	4.0	3.16	1.23	65	5
	43-60	Bs	4.4	1.02	0.75	10	0
	+60	C	4.5	0.20	0.53	11	27
KF	0-13	E	3.0	2.54	0.07	27	19
	13-29	Bhs	3.9	4.68	1.23	77	1
	29-66	Bs	4.5	0.54	0.56	12	25
GB	0-25	C	4.1	0.24	0.25	3	0
	25-75	C	4.4	0.22	0.30	2	0
	+75	C	4.3	0.08	0.39	2	0
GF	0-25	C	3.5	0.45	0.26	22	5
	25-75	C	4.0	0.24	0.40	9	0
	+75	C	4.0	0.33	0.29	5	0
HH	0-09	E	3.0	2.48	0.12	64	6
	9-15	Bhs	3.5	2.27	0.31	77	4
	15-33	Bs1	4.0	1.17	0.54	33	6
	33-73	Bs2	4.3	0.00	0.42	9	22
	+73	C	4.3	0.00	0.44	12	50
TF	0-20	E	3.0	1.60	0.06	27	4
	20-40	Bhs	3.6	4.54	1.17	127	0
	40-58	Bs1	4.2	0.62	0.70	14	0
	58-88	Bs2	4.3	0.02	0.31	4	0
	+88	Bs3	4.3	0.05	0.33	5	0

All six soils are acidic and low in CEC and exchangeable bases. A considerable fraction of the exchangeable acidity is Al (not shown). Base saturation is low in all soils, but highest in the two Danish Spodosols (KH and KF) and in the heathland soil in the Netherlands (HH). The variation in free Al content with depth (Table 3) in the four Spodosols, showing depletion in the E horizon and enrichment in the Bhs horizon, differs markedly from the noisy pattern of free Al concentrations with depth in the Upidsamments (GB and GF).

Both members of each soil-pair (KH and KF, GB and GF, HH and TF) are chemically similar (Tables 2 and 3). However, in comparison to the Dutch soils the two Danish soils (KH and KF) are somewhat richer in basic cations. This characteristic could result in a somewhat higher contribution of basic cation mobilization to acid neutralization at the Danish sites.

Monitoring Program

The monitoring program included (1) the determination of the quantity and chemical composition of precipitation and throughfall water, (2) chemical analysis of soil solutions, (3) estimation of waterfluxes in the soil, using a water transport model, and (4) estimation of the annual nutrient uptake by the vegetation. The monitoring program reported here was conducted from November 1982 to May 1987 (GB, GF, TF and HH) and from June 1985 to May 1987 (KH and KF).

Bulk precipitation was collected in open funnels mounted on 5- or 10-liter opake polyethylene bottles. Throughfall in the Dutch forests was collected in 2 PVC gutters (1800 cm^2),[18] and in the Danish forest by 1 plexiglass gutter (ca. 1000 cm^2),[15] At the Hasselsven site, throughfall was collected by means of 4 rectangular open top PVC boxes (6 cm high, 5 x 20 cm surface area). In the Klosterhede heathland no throughfall was collected. Meteoric water was sampled bi-weekly and pooled to monthly samples before analysis. Results of the monitoring of meteoric water were reported for the Dutch sites by Van Dobben and Mulder[18] and for the Danish sites by Rasmussen.[15] Note that we corrected the mean bulk deposition and throughfall Cl concentrations at the Danish sites, because (i) comparison of the analytical results showed that Cl concentrations reported by the Danish laboratory were consistently too low, and (ii) even in bulk deposition samples relatively high anionic charge deficits (10% of cationic charge) were reported. Corrected values for Cl concentrations, obtained by multiplying Na concentrations by the Cl/Na ratio in seawater, resulted in an acceptable anionic charge surplus in bulk deposition of 1%.

Soil solutions were sampled continuously, using acrylic copolymer filterplates and a permanent vacuum of ca. 150 mbar.[5,18] Solutions were collected bi-weekly and pooled to monthly samples before analysis. Lysimeters were installed in holes excavated laterally from a soil pit, by pressing the filterplates tightly to the undisturbed soil above. In order to prevent contamination of the forest floor leachates with water from the mineral soil, a segment of the forest floor was cut and placed in plate lysimeters with an extended PVC rim. Note that the use of a rim also prevents nutrient uptake by the vegetation. Single (KF) or duplicate lysimeters (all other sites) were installed at specific depths. Duplicate lysimeters were placed at a 2- to 3-m distance. Only at the Klosterhede heathland (KH) duplicate soil solutions were pooled before analysis. For details on lysimeters, procedures of installation and of collection of the soil solution reference is made to Van Dobben and Mulder.[18]

Soil moisture content in the field, necessary for calibrating the water transport model SWATRE, was measured monthly with a neutron probe. The neutron probe was calibrated gravimetrically. Estimates of the annual nutrient uptake by the vegetation were taken from de Visser[19] (TF and GF), Nieuwenhuyse[20] (HH) and Rasmussen[15] (KF). Nutrient uptake in the Danish heathland (KH) was assumed to equal that in the Dutch heathland (HH).

Analytical Procedures

Within 1 day of collection, all solution samples from the research sites in the Netherlands were analyzed for pH, specific conductance and total and inorganic dissolved C. Within 14 days of collection all samples, stored at room temperature in the dark, were analyzed for Ca, Mg (atomic absorption), Na, K (atomic emission), H_4SiO_4 (colorimetrically as the blue silica-molybdic acid complex), F^-, Cl^-, NO_3^-, SO_4^{2-} (ion chromatography), NH_4^+ (colorimetrically using Na-salicylate) and total Al (colorimetrically, using pyrocatechol violet). Soil solution samples from the two Danish sites were ca. 8 days in the mail before analyses. However, analytical results for the first year did not indicate significant changes in the major solute concentrations during this transport.

Total dissolved Al (monomeric plus polymeric forms) was determined after acidifying solutions to pH 2, at least 24 hours before analysis. In 4 monthly series of soil solutions from the sites in the Netherlands, monomeric Al, measured in non-acidified samples 6 min. after addition of pyrocatechol violet, was not significantly different from total dissolved Al, indicating that polymeric Al was negligible. In the Danish soil solutions only total Al concentrations were determined, because we anticipated a shift in the distribution of the various Al fractions during transport.

Organic monomeric Al concentrations, estimated from experimentally determined binding capacity values of dissolved organic acids, were only of quantitative importance in forest floor and E horizon leachates (ranging from 25 to 100% of total Al), which have elevated DOC concentrations. In soil solutions from the deeper horizons monomeric Al is almost exclusively (ca. 95%) in inorganic forms. The binding capacity of dissolved organic acids, present in our soil solutions, for Al (11 mmol Al/mol DOC at pH 4.0), determined from a Langmuir regression model, were of the same order of magnitude as those reported for organic acids in natural waters of the Southeastern U.S.[21] and for organic acids in soil leachates of the Solling area (FRG).[22] Driscoll's cation exchange method[23] for the direct analytical estimation of organic monomeric Al was used for forest floor and E horizon leachates only. This method was unsuitable for all other solutions, from which, due to high Al concentrations, inorganic monomer Al forms could not be removed completely.

Computational Procedures

Fluxes of solutes in meteoric water were calculated by multiplying monthly determined concentrations and measured amounts of rain and throughfall. Details and results are reported elsewhere.[15,18]

Total atmospheric input of the major seasalt components (K, Na, Ca, Mg, Cl and SO_4) was assumed to be equal to fluxes in bulk precipitation plus estimated amounts of additional dry deposition. The dry deposition of Cl was calculated by subtracting bulk deposition from throughfall fluxes. Dry deposition of the other seasalt components was calculated from Cl dry deposition, using K/Cl, Na/Cl, Ca/Cl, Mg/Cl, and SO_4/Cl ratios in seawater. The dry deposition of seaspray aerosol only appears to be of quantitative importance at the forested sites. The fluxes of Cl in throughfall and soil solutions at HH and GB suggest that the dry deposition of sea salt components is relatively unimportant here.

The total atmospheric input of SO_4 was assumed to equal fluxes in throughfall (or in the case of HH and GB to fluxes in the soil solution at 2 cm depth). Although SO_4 fluxes in throughfall did not differ significantly from those in bulk deposition at HH, soil solution fluxes did indicate a dry deposition term. Non-seasalt derived SO_4 in the atmospheric input was assumed to originate from atmospheric H_2SO_4.

Only in the Dutch forests, where N deposition is highest, total atmospheric inputs of NO_3 and NH_4 were assumed to equal throughfall fluxes. At all other sites the throughfall input may underestimate the total N deposition, due to foliar uptake. Therefore, the dry deposition of NO_3, non-seasalt SO_4, and NH_4 was estimated by multiplying dry deposition fluxes of non-seasalt SO_4 with the (NO_3/non-seasalt SO_4) and (NH_4/non-seasalt SO_4) ratios in bulk deposition (see Table 4).

At the Danish heathland (KH), where we did not measure throughfall deposition, we assumed the total atmospheric input of all major solutes to be equal to bulk deposition fluxes, because (i) additional dry deposition of NH_3, SO_2 and NO_x was probably small, like at the nearby forested KF site and (ii) additional dry deposition of seasalts at the other non-forested sites (GB, HH) was small relative to bulk deposition fluxes.

Atmospheric acid loads (Hdep) were calculated according to Van Breemen et al.:[24]

$$\text{Hdep} = (\text{H})\text{bu} + (\text{H})\text{dry} + (\text{NH}_4)\text{in} - (\text{NH}_4)\text{out} + (\text{NO}_3)\text{out} - (\text{NO}_3)\text{in} \quad (1)$$

where brackets denote fluxes, "bu" bulk precipitation, "dry" dry deposition, "in" atmospheric input and "out" output with drainage water. (H)dry is calculated according to Van Breemen et al.[25] as:

$$(\text{H})\text{dry} = (\text{SO}_4)\text{in,nss} - (\text{SO}_4)\text{bu,nss} + (\text{NO}_3)\text{in} - (\text{NO}_3)\text{bu} + (\text{NH}_4)\text{bu} - (\text{NH}_4)\text{in} \quad (2)$$

where "nss" indicates the non-seasalt component of SO_4.

Solute fluxes at various depths in the GB, GF, HH, and TF soils were calculated by multiplying simulated monthly water fluxes and measured monthly soil solution concentrations. Water fluxes were calculated using SWATRE, a one-dimensional simulation model for unsaturated water transport.[18,26] Fluxes were calculated for 4 hydrological years (between 1 April 1983 and 31 March 1987).

For the two Danish sites (KH and KF) annual soil water fluxes were estimated assuming Cl input equals Cl output. Solute fluxes in the Danish soils were estimated by multiplying annual water fluxes with mean annual solute concentrations. Solute fluxes and budgets were calculated for 1 hydrological year: 1 June 1985 - 31 May 1986 (KH and KF).

Although no error analysis has been conducted, we feel that uncertainties in the major solute fluxes in the Dutch soils will be similar to or slightly higher than the ones reported for the Hackfort loamy sands in the Netherlands. Errors in the major solute fluxes (Al, SO_4) below the rooting zone in these loamy sands are less than 50%[26] while flux errors at shallower depth are generally even lower. For the Danish sites the uncertainty in the annual solute fluxes are larger, because water fluxes, and associated with that solute fluxes, could only be estimated on an annual basis, rather than at monthly intervals. Errors in budget values (output minus input fluxes) can be larger (>100%) than errors in fluxes, especially when budget values are small (subtraction of two large numbers). For instance, large budget errors can be expected for Na, Mg, and Cl at KF, because both input and drainage water are dominated by seasalt components (see results).

Mineral equilibria in soil solutions were calculated for all solution samples from the mineral soil using the chemical equilibrium model ALCHEMI.[27] ALCHEMI includes temperature and activity correction (Debye-Hückel). Aluminum activities were calculated assuming all monomeric Al was in an inorganic form. For soil solutions from the surface 15 cm and for the KH site also for solutions at greater depth, this assumption leads to an overestimation of inorganic Al activities. Saturation indices (SI) with various secondary minerals were calculated according to:

$$SI = \text{Log}\,(Qp/Kp)$$

where Qp = ionic activity product and Kp = solubility product for the solid phase p. SI=0 indicates saturation, SI>0 oversaturation and SI<0 undersaturation with respect to the solubility of the mineral considered.

RESULTS AND DISCUSSION

Atmospheric Acid Loads

Atmospheric inputs differ markedly among the six sites studied (Table 4). The input of most solutes is considerably higher at forested sites than at non-forested ones. The Dutch sites differ from their counterparts in Denmark, because of lower inputs of seasalt components and considerably higher $(NH_4)_2SO_4$ loads. Seasalt inputs are high at the Danish sites, because of the short distance (16 km) from the North Sea. At the Dutch forested sites inputs of $(NH_4)_2SO_4$ are particularly elevated and mainly due to a large

Table 4. Estimated atmospheric deposition fluxes (input), output fluxes (at various depths below the rooting zone) and atmospheric acid loads (calculated according to equation 1) from April 1983 to March 1987 (NL) and from June 1985 to May 1986 (DK)

Site	Depth (cm)	Water (mm)	H	NH_4	BC[a]	Al	Cl	NO_3	SO_4	Si	Acid Load
			kmol$_c$/ha·yr[b]								kmol/ha·yr
KH	input	992	0.39	0.99	3.96	0.10	3.45	0.84	1.25	-	0.5
	49	693	0.22	0.01	4.26	0.46	3.45	0.01	1.36	0.55	
KF	input	992	0.60	1.28	9.82	0.10	8.72	1.08	2.06	-	0.8
	31	650	0.30	0.01	8.57	3.21	8.72	0.01	2.99	0.39	
GB	input	839	0.46	1.45	1.13	0.02	0.77	0.70	1.48	0.06	2.2
	100	677	0.22	0.03	1.87	1.44	0.75	0.99	1.96	0.48	
GF	input	839	-0.02[c]	3.02	2.13	0.02	1.66	0.80	2.61	0.06	4.1
	100	283	0.20	0.04	2.12	4.65	1.98	1.93	3.16	0.43	
HH	input	785	0.98	2.27	1.01	0.03	0.60	1.08	2.49	0.05	2.7
	70	354	0.21	0.01	1.12	2.05	0.59	0.54	2.20	0.29	
TF	input	800	-0.24[c]	3.68	1.67	0.03	1.20	0.55	3.21	0.07	4.3
	57	237	0.22	0.69	1.33	4.82	1.18	2.05	4.35	0.40	

Note: All values are annual means for 4 hydro years.

[a] Sum of base cations (K, Na, Ca, Mg).

[b] Kmol$_c$/ha·yr refers to kmoles of unit charge per ha and year.

[c] Negative number indicates net alkalinity (HCO_3) input.

contribution of dry deposition. Total deposition of NO_3 is similar at all sites and less than that of non-seasalt derived SO_4. Atmospheric inputs of free H^+ are relatively low at all sites and range from 0 to 0.9 kmol/ha·yr (Table 4). At the forested TF and GF sites, where dry deposition of NH_3 exceeds that of H_2SO_4, we estimated the input of HCO_3^- to be even higher than the input of free H^+.

Atmospheric acid loads, calculated from equation 1, strongly depend on the fate of N in the ecosystem considered.[10,24] They are highest in the Netherlands and considerably lower in Denmark (Table 4). In the Netherlands the atmospheric acid load is primarily due to transformations of atmospherically derived N (assimilation, nitrification), while in Denmark this load is predominantly caused by the input of free H^+ from the atmosphere. The contribution of N transformations to the acid load at the Danish sites is small, because removal of NH_4 (acid production) and NO_3 (acid consumption) from the percolating soil solutions nearly balance. Due to increased dry deposition of (potentially) acidic compounds, atmospheric acid loads at forested sites invariably are higher than at their non-forested counterparts.

Al Concentrations in Soil Solutions

Aluminum is the major dissolved cation (on an equivalent basis) in all sub-soil solutions of the Dutch sites (Table 5a). In the Danish soils, Na concentrations, included in BC (basic cations), are higher than the concentrations of Al, due to elevated inputs of seaspray. The higher deposition of seasalt at the Danish sites is also reflected by the relatively high dissolved Cl concentrations.

Table 5. Soil solutions below the rooting zones

A. Mean solute concentrations

Site	Depth cm	pH[a]	BC[b]	Al	F	Cl	NO_3	SO_4	Si
			mmol$_c$/m^3 [c]						mmol/m^3
KH	49	4.50	615	67	0	498	1	196	80
KF	31	4.34	1319	494	0	1341	2	460	60
GB	100	4.51	301	188	3	114	128	298	79
GF	100	4.15	905	1751	11	771	799	1173	171
HH	70	4.25	339	488	6	149	103	583	75
TF	57	4.03	563	2065	6	469	698	1757	187

B. Coefficients of variation (100% * standard deviation/mean) of major solute concentrations

Site	Depth cm	H	BC	Al	Cl	NO_3	SO_4	Si
KH	49	60	25	52	40	283	22	33
KF	31	39	36	66	56	214	23	28
GB	100	27	54	42	45	67	29	60
GF	100	28	52	51	61	68	41	39
HH	70	40	48	65	40	246	24	23
TF	57	40	40	22	31	65	17	25

Note: To facilitate comparison between sites, soil solutions below the rooting zone were used, so that effects of biocycling on solute concentrations (which will be significantly different for the various vegetations) will be small. Data are from duplicate lysimeters for the complete monitoring period.

[a] Calculated from mean H^+ concentrations.

[b] Sum of base cations (K, Na, Ca, Mg).

[c] Mmol$_c$/m^3 refers to mmol of unit ionic charge per m^3.

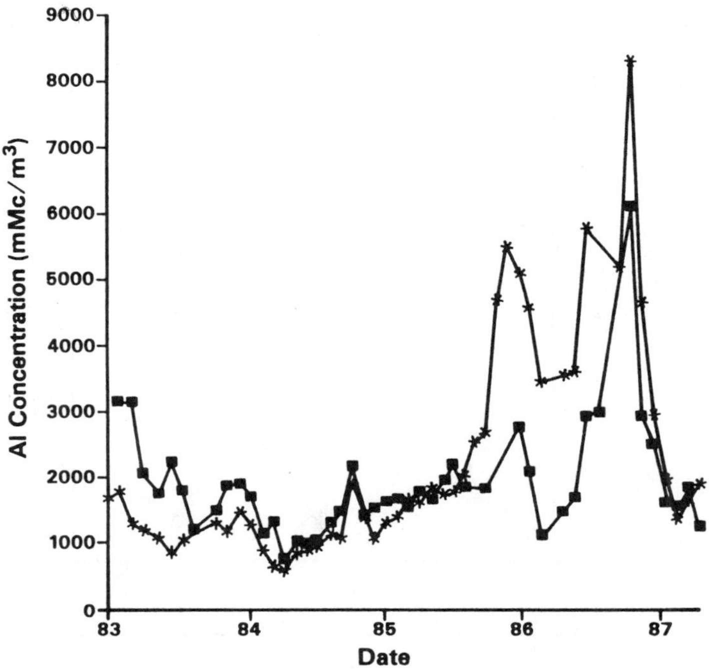

Figure 1. Total dissolved Al concentrations ($mmol_c/m^3$) in duplicate lysimeters at GF at 40 cm depth.

Coefficients of variation are small for all major solute concentrations (Table 5b) and are mainly due to temporal variations. The contribution of spatial variability to CV values at our 3 x 3 m^2 plots is small as is illustrated by the Al concentrations in duplicate lysimeters in Figure 1.

Inorganic monomeric Al is primarily (>80%) free aquo-Al (Al^{3+}) in solutions from all mineral soils. Other dissolved Al species (e.g., $AlSO_4^+$, $AlOH^{2+}$) are only of minor significance.

Total dissolved Al concentrations are positively correlated with the sum of NO_3 and SO_4 concentrations and not with Cl concentrations (Table 5a), suggesting that these anions rather than Cl, are associated with solubilization of Al. Nitrate concentrations are distinctly higher in forested than in non-forested soils, but virtually absent in both Danish soils. This reflects the luxurious N supply at the Dutch sites. Lowest Al concentrations, observed in the Danish heathland soil (KH), are similar to those reported for Spodosol solutions in the U.S. and Scandinavia.[4,5,6,28] Dissolved Al concentrations are extremely high in the forested GF and TF soils. Similarly high values are known to us only from other sandy to loamy soils with high atmospheric acid inputs (>2.5 kmol/ha·yr) in the Netherlands[11] and in the F.R.G.[29]

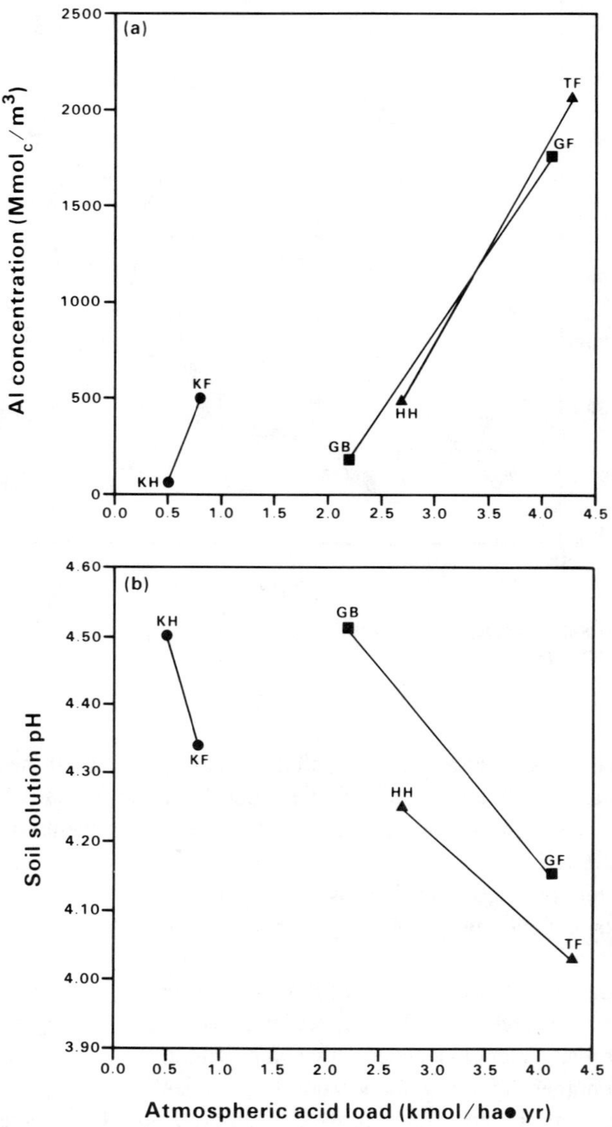

Figure 2. Mean dissolved Al concentrations (mmol$_c$/m^3) (a) and soil solution pH (b) as a function of the atmospheric acid load (kmol$_c$/ha yr). Soil solution data are from below the rooting zone. Lines connect data for the forested site and their non-forested counterpart.

Increased mean dissolved Al concentrations and decreased mean pH in our sub-soils are associated with increased atmospheric acid loads (Figure 2). This trend is also true for the Danish soils, even though they have a slightly higher basic cation content than Dutch soils. Our data support the hypothesis that strong acid inputs play a prominent role in the current increase in soil acidification rate and Al solubility in acidic sandy soils.

Al Chemistry

Dissolved Al correlates positively with both NO_3 and SO_4 concentrations in almost all Dutch soil horizons (Table 6). The prominent covariation of dissolved Al and NO_3 is also illustrated for the GF and TF sites in Figure 3. Our findings suggest the importance of H_2SO_4 and HNO_3 inputs for Al mobilization in the Dutch soils. In contrast, no significant correlations between dissolved Al and NO_3 and SO_4 were observed in the Danish soils, indicating that other processes may even be more relevant in Al mobilization there. Dissolved Al also correlates positively with Cl concentrations in almost all soil horizons (Table 6). However, this correlation must be due to evaporative concentration effects resulting in an increase in all solute concentrations or to cation exchange, where Al is exchanged for Mg plus Na.

Soil solutions from most individual mineral soil layers of all sites exhibit significant negative pH-Al correlations (Table 6 and Figure 4), as is expected on the basis of chemical equilibrium between solutions and Al-bearing minerals. In contrast, the pH-Al relationship with depth is positive; both concentrations of dissolved Al and pH values increase in the upper 40 cm of nearly all soils (Table 7). Similar trends with depth were reported for acidic loamy sands in the Netherlands, which receive high atmospheric acid inputs.[11]

Chemical equilibrium calculations confirm that saturation of soil solutions with secondary Al-bearing minerals increase with depth (Table 8). Sub-soil solutions at all sites reach saturation with gibbsite solubility, indicating that this mineral may serve as an Al solubility control in these soils. Other researchers also found (sub)soil solutions close to saturation with respect to the solubility of gibbsite.[4,5,28]

In the KH, KF, and HH soils, with relatively low acid loads, Bhs horizon leachates attain equilibrium with gibbsite solubility, while in the TF soil, with a higher acid load, saturation with gibbsite does not occur until the soil solution reaches the Bs horizon. In driftsands with relatively low acid loads (GB) even the surface soil solutions are close to saturation with respect to gibbsite solubility, while at GF, with the high acid load, saturation is not evident until the soil solution reaches a depth between 40 and 100 cm.

In an earlier paper[11] we hypothesized that in the surface layers of the mineral soils, Al dissolution could not keep pace with the addition of strong acids, so that chemical equilibrium was not reached until in the subsoil. In this hypothesis, kinetic constraints were the major cause of the strong under saturation with respect to most Al-bearing minerals in the surface soil. However, batch studies,[30] where we added strong acid to

Table 6. Correlation coefficients for Al and pH and for Al and various anions in soil solutions over the complete monitoring period

Site	depth cm	Al/pH	Al/NO_3	Al/SO_4	Al/Cl
KH	0	-0.77*	0.45	0.96**	0.89
	13	-0.46	-0.37	-0.41	0.03
	49	-0.65*	0.04	-0.77*	0.80*
KF	0	-0.08	-0.23	0.36	0.40
	12	-0.41	0.13	0.13	0.77*
	31	-0.71*	-0.17	-0.05	0.96*
	80	-0.58	0.62*	0.35	0.92*
GB	0	-0.54***	0.59***	0.55***	0.31**
	40	-0.41	0.76***	-0.29*	-0.05
	100	-0.53	0.46***	-0.58***	-0.17
GF	0	-0.12	-0.01	0.19*	0.09
	10	-0.46*	0.94***	0.56**	0.70**
	40	-0.46***	0.95***	0.65***	0.58***
	100	-0.51***	0.88***	0.81***	0.88***
HH	0	-0.33**	0.30**	0.05	-0.04
	7	-0.55***	-0.30*	0.85***	0.32*
	22	-0.34**	0.50***	0.80***	0.60***
	70	-0.55***	0.62***	0.02	0.09
	110	-0.34	0.70**	-0.17	-0.52**
TF	0	-0.10	0.01	0.14	0.07
	12	-0.60***	0.02	0.69***	0.54***
	35	-0.40***	0.40***	0.40***	0.52***
	57	-0.43**	0.75***	0.41**	0.45**
	100	-0.80**	0.36	-0.16	0.50*

* $P<0.05$
** $P<0.01$
*** $P<0.001$

mineral soil samples, showed that chemical equilibrium was rapidly reached, even in surface soils. In these experiments chemical equilibrium in the surface soil was characterized by solutions that were highly under saturated with respect to $Al(OH)_3$, solubility. Therefore, the observed undersaturation with $Al(OH)_3$ on surface soil horizons are not necessarily caused by kinetic constraints, but alternatively may be due to equilibrium with low solubility soil Al (e.g., organically bound soil Al).[30]

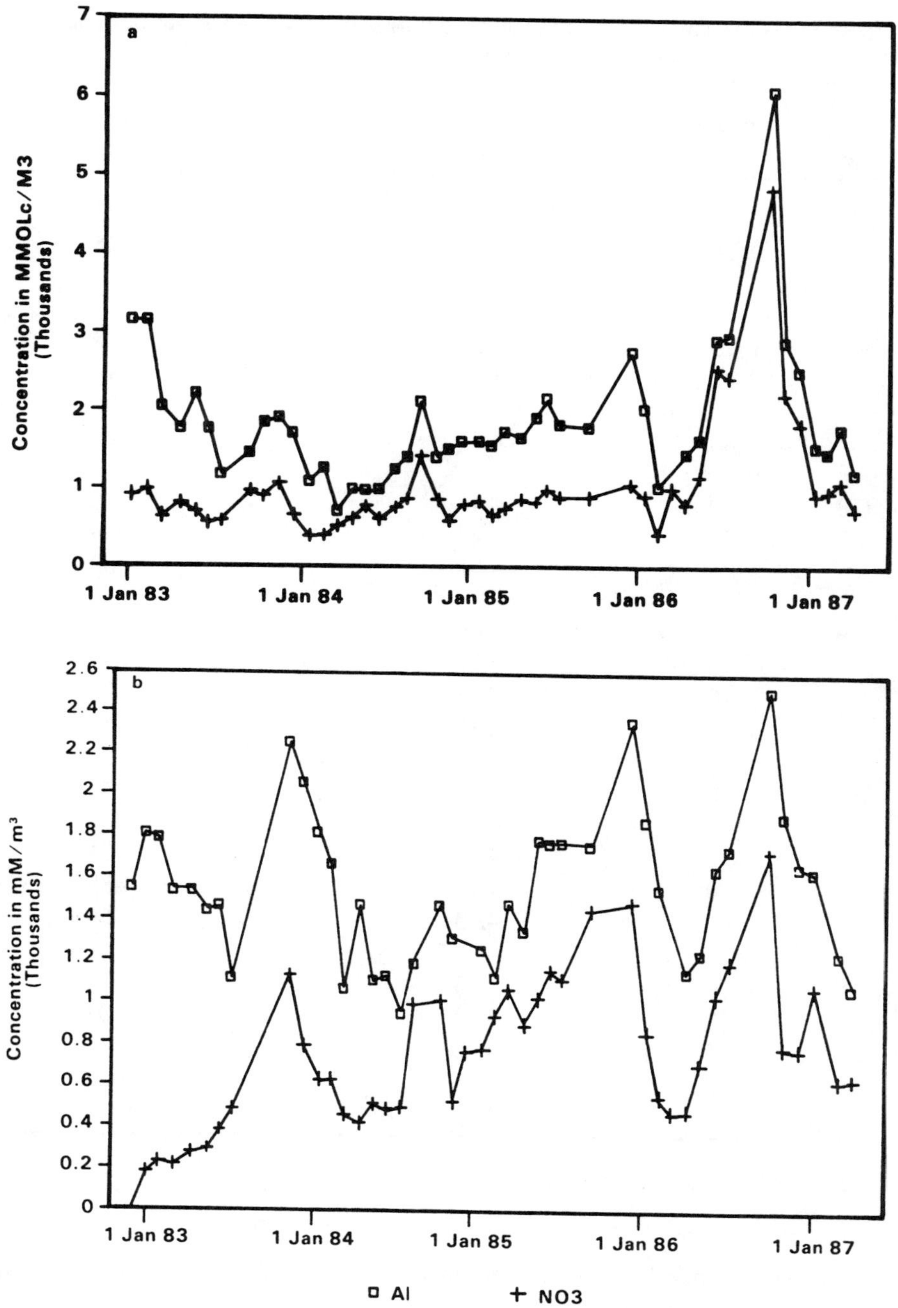

Figure 3. Concentrations of Al and NO_3 in GF, 40 cm (a) and TF, 35 cm (b) soil solutions as a function of time. Concentrations are in $mmol_c/m^3$.

Table 7. Mean soil solution pH values and total dissolved Al at various depths for the complete monitoring period

Site	depth (cm)	pH	Al(tot) ($mmol_c/m^3$)[a]
KH	0	4.04	20
	13	4.11	65
	49	4.50	67
KF	0	3.60	32
	12	3.62	194
	31	4.34	494
	80	4.57	312
GB	0	4.41	178
	40	4.40	239
	100	4.51	188
GF	0	4.08	43
	10	3.91	839
	40	3.97	2073
	100	4.15	1751
HH	0	3.83	43
	7	3.53	372
	22	4.21	551
	70	4.25	488
	110	4.13	364
TF	0	3.95	33
	12	3.38	107
	35	4.02	1515
	57	4.03	2065
	100	4.03	1821

[a] mmol of unit ionic charge per m^3.

Although chemical equilibrium calculations indicate gibbsite may be present in the sub-soils, this mineral was not detected by X-ray and optical mineralogical analysis. Gibbsite may form more readily from soil solutions with elevated concentrations of dissolved inorganic Al than from solutions with relatively high amounts of organic acids, where dissolved Al is primarily in an organic form. This latter condition is true in natural podzolic soils, where acidification is mainly driven by production and transport of organic acids. Gibbsite formation, if it takes place at all, will most likely be a recent phenomen, associated with elevated acid inputs, and its concentration will be small. The "gibbsite-like" Al solubility control may also be caused by Al-containing secondary minerals, which have solubility characteristics similar to gibbsite, but which are mineralogically different, e.g., Al interlayers in clay minerals, or "free" Al.

Table 8. Saturation Index (SI) with respect to gibbsite (pAl-3pH=-9.00 at 10°C) and jurbanite ($pAl+pSO_4-pH=3.80$)[31]

Site	Depth (cm)	Horizon	Gibbsite		Jurbanite		
			mean	sd	mean	sd	n
KH	13	E*	-1.36	0.64	-1.09	0.26	8
	49	Bhs*	-0.26	0.46	-0.64	0.16	9
KF	12	E*	-2.43	0.88	-0.83	0.27	10
	31	Bhs	-0.02	0.44	0.23	0.23	10
	80	C	0.36	0.32	0.34	0.08	7
GB	0	C	-0.26	0.48	-0.37	0.51	95
	40	C	0.03	0.36	-0.12	0.20	54
	100	C	0.05	0.26	-0.08	0.31	55
GF	10	C*	-1.12	0.66	0.16	0.31	27
	40	C	-0.64	0.30	0.67	0.22	86
	100	C	-0.15	0.27	0.84	0.21	69
HH	7	E*	-2.48	0.78	-0.35	0.28	57
	22	Bhs	-0.38	0.34	0.39	0.25	69
	70	Bs	-0.15	0.30	0.36	0.18	41
	110	C	-0.10	0.27	0.42	0.12	13
TF	12	E*	-3.52	0.41	-1.06	0.26	54
	35	Bhs	-0.60	0.34	0.78	0.20	83
	57	Bs	-0.39	0.28	1.00	0.10	45
	100	BC	-0.04	0.44	1.00	0.13	8

Source: Values are means, standard deviations and number of observations for the complete monitoring period.

*Values are overestimated, because organic monomeric Al was assumed to be negligible.

Jurbanite, a basic Al-hydroxy-sulfate,[31] is also often proposed to control Al solubility in mineral soils receiving high inputs of sulfuric acid.[29] However, only GB sub-soil solutions are in equilibrium with respect to jurbanite solubility; at the other sites solutions range from undersaturated (SI=-0.64; KH) to oversaturated ($0.34 < SI < 1.00$; KF, HH, GF, TF conditions). Because in most soils conditions of oversaturation with respect to solubility of jurbanite and net SO_4 release (Table 4) occur simultaneously, the formation and occurrence of jurbanite seems unlikely.

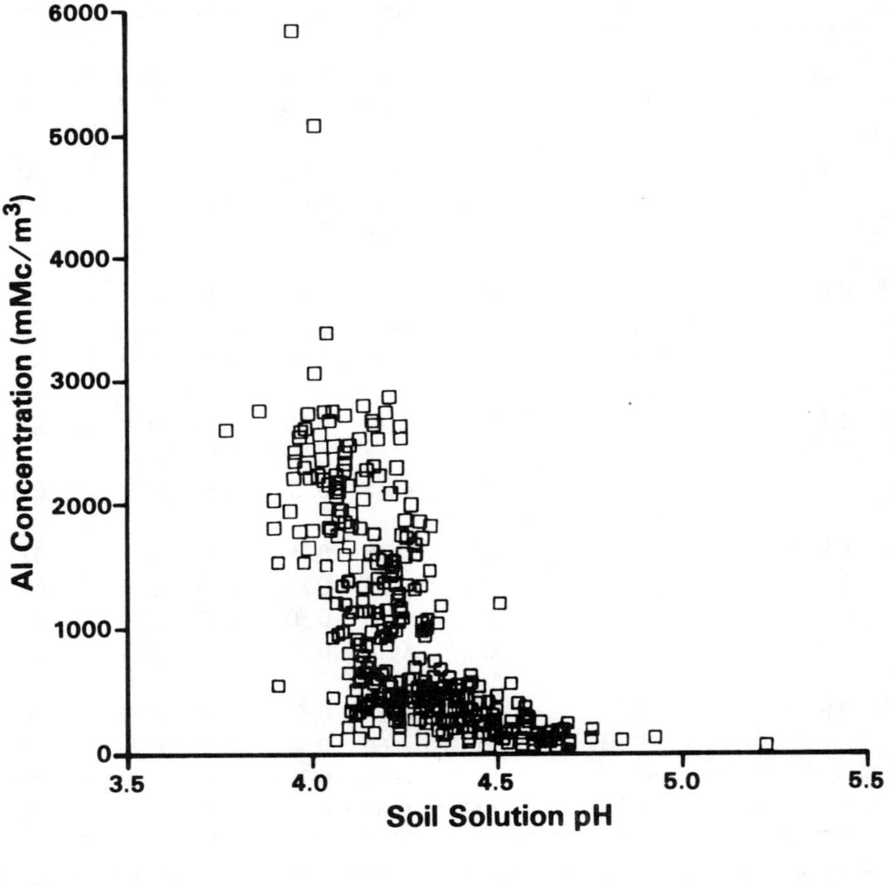

Figure 4. Total dissolved Al as a function of soil solution pH in all sub-soil solutions (below 50 cm depth) of the six study sites. Concentrations are in $mmol_c/m^3$ (all Al was assumed to be trivalent).

Al Budgets

Proton budgets[24] indicate that atmospheric acid inputs (H(inp) plus N(tr.) in Table 9) are by far the major source of the total acid load in the Dutch soils (Table 9). In the Danish soils, atmospheric acid inputs are relatively low and about equal to internally produced acidity (basic cation assimilation by the biomass, deprotonation of dissolved organic acids). Dissolution of Al is the major acid sink in all soils, neutralizing 30 to 95% of the total acid load. In the KF soil Al mobilization even exceeds the total acid load, which may be due to exchange of Al for Mg plus Na, as is also suggested by the negative value for basic cation mobilization (Table 9). Possibly, seasalt events, which commonly occur at KF, mobilize exchangeable Al and cause a (short term) increase in base saturation. It should be stressed that neutral seasalts can never dissolve Al from the soil minerals, but can only mobilize previously dissolved Al, stored in an exchangeable form. Strong or weak acids are necessary for the actual dissolution of Al from soil minerals. So, continued acidification after the pulse will reduce base saturation and restore exchangeable acidity again.

Table 9. Proton budgets for all sites for 1 hydrological year (KH, KF) and 4 hydrological years (GB, GF, HH, TF), respectively

	Acid Sources						Acid Sinks			
Site	H inp	N tr.	Org Ac.[c]	Biom assi	Anion mob.	SUM	SUM	Base mob.	Al mob.	H out
KH	0.4	0.2	0.2	0.3	0.1	1.2	1.2	0.6	0.4	0.2
KF	0.6	0.2	0.4	0.7	0.9	2.8	2.8	-0.6	3.1	0.3
GB	0.5	1.7	-0.2	0.0	0.5	2.5	2.3	0.7	1.4	0.2
GF	0.0	4.1	-0.2	0.4	0.9	5.2	5.3	0.4	4.7	0.2
HH	1.0	1.7	0.0	0.3	-0.3	2.7	2.7	0.4	2.1	0.2
TF	-0.2[b]	4.5	-0.2	0.5	1.1	5.7	5.6	0.1	5.3	0.2

Note: Values are in kmolc/ha·yr. The respective categories of acid sources and sinks are: free H+ input, N transformations[a], deprotonation of organic acids, biomass assimilation, anion mobilization, sum of acid sources, sum of acid sinks, base cation mobilization, Al mobilization, H^+ in drainage water

[a] Acid production due to N transformation is essentially caused by net removal of atmospherically derived NH_4 in the percolating soil solution and its partial transformation to NO_3. Therefore H^+ input plus N-transformations can be assigned as atmospheric acid input.

[b] Negative number indicates net alkalinity (HCO_3) deposition.

[c] Calculated from charge balance discrepancies.

Rates of basic cation mobilization (weathering plus cation exchange), which are low relative to total acid loads, are similar in all soils (-0.6 to 0.7 $kmol_c/ha \cdot yr$; Table 9) and hence do not seem to vary much with acid input. Although cation exchange probably modifies rates of mobilization of basic cations, our data indicate that rates of basic cation weathering are not greatly increased (1) by elevated acid loads or (2) in soils with a somewhat higher basic cation content (KH and KF). The suggestion that weathering rates do not strongly depend on acid loads in Dutch Spodosols is supported by similar rates, estimated for the whole period of soil formation (ca. 10.000 yr).[32] Also, the rates of mobilization of Si, which are similar in all soils, do not seem to depend on acid load (Table 4).

The apparent minor effect of acid load on the weathering rates of primary minerals implies that any additional acid input will be neutralized almost exclusively by Al solubilization in these soil types. Relative to the "natural" rate of Al mobilization (i.e., the mean Al weathering rate from 10,000 yr BP until now, estimated for similar soils)[32] current rates have increased from only slightly (KH) to ca. 10 times at GF and TF.

Chemical budgets for individual soil horizons (Table 10) indicate that Al mobilization is limited in the forest floor (0.10 to 0.16 $kmol_c/ha \cdot yr$), and increases markedly in the surface mineral soil. Solubilization of Al in the Spodosol E horizons ranges from 0.16 to 1.50 $kmol_c/ha \cdot yr$. In the two forested Spodosols, Al mobilization continues in the Bhs horizon (1.95 and 3.44 $kmol_c/ha \cdot yr$ at KF and TF, respectively). Some Al retention was observed in four of the sub-soils, but most of the mobilized Al is currently leached and lost to the groundwater at all sites.

Implications for Pedogenesis

The amounts of Si mobilized or immobilized in the various soil horizons are small compared to those of Al (Table 10). Only at KH mobilization rates of Al and Si are similar. Like Al, Si is mostly mobilized in the surface soil and partly retained in the sub-soil. The ratio of Al/Si mobilized provides information on the source of mobilized Al. Congruent dissolution of aluminosilicates would result in molar Al/Si ratios between 0.3 and 1.0. Because the observed ratios generally exceed 1.0 (Table 10), non-silicate bound, free Al rather than Al-silicates is apparently the predominant source of mobile Al. This observation is supported by laboratory experiments on the same soils.[30] Free Al is present in all soils and peaks in the Spodosol Bhs horizons (Table 3).

Relative to the current Al mobilization rate, the soils' free Al pools are small, so that they may be exhausted in a short period of time, even in Spodosols. From free Al content, bulk density, and the thickness of the Bhs it can be calculated that at the present mobilization rate all free Al can be completely removed from this horizon in ca. 200 yr. In forested driftsands, with low free Al contents, the free Al depleted soil layer may currently even deepen at a rate of 0.5 cm/yr.

Once free Al has been depleted, acid buffering of the soil will be limited, so that, upon continuation of the present deposition rates, soil solutions low in pH and dissolved Al will occur at increasing depth. Currently, this condition is only evident in the Spodosol E horizons.

Table 10. Mean annual Al and Si budgets and molar ratios of mobilized Al and Si per soil layer

Site	Depth cm	Horizon	Al $kmol_c$/ha·yr	Si kmol/ha·yr	Al/Si
KH	surf- 0	O	0.06	0.16	0.13
	0- 13	E	0.34	0.39	0.29
	13- 49	Bhs	-0.04	-0.00	...
KF	surf- 0	O	0.10	0.29	0.11
	0- 12	E	1.06	0.62	0.57
	12- 31	Bhs	1.95	-0.52	...
GB	surf- 2	C	0.98	0.14	2.33
	2- 40	C	0.70	0.14	1.67
	40- 100	C	-0.26	0.14	...
GF	surf- 0	O	0.16	0.11	0.48
	0- 40	C	5.46	0.48	3.79
	40- 100	C	-0.99	-0.22	...
H	surf- 0	O	0.20	0.41	0.16
	0- 7	E	1.50	0.07	7.14
	7- 22	Bhs	0.44	-0.28	...
	22- 70	Bs	-0.13	0.02	...
TV	surf- 0	O	0.11	0.05	0.73
	0- 12	E	0.16	0.20	0.27
	12- 35	Bhs	3.44	0.10	11.47
	35- 57	Bs	1.58	0.04	13.17

Note: Negative values denote immobilization.

CONCLUSIONS

Atmospheric acid inputs have a strong impact on the Al chemistry of acidic, sandy soils, with low concentrations of basic cations. Such soils, which cover large areas of Belgium, the Netherlands, Northwestern Germany and Denmark, are highly vulnerable to acidic deposition. Their base saturation is low and the rate of basic cation weathering does not increase with the rate of acid input. Therefore, any additional strong acid deposition in these soils is almost exclusively neutralized by Al solubilization. Aluminum concentrations in the surface soil layers are relatively low and may be in equilibrium with free (organically-bound) soil Al. Soil bound Al is probably least soluble in the surface horizons, where most of the Al leaching has occurred. Upon percolation, the soil solution probably attains equilibrium with more soluble Al compounds. Due to continued Al mobilization plus evaporative concentration, subsoil solutions reach equilibrium with respect to the solubility of gibbsite.

The source of mobilized Al is mostly free, non-silicate bound Al. The pools of free Al will be exhausted rapidly at the present high Al leaching rate. Our data indicate that the strong impact of atmospheric acid input on the Al chemistry of the soil may not only result in negative effects on nutrient solutions in the rooting environment, and on ground and surface water quality, but may also dramatically change pedogenetic processes.

ACKNOWLEDGMENTS

Thanks are due to E. J. Velthorst, N. Nakken, H. Sliepenbeek, A. Van Osch, A. Vallent, and H.W.G. Booltink for help during sampling and chemical analysis. Calculation of the waterfluxes was carried out by H.W.G. Booltink. We gratefully acknowledge Staatsbosbeheer (Gerritsfles, Hasselsven), Mr. W.A.M. Rijken (Tongbersven) and Mr. J. E. Handberg (Klosterhede) for making the sites available for our research. This research was supported in part by the EEC (contract ENV-650-NL), the Netherlands Technology Foundation (STW), the (Danish) Agency of the Environmental Protection and the (Danish) Ministry of Energy.

REFERENCES

1. Duchaufour, P. *Pedology* (translated by T.R. Paton; London: George Allen and Unwin, 1982).

2. Farmer, V. C., M. L. Berrow and J. D. Russel. "Imogolite and protoimogolite in spodic horizons evidence for a mobile aluminum silicate complex in podzol formation," *J. Soil Sci.* 31:673-684 (1980).

3. Childs, C. W., R. L. Parfitt and R. Lee. "Movement of aluminum as an inorganic complex in some podzolized soils, New Zealand," *Geoderma* 29:139-155 (1983).

4. David, M. B. and C. T. Driscoll. "Aluminum speciation and equilibria in soil solutions of a Haplorthod in the Adirondack Mountains," *Geoderma* 33:297-318 (1984).

5. Driscoll, C. T., N. Van Breemen and J. Mulder. "Aluminum chemistry in a forested Spodosol," *Soil Sci. Soc. Am. J.* 49:437-444 (1985).

6. Ugolini, F.C., R. Minden, H. Dawson and J. Zachara. "An example of soil processes in the *Abies amabilis* zone of Central Cascades, Washington. *Soil Sci.* 124:291-302 (1977).

7. Manley, E. P., W. Chesworth and L. J. Evans. "The soil solution chemistry of some podzolic soils from the eastern Canadian shield: A thermodynamic interpretation of the mineral phases controlling the content of Al^{3+} and H_4SiO_4," *J. Soil Sci.* 38:39-51 (1987).

8. Rosenqvist, I. T. *Sur Jord - surt vann: bidrag til en analyse af geologiske materialers buffervirkning overfor sterke syrer i nedbor* (Oslo: Ingenior Forlaget, 1977).

9. Krug, E. C. and C. R. Frink. "Acid rain on acid soil: a new perspective," *Science* 221:520-525 (1983).

10. Van Breemen, N., C. T. Driscoll and J. Mulder. "Acidic deposition and internal proton sources in acidification of soils and waters," *Nature* 307:599-604 (1984).

11. Mulder, J., J.J.M. Van Grinsven and N. Van Breemen. "Impacts of acid atmospheric deposition on woodland soils in the Netherlands: III. Aluminum chemistry," *Soil Sci. S oc. Am. J.* 51:1640-1646 (1987).

12. Brown, K. A. "Chemical effects of pH 3 sulfuric acid on a soil profile," *Water Air Soil Pollut.* 32:201-218 (1987).

13. Rasmussen, L., N. E. v. Freisleben and P. Jorgensen. "Leaching of ions from a forested Typic Udipsamment by acidified throughfall," *Geoderma* 43:1-15 (1988).

14. Van Breemen, N., P. A. Burrough, E. J. Velthorst, H. F. Van Dobben, Toke de Wit, T. B. Ridder and H.F.R. Reijnders. "Soil acidification from atmospheric ammonium sulfate in forest canopy throughfall," *Nature* 229:548-550 (1982).

15. Rasmussen, L. "Sur nedbors effekt pa ionbalancen og udvaskningen af metaller og anioner i danske naleskovsokosystemer iperioden 1983-1987," Unpublished report *Lab. Environ. Sci. Ecol.*, Techn. Univ., Lyngby, Denmark (1988).

16. Begheijn, L. Th. "Methods of chemical analyses for soils and waters," Internal Report, Dept. Soil Sci. Geol., Agric. Univ. Wageningen, the Netherlands (1980).

17. Bascomb, C. L. "Rapid method for the determination of cation exchange capacity of calcareous and non-calcareous soils," *J. Sci. Food Agric.* 12:821-823 (1964).

18. Van Dobben, H.F. and J. Mulder, Eds. *Impacts of Acid Atmospheric Deposition on Oligotrophic Ecosystems in the Netherlands: II. Biogeochemistry of Moorland Pools a nd Surrounding Terrestrial Area* (in prep.).

19. de Visser, P.H.B. "Interaction between soil, vegetation and atmospheric deposition," Report 02-01, Dutch Priority Programme on Acidification, Dept. Soil Sci. Geol., Agric. Univ., Wageningen, the Netherlands (1986).

20. Nieuwenhuyse, A. "Protonenbudgetten en nutrientencycli van twee grove den-opstanden en een struikheide vegetatie over het jaar 1983-1984," MSc Thesis, Dept. Soil Sci. Geol., Agric. Univ., Wageningen, the Netherlands (1986).

21. Pott, D.B., J.J. Alberts and A.W. Elzerman. "The influence of pH on the binding capacity and conditional stability constants of aluminum and naturally occurring organic matter," *Chem. Geol.* 48:293-304 (1985).

22. Dietze, G. "Bindungskapazitat und Bindungsstabilitatskonstanten von Al und gelosten organischen Sickerwasserinhaltsstoffen in Abhangigkeit von der Wasserstoffionenkonzentration," in *Berichte des Forschungszentrums Waldoekosysteme/Waldsterben* Reihe B, Bd. 3:25-83 (1986).

23. Driscoll, C.T. "A Procedure for the fractionation of aqueous aluminum in dilute acidic waters," *Int. J. Environ. Anal. Chem.* 16:267-284 (1984).

24. Van Breemen, N., J. Mulder and C.T. Driscoll. "Acidification and alkalinization of soils," *Plant and Soil* 75:283-308 (1983).

25. Van Breemen, N., P.H.B. de Visser and J.J.M. Van Grinsven. "Nutrient and proton budgets in four soil-vegetation systems underlain by pleistocene alluvial deposits. *J. Geol. Soc.* London 143:659-666 (1986).

26. Van Grinsven, J.J.M., N. Van Breemen and J. Mulder. "Impacts of acid atmospheric deposition on woodland soils in the Netherlands: I. Calculation of hydrologic and chemical budgets," *Soil Sci. Soc. Am. J.* 51:1629-1634 (1987).

27. Schecher, W. D. and C. T. Driscoll. "An evaluation of uncertainty associated with aluminum equilibrium calculations," *Water Resour. Res.* 23:525-534 (1987).

28. Nilsson, S.I. and B. Bergkvist. "Aluminum chemistry and acidification processes in a shallow podzol on the Swedish west coast," *Water Air Soil Pollut.* 20:311-329 (1983).

29. Ulrich, B. und E. Matzner. *Abiotische Folgewirkungen der weitraumigen Ausbreitung von Luftverunreinigungen* (Luftreinhaltung Forschungsbericht 104 02 615, Dept. Soil Sci. Forest Nutr., Univ. Gottingen, FRG, 1983).

30. Mulder, J., N. Van Breemen and H. C. Eijck. "Depletion of soil aluminum by acid deposition and implications for acid neutralization," *Naure.* 337:247-249 (1989).

31. Nordstrom, D. K. "The effect of sulfate on aluminum concentrations in natural waters: Some stability relations in the system Al_2O_3-SO_3-H_2O at 298°K," *Geochim. Cosmochim. Acta* 46:681-692 (1982).

32. Breeuwsma, A. and W. de Vries. "Gevolgen van zure regen voor de bodem 2. Aandeel in de bodemverzuring in Nederland," Report 1787, Stichting voor Bodemkartering, Wageningen, the Netherlands (1984).

33. "Soil taxonomy: A basic system of soil classification for making and interpreting soil surveys," USDA-SCS Agric. Handbook 436, U.S. Government Printing Office (1975).

CHAPTER 11

ALUMINUM SPECIATION AND ORGANIC CARBON IN WATERS OF CENTRAL ONTARIO

Bruce D. LaZerte

INTRODUCTION

The Muskoka-Haliburton region of Central Ontario is currently receiving about 62 meq/m^2 of sulphate deposition per year with an equivalent deposition of hydrogen ions[1]; even more sulfate and acidity was deposited annually 10 years ago.[1] This translates into acidified lakes[2] and streams[3], with relatively high levels of aluminum in the acidified streams.[4]

As part of an ongoing project to determine sources, sinks and forms of aqueous aluminum, we have monitored over the last two years a series of 26 different streams in two catchments, as well as 60 lysimeters at different locations in the catchments and soil profiles. Using these data sets, we will demonstrate where in the catchments aluminum is coming from, where it is lost and what form it is in. We will also attempt to identify the mineralogical sources and sinks for aluminum by constructing solubility diagrams. Finally, current models that predict the speciation of aluminum as a function of dissolved organic carbon (DOC) will be tested and an alternative developed from our data.

SITE DESCRIPTION

The data presented in this study comes from Harp and Plastic Lake catchments which have been intensively monitored over the last few years by the Acid Precipitation In Ontario Study (APIOS) of the Ontario Ministry of the Environment. The Plastic Lake catchment has a primarily coniferous forest on thin (< 1 m) podzolic soils, whereas the Harp Lake catchment has a primarily deciduous forest on much deeper and less acidic podzols.[5]

Both areas are characterized by extensive well drained upland soils, interspersed with wetlands, beaver ponds and conifer swamps in the lowland drainage areas. Details of the stream sampling stations and soil water samplers (lysimeters) will be presented elsewhere. The soil water data comes from both plexiglass zero-tension and ceramic-plate constant-tension lysimeters placed under the organic (LFH), elluviated mineral (Ae) and deeper mineral (Bf, and BC, C when present) horizons. We use here the Canadian system of soil classification.

METHODS

All water samples were analyzed for "total monomeric" and "organic monomeric" aluminum[6] using an automated catechol violet ion-exchange procedure[7] that agrees well with an older dialysis method[4,7] when estimating inorganic monomeric aluminum. In the automated system, inorganic monomeric aluminum is estimated by the difference between total and organic monomeric aluminum. In addition, a "total acidified" fraction was obtained by acidification to 0.1% (v/v) with concentrated nitric acid and subsequent analysis with graphite furnace atomic absorption spectroscopy.

From inorganic monomeric aluminum, pH, sulfate, total fluoride and ionic strength[4], we estimated $\{Al^{3+}\}$, where the {} indicate activities. Our calculation method[4] agrees well with other commonly available programs (e.g., ALCHEMI).

RESULTS AND DISCUSSION

Spatial Variability

We will first describe the sequence of changes in aluminum chemistry as precipitation passes through the well-drained upland soils of Plastic Lake catchment into a stream channel, from the stream through an organic rich conifer swamp to Plastic Lake, and finally out of the lake itself (Figures 1-3). In general, the spatial pattern in Harp Lake catchment is less pronounced because of the less acidic and deeper soils. The lysimeter data used in these figures comes only from the zero-tension lysimeters, as these concentrations should be more appropriate for direct comparisons with surface water. The data is presented as annual (1986-87) un-weighted means with 95% confidence intervals.

Organic aluminum and DOC concentrations are highest after passing through the LFH soil horizon (Figures 1,2). They then drop in the mineral horizons, primarily in the Bf, and remain low through the up- and downstream stations to the swamp. Levels of DOC and organic aluminum increase in the swamp, but then both decrease in the lake.

Inorganic monomeric aluminum (the difference between organic and total monomeric aluminum in Figure 1) is a nonsignificant fraction of total monomeric in the LFH horizon. However, it begins to dominate the total in the mineral horizons, and both the inorganic and total monomeric aluminum reach their highest values at the bottom of the Bf. Conversely, from the upland soil Bf horizons down through the stream, swamp and lake, inorganic monomeric aluminum is steadily lost.

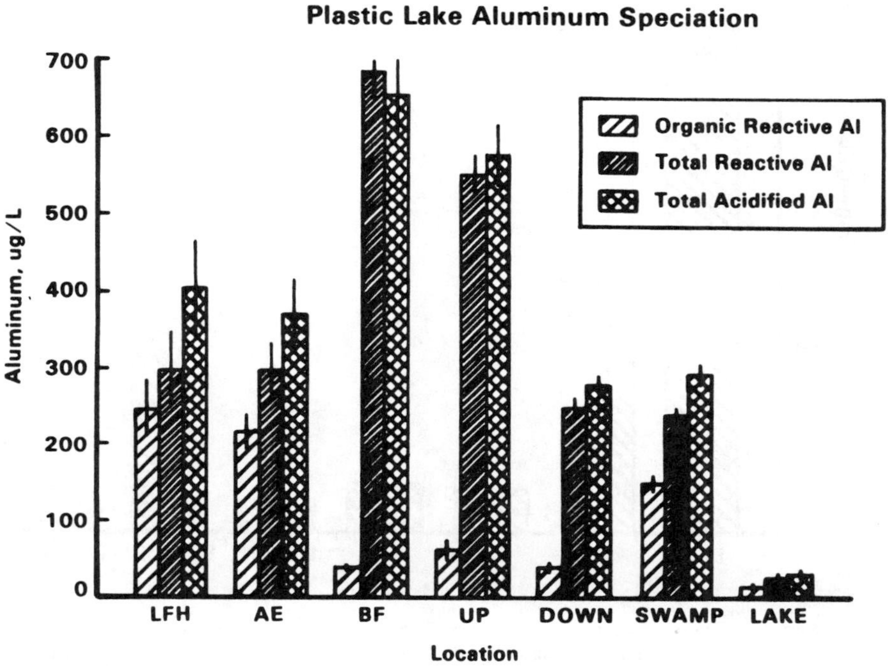

Figure 1. A comparison of three aluminum fractions across upland soils sites (LFH, Ae, Bf), two stream sites (UP-stream and DOWN-stream), a downstream swamp outflow and finally Plastic Lake outflow.

The difference between total acidified and total monomeric aluminum is usually insignificant, although there is a trend toward greater differences with higher DOC and organic aluminum levels. This may indicate that tightly bound or occluded organic aluminum is responsible. As noted elsewhere, polymeric, colloidal and particulate aluminum does not appear to be important intermediaries in the loss of aluminum from these waters.[4]

Sources and Sinks

The primary source of inorganic monomeric aluminum is the Bf horizon of upland soils (Figure 1). As has been observed elsewhere[8], Hem and Robertson's[9] microcrystalline gibbsite provides an upper limit on aluminum solubility over much of the soil water pH range as shown by plotting pAl (-LOG$\{Al^{3+}\}$) versus pH (Figure 4), whereas May et al.'s[10] natural gibbsite is usually oversaturated. The interlayer aluminum hydroxide line[11] falls between these two. This is consistent with the

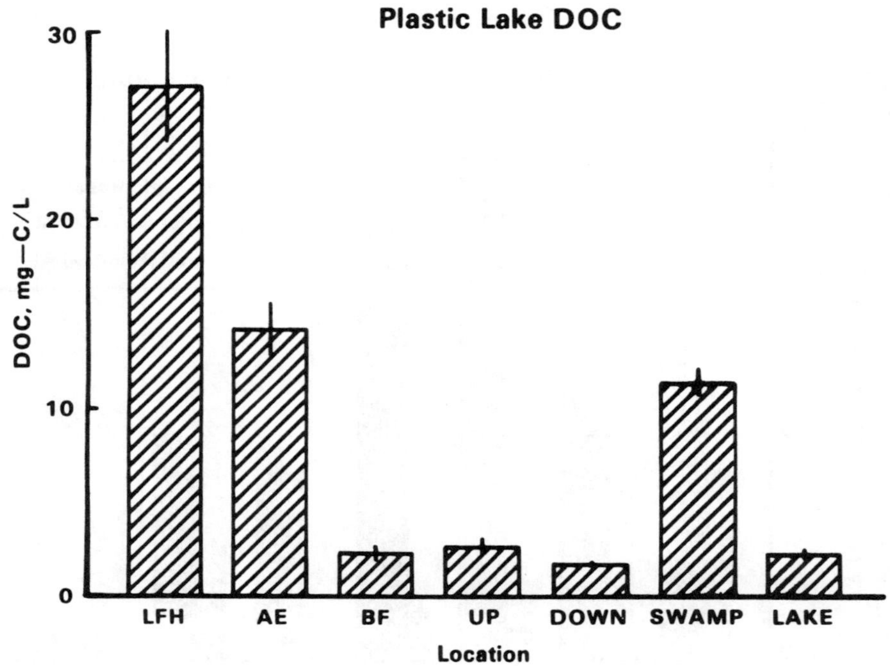

Figure 2. A comparison of DOC levels across upland soil sites (LFH, Ae, Bf), two stream sites (UP-stream and DOWN-stream), a downstream swamp outflow, and finally Plastic Lake Outflow.

hypothesis that the source of inorganic monomeric aluminum in the Bf horizon is some form of aluminum trihydroxide.

However, it is necessary to be cautious in interpreting these figures, especially at the higher pH's (>5.5), as Neal et al.[12] have shown a high degree of spurious correlation between pAl and pH in this range. The problem occurs because pAl is not directly measured, rather inorganic monomeric aluminum is measured and pAl calculated using measured pH. It is this calculation that introduces the spurious correlation at higher pH's.

High DOC samples (> 5 mg-C/L) primarily from the LFH and Ae soil horizons, have been deliberately excluded from Figure 4. Little inorganic monomeric aluminum is leached from the LFH and Ae horizons, and as might be expected,[8,13] the organic soil layer is often undersaturated with respect to the trihydroxide mineral phases (Figure 5).

The *loss* of inorganic monomeric aluminum in the low DOC lakes and streams is also associated with aluminum trihydroxide solubility (Figure 6) as has been shown frequently elsewhere,[14] but high DOC swamp waters are often undersaturated (Figure 7).

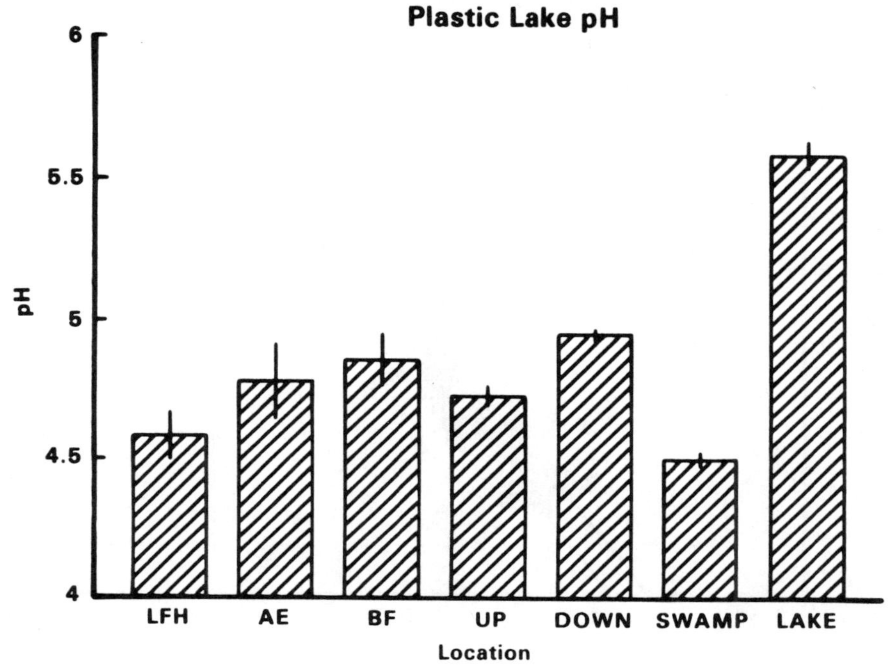

Figure 3. A comparison of pH levels across upland soil sites (LFH, Ae, Bf) two stream sites (UP-stream and DOWN-stream), a downstream swamp outflow, and finally Plastic Lake Outflow.

The undersaturation observed at higher pH's in Figure 6 may represent a methodological error. The ion-exchange procedure only separates positively charged aluminum species from the neutral or negatively charged. At pH's greater than 6.5, $Al(OH)_4^-$ begins to dominate the inorganic monomeric fraction and would not be measured as such by the cation exchange procedure.[7]

Alternatively, at these pH's the difference between total and inorganic monomeric aluminum is slight, and small systematic discrepancies could be causing the trend. In fact a substantial quantity of data that was high in pH and DOC had to be discarded because the measured organic monomeric aluminum was slightly greater than or equal to the total monomeric aluminum, making the calculation of pAl impossible.

It is doubtful whether other minerals are responsible for the inorganic monomeric aluminum levels observed in both the surface and soil waters. Only a few samples have ever approached equilibrium with respect to halloysite,[15] alunite[16] or other possible phases [see also ref. 14], and there is no observed relationship with silica or sulfate.

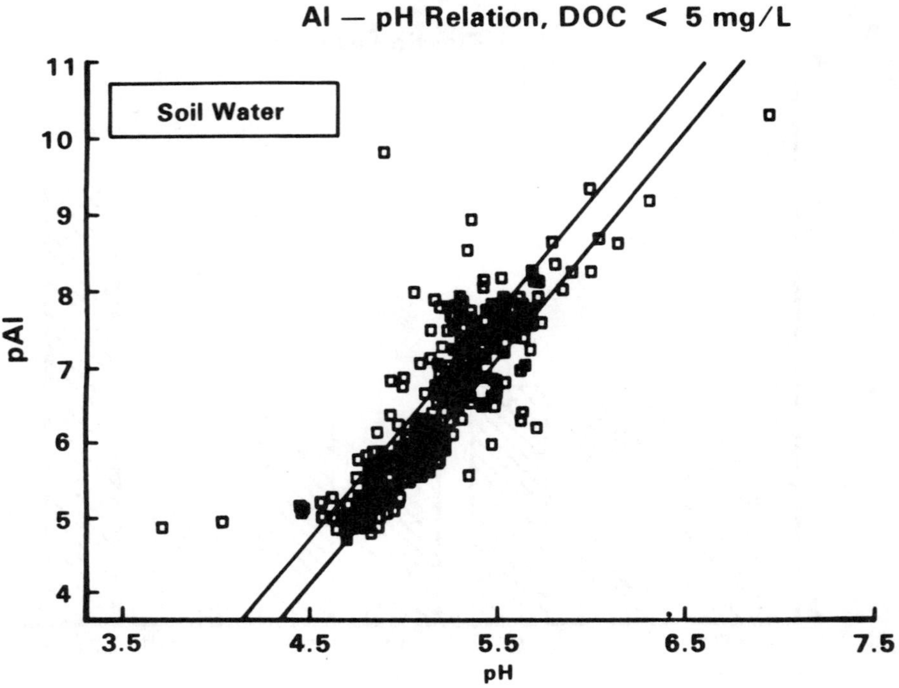

Figure 4. Solubility diagram for low DOC surface water. The lower line represents equilibrium with Hem and Robertson's[11] microcrystalline Gibbsite, whereas the upper line is for May et al.'s[10] natural Gibbsite.

DOC - Al Relations

There are two widely separated sources of organic aluminum and DOC: the upland soil LFH horizon and wetland conifer swamps. We do not know if the DOC associated with these two environments is chemically different or behaves differently. However, as noted above, neither of these environments appear to be in equilibrium with any aluminum trihydroxide mineral phase. For this reason, attempts have been made recently to predict the separation of organic from inorganic monomeric aluminum in high DOC soil and surface waters.[13,17]

Cronan et al.,[13] eqn. 2, predict pAl as a function of solution pH and the fractional loading of the solid phase carboxyl groups with aluminum. If it is assumed that the fractional loading of the solution phase DOC is the same as that of the solid phase, then the Cronan et al. model can be applied to aqueous phase measurements. We estimated the concentration of aqueous phase carboxyl groups with the Oliver et al. procedure[18] which has been previously verified for our waters.[3] Dividing this into our measurement of organic aluminum gives a rough estimate of the bound aluminum ratio used by Cronan et al.[13]

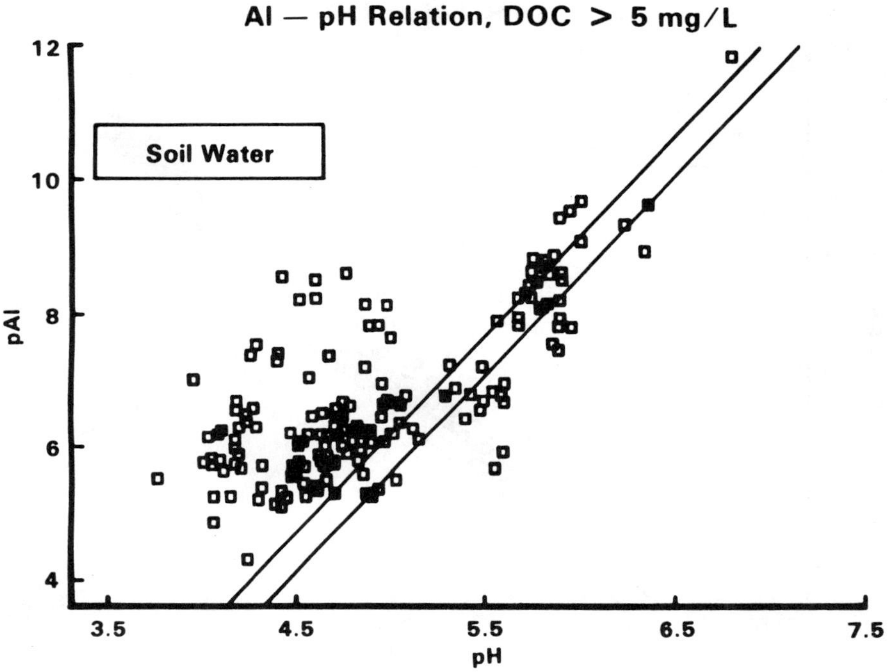

Figure 5. Solubility diagram for high DOC soil water. The lower line represents equilibrium with Hem and Robertson's[11] microcrystalline Gibbsite, whereas the upper line is for May et al.'s[10] natural Gibbsite.

Although the model does fit around pAl = 5 (which is the region originally verified by Cronan et al.,[13] at higher observed pAl's the model underestimates severely in high DOC (>5 mg/L) soil waters. For most of our data $5 < pAl < 10$. Similar or worse relationships were obtained for high DOC surface waters and low DOC surface and soil waters.

The Backes and Tipping model[17] is a more general chemical equilibrium model. It predicts organically complexed aluminum in solution as a function of pH, DOC and pAl. Cronan et al.[13] also present an alternative representation of their model [their eqn. 3] which is similar in form, but with more constraints on the parameters. The generalized form of the two models are:

$$p(\text{Org-Al}) = A + (B * \text{pDOC}) + (C * \text{pAl}) + (D * \text{pH})$$

where $\text{pDOC} = -\log(\text{g-C/L})$

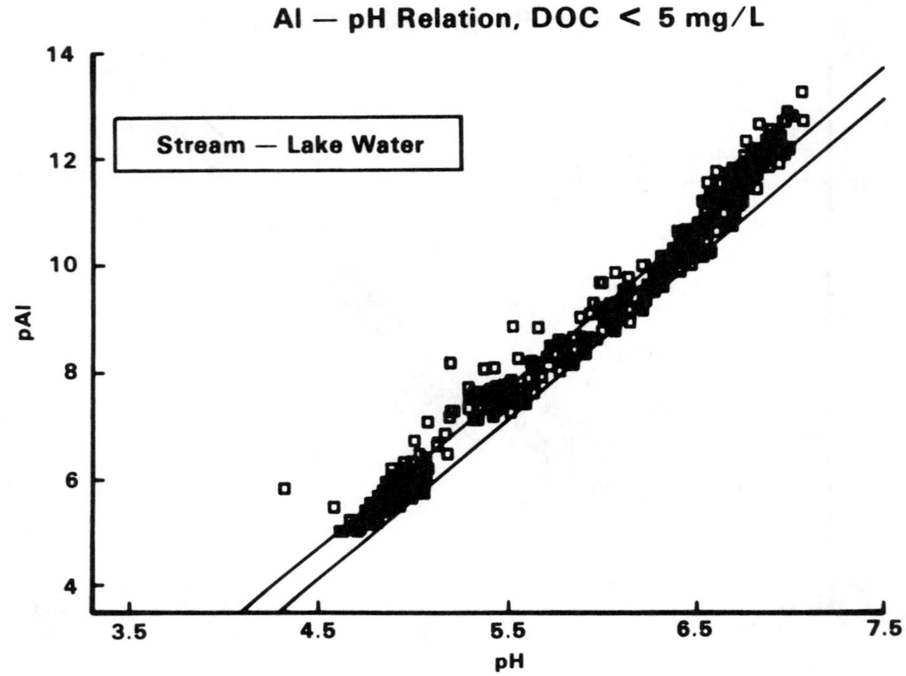

Figure 6. Solubility diagram for low DOC soil water. The lower line represents equilibrium with Hem and Robertson's[11] microcrystalline Gibbsite, whereas the upper line is for May et al.'s[10] natural Gibbsite.

Backes and Tipping[17]:

B = 1, all other terms estimated

Cronan et al.[13]:

A = -pK/n + pX, where X = Eq. of exchange sites/g - C of DOC, n is an estimated term, and pK is from Cronan et al.[13]

B = 1
C = l/n
D = -3/n

The similarity in general structure of the two models and their apparent success suggested that the additive log-linear chemical equilibrium model would be the most successful with our database. But even the simple B = 1 constraint in the Backes and Tipping model restricts its success with our database (Table 1). Only by letting all parameters "float" was a good fit ("Total" $R^2 > 0.9$) obtained (Table 2).

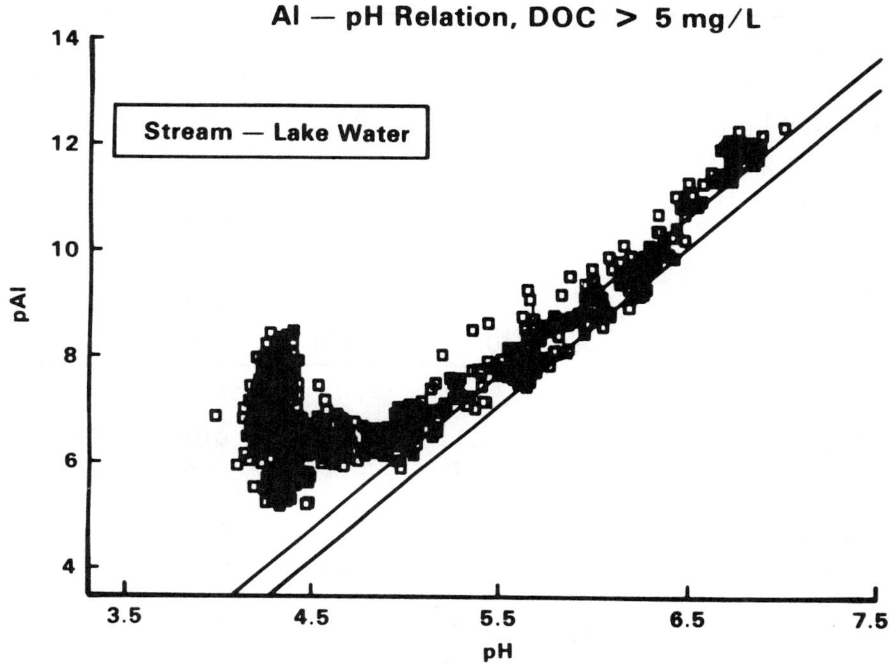

Figure 7. Solubility diagram for high DOC surface water. The lower line represents equilibrium with Hem and Robertson's[11] microcrystalline Gibbsite, whereas the upper line is for May et al.'s[10] natural Gibbsite.

In all four data sets as well as the total, p(Org-Al) was best predicted by pDOC and pAl (Table 2). Although occasionally significant, pH contributed little to the regressions (adding <.01 to R^2) and was found through collinearity diagnostics to seriously degrade the parameter estimates. As a result, pH was dropped. Other regression models were explored (e.g., strictly linear models) but none performed as well as those in Table 2. The range of variables in the total data set is given in Table 3.

The "Total" estimate of the A coefficient (3.20) is reasonably close to the original estimate of Backes and Tipping (3.58).[17] However, it is even closer to the revised estimate Tipping et al.[19] obtained from some of our data from Plastic Lake Inflow No. 1 (3.28).

While the "Total" model has B = 0.84, the low DOC data sets had B coefficients closer to 1.0 (Table 2) as used in the Backes and Tipping[17] and Cronan et al.[13] models, whereas the high DOC data sets had estimated B values substantially lower. This could indicate that in high DOC waters we are underestimating organically bound aluminum with the catechol violet procedure (as noted above).

Table 1. Estimated parameters (std. error) for the Backes and Tipping model

Database	N	A	C	D	R^2
Backes and Tipping	-	3.58[a]	0.72	-1.05	0.96[b]
Soil < 5 mg-C/L	434	3.64 (0.13)	0.07 (0.01)	-0.18 (0.04)	0.09
Soil > 5	169	3.12 (0.10)	0.13 (0.02)	-0.14 (0.03)	0.33
Surface < 5	417	2.73 (0.19)	0.09 (0.02)	-0.03[c] (0.06)	0.69
Surface > 5	634	3.09 (0.02)	0.12 (0.00)	-0.12 (0.01)	0.63
Total	1654	3.22 (0.02)	0.11 (0.00)	-0.14 (0.01)	0.51

[a]Calculated for DOC in units of mg-C/L, not mg/L of humic acid, assuming 2 g-HA = 1 g-C/DOC.
[b]The fit to laboratory data in Backes and Tipping.[17]
[c]Not significant, $p = 0.62$.

Our estimates of the C coefficient differ from Backes and Tipping's by almost an order of magnitude. At present, we have no explanation for this.

Despite the wide range in observed pH (Table 3), the high correlation between pH and pDOC and pAL effectively made pH a redundant variable in our models. An experimental protocol with crossed gradients of pH, pDOC and pAl could more effectively estimate the pH effect. However, in the systems we are studying here, this does not appear necessary.

SUMMARY

Soils and surface waters in the Muskoka-Haliburton region of Central Ontario exhibit well defined regional sources and sinks of inorganic and organic monomeric aluminum. For inorganic monomeric aluminum, the source is the Bh soil horizon and the sink is all downstream locations. Organic monomeric aluminum and DOC is obtained from two widely separated sources: the soil LFH horizon and wetlands.

Table 2. Estimated parameters (std. error) for the reduced model (no pH term)

Database	N	A	B	C	R^2
Backes and Tipping	-	3.58	1.0	0.72	-
Soil < 5 mg-C/L	434	3.30 (0.13)	0.90 (0.05)	0.028 (0.006)	0.47
Soil > 5	169	3.09 (0.10)	0.78 (0.05)	0.092 (0.011)	0.71
Surface < 5	417	2.83 (0.10)	0.93 (0.04)	0.075 (0.003)	0.74
Surface > 5	634	3.41 (0.03)	0.65 (0.02)	0.084 (0.002)	0.89
Total	1654	3.20 (0.02)	0.84 (0.01)	0.063 (0.002)	0.91

Table 3. Descriptive statistics for the total dataset (n=1654)

	pAl	p(Org-Al)	pDOC	pH
Min	4.29	4.21	1.09	3.71
Max	13.27	6.53	3.00	7.07
Mean	7.34	5.57	2.27	5.20
Std. Dev.	1.91	0.39	0.41	0.77

The mineralogical source and sink of inorganic monomeric aluminum appears to be some aluminum trihydroxide solid phase, and it can be effectively predicted in low DOC waters by assuming equilibrium with this phase. In waters higher in DOC, mineralogical control does not always occur. Nonetheless, the separation between free and organic aluminum can be effectively predicted in high DOC waters using a Backes and Tipping type of chemical equilibrium model. There does not appear to be a universal set of coefficients that apply to all environments.

ACKNOWLEDGMENTS

The work presented here is the result of the combined efforts of many individuals. In particular, Lem Scott supervised the field sampling, while Peter Grauds and Charlie Chun performed the chemical analyses.

REFERENCES

1. Dillon, P. J. Unpublished results (1987). Ontario Ministry of the Environment, Dorset Research Centre, Dorset, Ontario, Canada (1987).

2. Dillon, P. J. "The rate of acidification of aquatic ecosystems in Ontario, Canada," *Nature.* in press (1987).

3. LaZerte, B. D., and P. J. Dillon. "Relative importance of anthropogenic versus natural sources of acidity in lakes and streams of central Ontario," *Can. J. Fish. Aquat. Sci.* 41(11):1664-1677 (1984).

4. LaZerte, B. D. "Forms of aqueous aluminum in acidified catchments of central Ontario: A Methodological Analysis," *Can. J. Fish. Aquat. Sci.* 41(5):766-776 (1984).

5. Neary, A., E. Mistry and L. Vanderstar. "Sulphate relationships in some central Ontario forest soils," *Can. J. Soil. Sci.* 67:341-352 (1987).

6. Driscoll, C. T. "A procedure for the fractionation of aqueous aluminum in dilute acidic waters," *Intern. J. Environ. Anal. Chem.* 16:267-283 (1984).

7. LaZerte, B. D., C. Chun, D. Evans, and F. Tomassini. "Measurement of aqueous aluminum species: Comparison of dialysis and ion exchange techniques," *Environ. Sci. Technol.* 22:1106-1108 (1988).

8. David, M. B. and C. T. Driscoll. "Aluminum speciation and equilibria in soil solutions of a haplorthod in the Adirondack Mountains (New York, U.S.A.)," *Geoderma* 33:297-318 (1984).

9. Hem, J. D. and C. E. Robertson. "Form and stability of aluminum hydroxide complexes in diluted solution," U.S. Geol. Surv. Water Supply Paper 1827-A (1967).

10. May, H. M., P. A. Helmke, and M. L. Jackson. "Gibbsite solubility and termodynamic properties of hydroxyaluminum ions in aqueous solution at 25°C," *Geochim. Cosmochim. Acta* 43:861-868 (1979).

11. Bloom, P. R., M. B. McBride, and B. Chadbourne."Adsorption of aluminum by a smectite: I. Surface hydrolysis during Ca^{2+}-Al^{3+} exchange," *Soil Sci. Soc. Amer. J.* 1:1068-1073 (1977).

12. Neal, C., R. A. Skeffington, R. Williams, and D. J. Roberts, "Aluminium solubility controls in acid waters: The need for a reappraisal," submitted manuscript (1988).

13. Cronan, C. S., W. J. Walker, and P. R. Bloom. "Predicting aqueous aluminum concentrations in natural waters," *Nature* 324:140-143 (1986).

14. Driscoll, C. T., J. P. Baker, J. J. Bisogni, and C. L. Schofield. "Aluminum speciation and equilibria in dilute acidic surface waters of the Adirondack region of New York State," In *Geological Aspects of Acid Deposition*, O. P. Bricker, Ed. (Boston: Butterworth Publishers, 1984), p. 55.

15. Hem, J. D., C. E. Robertson, C. J. Lind, and W. L. Polzer. "Chemical interactions of aluminum with aqueous silica at 25°C," U.S. Geol. Surv. Water Supply Paper 1827-E (1973).

16. Nordstrom, D. K., "The effects of sulphate on aluminum concentrations in natural waters: Some stability relations in the system Al_2O_3-SO_3-H_2O at 298°K," *Geochim. Cosmochim. Acta* 46(4):681-692 (1982).

17. Backes, C. A. and E. Tipping, "An evaluation of the use of cation exchange resin for the determination of organically complexed Al in natural acid waters," *Int. J. Environ. Anal. Chem.* 30:135-143 (1987).

18. Oliver, B. G., E. M. Thurman, and R. L. Malcolm, "The contribution of humic substances to the acidity of colored waters," *Geochim. Cosmochim. Acta* 47:2031-2035 (1983).

19. Tipping, E., C. Woof, C. A. Backes, and M. Ohnstad, "Aluminum speciation in acidic natural waters: Testing of a model for Al-humic complexation," *Wat. Res.*, 22(3):321-326 (1988).

CHAPTER 12

ALUMINUM SPECIATION AND TOXICITY IN UPLAND WATERS

A.J. Dobbs, P. French, A.M. Gunn, D.T.E. Hunt, D.A.Winnard

INTRODUCTION

The subject of metal speciation in relation to environmental effects is very technically demanding and complex. One of the reasons for this complexity is the reactivity of many of the metal species involved. It is interesting to compare metal speciation with carbon speciation, which is basically organic chemistry. The carbon "species" are organic chemicals, and their general lack of kinetic lability allows extraction and analysis to be routinely undertaken. Not so, in general, with metals, with the exception of some organometallic species, where attempts to extract a particular metal species generally result in rapid re-equilibration in the sample. When this lability feature is combined with the requirement for speciation measurements at low, environmentally realistic concentrations, it usually turns out that the available analytical techniques are not capable of determining specific metal forms. This is certainly true of aluminum. All that can be achieved is some form of empirical separation based, for example, on differences in molecular size or reaction kinetics. It is, in our view, important to bear in mind that such separation procedures are operationally defined. Results from size based separations, such as dialysis or ultrafiltration, and kinetic separations based on ion exchange or kinetic colorimetry depend critically on the way the experiments are conducted. It is therefore essential that the methodologies used are well documented and consistently applied. We also believe that the operational nature of such separations should be clearly reflected in the names of the fractions, and the tendency to assign biologically or chemically more "relevant" names to them (such as "toxic", "bioavailable" or "monomeric") should be resisted unless there is good evidence to equate the measurements to the concepts. Since the main objective of speciation studies in natural waters is to determine the effects, actual or potential, on biota it is essential that experimental analytical speciation studies are conducted in concert with toxicity or bioavailability studies. This approach is followed in the studies described in this chapter.

Aluminum is known to be toxic to a wide range of aquatic organisms under conditions of low pH and hardness, and its mobilization in acidic upland waters is thought to be an important contributory factor to the decline of certain fisheries in Europe and North America. The toxicity is, however, strongly influenced by the forms in which the aluminum is present. Measurements of total aluminum levels alone are, therefore, of limited use in investigations of the role of aluminum in surface water acidification and its biological effects. Two approaches are commonly employed for studying aluminum speciation: (i) thermodynamic modeling, and (ii) analytical speciation techniques. The thermodynamic modeling approach suffers from the disadvantage that constants for important species may be either unavailable or of questionable accuracy, in addition to which waters of interest may not be at thermodynamic equilibrium. However, the approach can provide information on simple inorganic complexes - such as monomeric aluminum hydroxides - which cannot readily be measured by analytical techniques. It can also give accurate data on artificial samples prepared for toxicity studies or for performance tests on analytical speciation techniques.

It is clearly not possible to select a single toxicity or bioavailability test procedure that is representative for all aquatic organisms. However, convenience and costs are often important considerations that must be taken into account. For the purposes of this study we selected the Microtox system[1] based on the bioluminescent response of reconstituted freeze-dried bacteria, *Photobacterium phosphoreum*. From the environmental point of view this system is not ideal since the bacteria are of marine origin and aluminum toxicity is not generally recognized as a problem for marine bacteria. However, Microtox is very widely used as a general water quality indicator and is also very convenient and relatively cheap. Additionally, we have developed an on-line field system to monitor aluminum speciation and pH in conjunction with fish stress measurements. Thermodynamic calculations have been used both to aid interpretation of the speciation/toxicity experiments and to model aluminum speciation and solubility in typical natural waters.

THERMODYNAMIC MODELING

A computer program developed at the Water Research Centre[2] was used to model aluminum speciation in natural upland waters[3] using available equilibrium data. Complexation by hydroxide, fluoride, sulfate, phosphate and by model organic ligands was considered. Predictions of monomeric hydroxide distributions in the pH range relevant to acid waters were found to be influenced strongly by the choice of stability constants. In particular, there was considerable uncertainty concerning $AlOH^{2+}$, $Al(OH)_2^+$ and $Al(OH)_3^0$ (Figure 1) which are important species between pH 4.5 and 6, but Al^{3+} was always the dominant form at pH values below 5. Sulfate, phosphate and chloride levels are normally low in upland waters and their complexes with aluminum were predicted to be unimportant. Calculations for fluoride, when present at concentrations equal to or greater than that of aluminum, showed its complexes to dominate aluminum chemistry at pH values below 6. During acid events in upland waters total aluminum levels are usually elevated to such an extent that the fluoride-aluminum ratio is relatively low and under such conditions fluoride complexes were predicted to be of only minor importance, with the main inorganic species being the free ion and aluminum hydroxides.

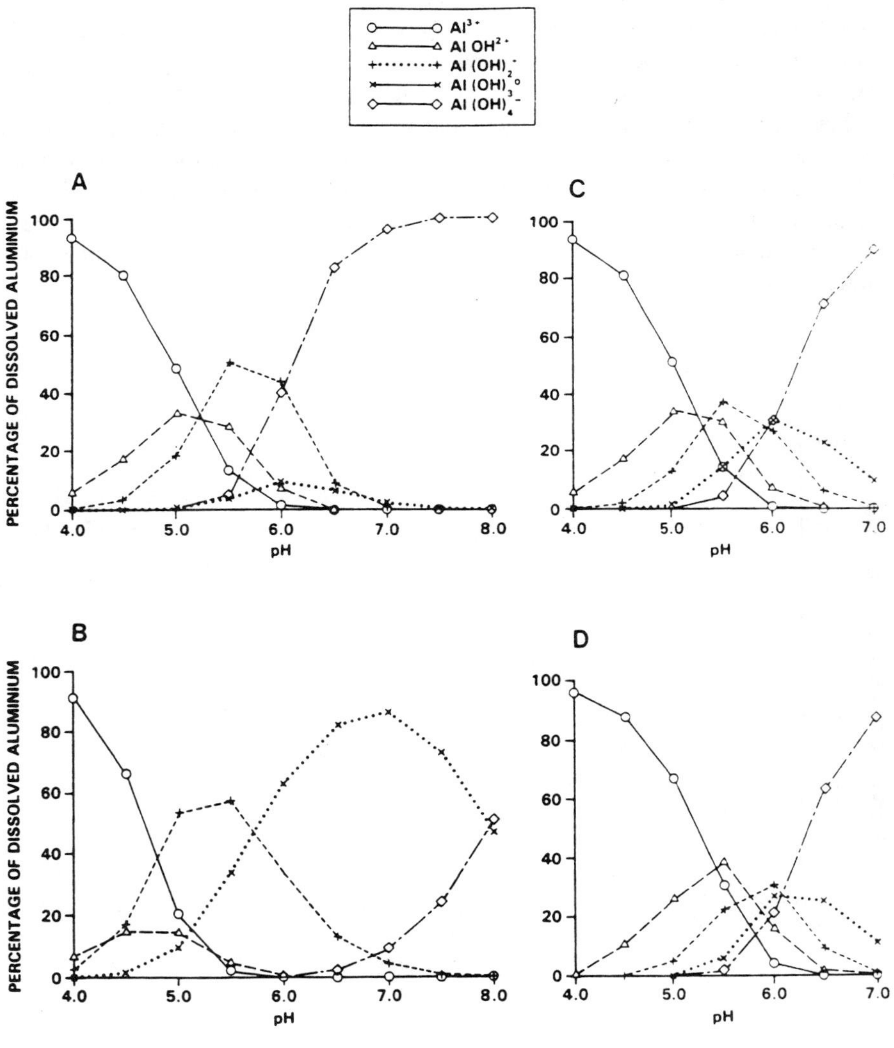

Figure 1. Distribution of free aluminum and monomeric hydroxide complexes, using four different sets of stability constants. (A) May, Helmke and Jackson,[4] (B) Baes and Mesmer,[5] (C) Dryssen,[6] (D) Helliwell et al.[7]

The likely importance of complexation by organic substances was assessed using equilibrium data for the model organic ligands salicylate and oxalate. These have similar functional groups to those thought to be responsible for complexing in fulvic and humic acids, and a recent study has indicated the likely presence of salicylate-type aluminum binding sites in fulvic acid.[8] Predictions for salicylate and oxalate indicated that their complexes were important when present at molar ratios approaching or exceeding that of aluminum. Figure 2 shows the calculated distribution of salicylate complexes in the absence of fluoride, for an aluminum level of 1.6×10^{-5} *M* and a range of salicylate concentrations (10^{-6} *M* to 10^{-4} *M*) considered broadly representative of that at which organic ligands could be found. Oxalate binds more strongly than salicylate and, when present in excess, complexes virtually all aluminum in the pH range 4 - 5. It has been suggested that fulvic acid and fluoride react with aluminum by similar mechanisms, having the same pH dependence, and thus are in direct competition.[8] Calculations with fluoride and either salicylate or oxalate included in the model suggested that organic complexation would only be important when ligand levels greatly exceeded that of fluoride. Of course, such conditions do often exist in many upland waters.

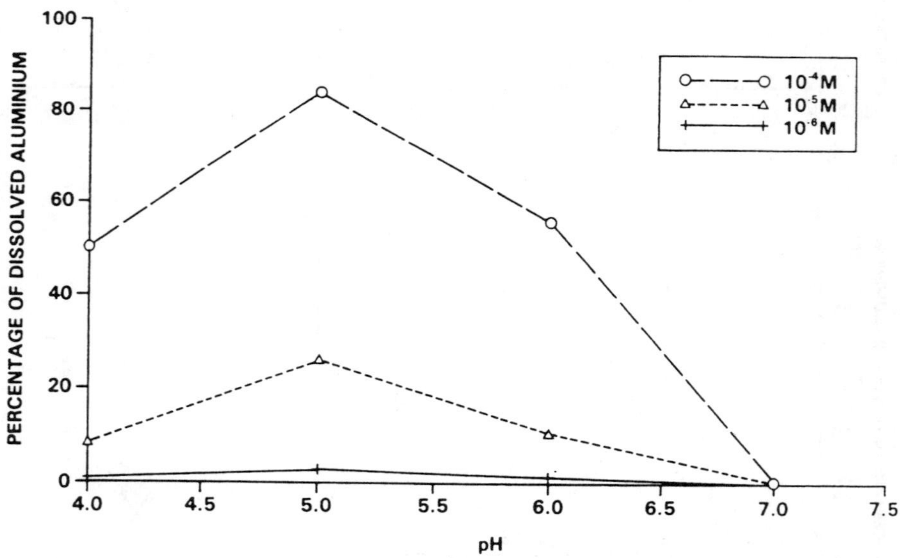

Figure 2. Percentage of aluminum complexed with salicylic acid as a function of pH, at an aluminum concentration of 1.6×10^{-5} *M* and three different levels of salicylic acid.

Speciation/Toxicity Relationships

The effect of complexation on aluminum toxicity was investigated using the Microtox system[1] based on measurement of the light output of a luminescent bacterium. Initial work revealed the presence of a pH 7.2 buffer in the reconstituted bacterial medium. In order to operate under acid conditions (pH 5.5) this was removed by centrifugation and the bacteria resuspended in a zwitterionic "biological" buffer (MES: 2-(*N*-morpholino)ethane sulfonic acid) at the minimum concentration commensurate with adequate pH control. Buffers of the MES type have been shown to exhibit low metal binding strengths and low toxicities.[9] In order to avoid changes in speciation during the experiment, the dilution effect of the bacterial suspension on the toxicant medium was kept to a minimum, representing only 9.1% of the toxicant concentration, as compared with 50% in the standard Microtox assay. Smaller dilution than this was not possible as a minimum of 100 μL of bacterial suspension was necessary to provide a reproducible initial light output when added to sample and control solutions. The bacteria were also very sensitive to solution conditions such as pH (5.5), salinity (2.1%) and temperature (15°C), and these were strictly controlled throughout the procedure.

For studies at aluminum concentrations of 100 - 1000 μg/L, sufficient sensitivity was achieved using relatively long exposure times (about one hour); longer times than this were not possible because of the rapid natural decay of the light output of the organism at the test pH. The high ionic strength resulting from the use of sodium chloride to provide osmotic protection for the marine bacterium was taken into account in equilibrium speciation calculations. Toxicity was calculated as the ratio of light lost to light remaining, after correction for decay in the control solutions.

Concurrent chemical characterization of test solutions was achieved using a modified catechol violet colorimetric procedure[10] with a short reaction time. This approach was preferred to an ion-exchange separation, as it provides a kinetic split of hydroxy monomeric aluminum from polymeric species.[11,12] The polymeric aluminum species were expected to have limited toxicity to the Microtox bacteria, and this was borne out by initial tests in which a pronounced flattening in both Microtox response and catechol violet reactivity was observed at high aluminum concentrations. For the complexation tests the concentrations of aluminum used were therefore kept low, and all solutions were aged for at least 24 hours to allow monomeric aluminum concentrations to stabilize.

Citrate, which complexes aluminum very strongly, was examined first as a model ligand at a series of concentrations in the range 0 - 0.001 *M*. These tests indicated that the Al-citrate complex was essentially non-toxic to levels of at least 25 mg/L (as Al); this contrasts with a one hour EC50, defined as the concentration required to induce a 50% reduction in light output relative to controls, of 300 μg/L in the absence of added citrate. Results for both the toxicity and catechol violet reactivity tests are shown in Figure 3. Catechol responses were essentially flat in all cases where aluminum was present at lower levels than citrate, while "breakthrough points", corresponding to abrupt increases in uncomplexed aluminum, occurred at levels in good agreement with those predicted on the basis of a 1:1 Al-citrate complex. Breakthrough points for the

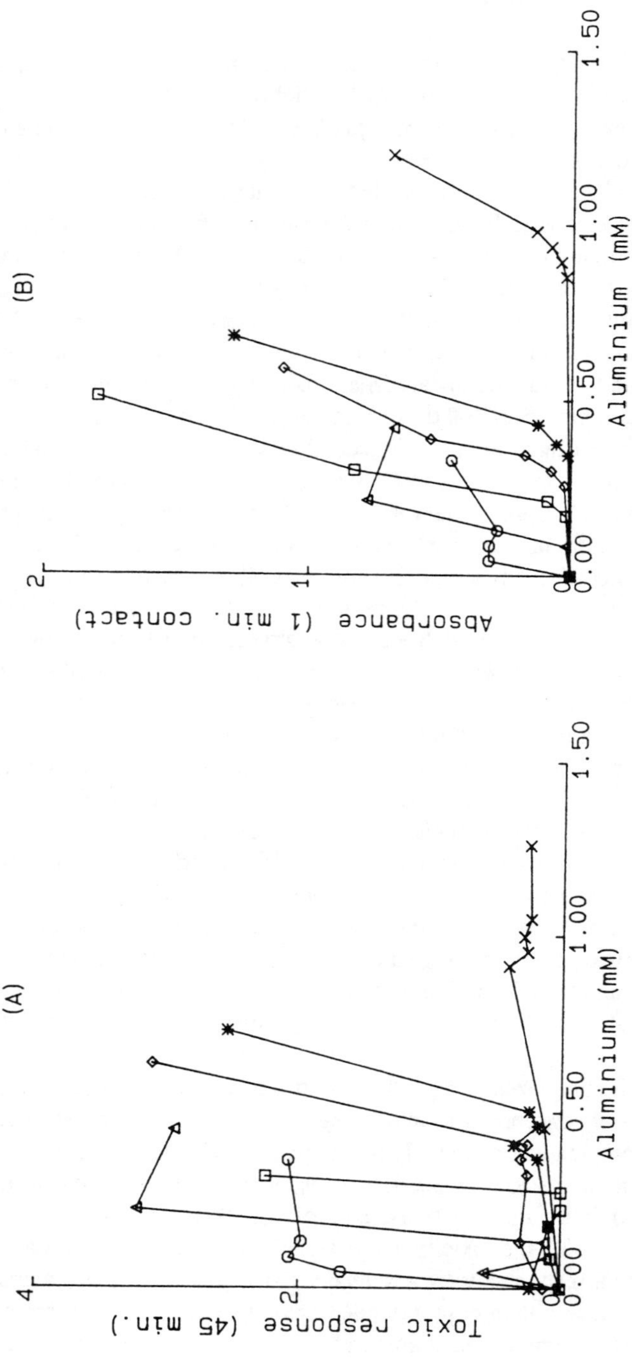

Figure 3. Effect of citrate (o = 0.0 m*M*, Δ = 0.091 m*M*, □ = 0.182 m*M*, ◇ = 0.273 m*M*, * = 0.364 m*M*, X = 0.910 m*M*) on toxicity (A) and catechol violet reactivity, (B).

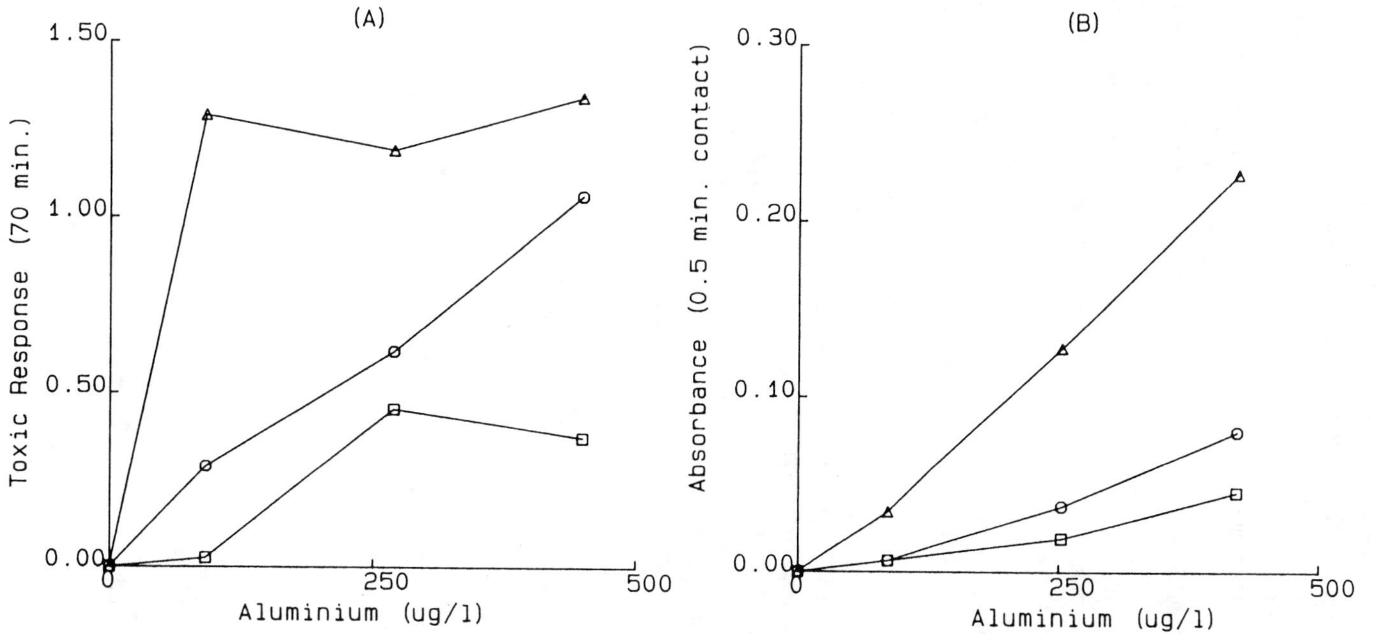

Figure 4. Effect of flouride (Δ = 0 mg/L, o = 0.9 mg/L, □ = 1.8 mg/L on toxicity (A) and catechol violet reactivity (B).

data also showed a clear linear relationship to citrate concentration, although in this case occurring at slightly higher total aluminum levels than expected.

Complexation by fluoride was also found to influence aluminum toxicity, but here interpretation of the data was complicated by toxic response to fluoride itself. Free fluoride concentrations in each test solution were therefore estimated (a) by calculation using available thermodynamic data, and (b) by measurement with an ion-selective electrode. Results of the two procedures showed excellent agreement and appropriate toxicity corrections for free fluoride levels were determined by extrapolation from control solution responses. This procedure of course implies simple independent action for the toxicity of free fluoride and that of other species, and makes no allowance for possible non-additive effects. With this reservation, the corrected responses, shown in Figure 4A, are much higher than can be explained by calculated levels of "free" aluminum in the test solutions and suggest appreciable toxicity of the Al-fluoride complexes themselves. Colorimetric data (Figure 4B) also indicate a relatively high lability of the complexes to the catechol violet reagent.

The effect of aquatically derived fulvic acid (extracted - Amberlite XAD4 + XAD2 - from the regenerant solution of a power station cooling water clean-up tower) is shown in Figure 5. Here the toxic response in the presence of 9 mg/L and 36 mg/L fulvic acid was reduced by 24% and 61%, respectively, and there was again good agreement with colorimetric data. As total aluminum levels, and hence toxicities, were relatively low in this test the number of replicates examined was increased from two to four, enabling a statistical comparison of the data. Significances of the differences between relevant mean responses are shown in Table 1.

Table 1. Significance tests on data in figure 5A

	B2	B3	C2	C3
A2	**	-	*	-
A3	NS	-	**	
B2	NS	-		
B3	**			

NS - Not significant
* - Significant at $p = 0.05$ level
** - Significant at $p = 0.01$ level

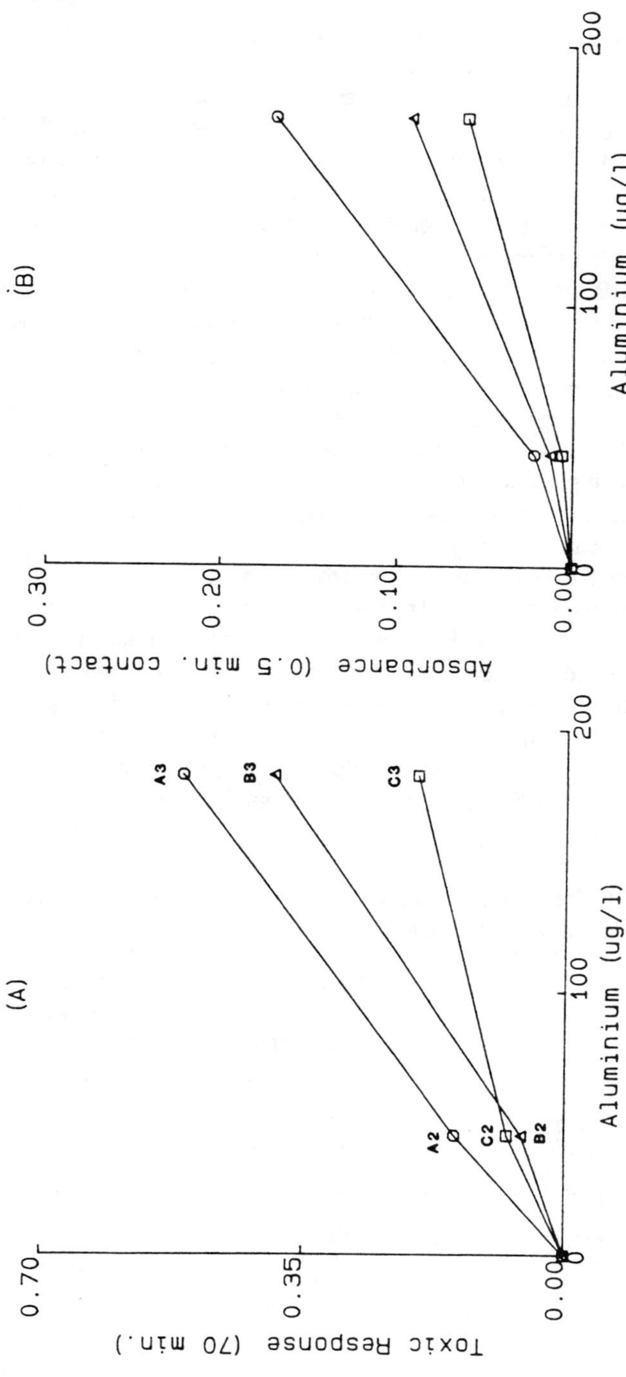

Figure 5. Effect of fulvic acid (o = 0 mg/L, Δ = 9 mg/L, □ = 36 mg/L) on toxicity (A) and catechol violet reactivity (B).

In order to examine amelioration of aluminum toxicity in a real sample of colored upland water, a suitable sample of low natural aluminum content and toxicity was selected from a number of highly colored low hardness waters collected from the Yorkshire Peak District. The selected water was analyzed for a range of determinands after filtration to 0.45 μm and found to contain 0.6 mg/L Ca, 0.6 mg/L Mg, 4.2 mg/L Na, 1.2 mg/L K, 0.1 mg/L F and 41 mg/L TOC. Total aluminum content was determined as 170 μg/L using graphite furnace atomic absorption spectrometry. When spiked with aluminum, and subjected to Microtox assay at pH 5.5, concentrations of up to 500 μg/L could be added without increase in toxic response (Figure 6) and the EC50 was raised to 2100 μg/L. Reactivity to the colorimetric procedure was also substantially reduced. To confirm that the observed amelioration in toxicity was due primarily to organic complexation, the water sample was examined after treatment by ozone/UV digestion. Unfortunately this met with limited success, with long periods of time being required to decolorize the sample and the toxic response at pH 5.5 still being largely unaltered. Whether this was due to incomplete destruction of organic matter, an indication of complexation of the aluminum by inorganic species, or adsorption on colloidal material, was not clear. However, a subsequent test was carried out after filtration of the sample through a 20,000 MW cut-off ultrafilter. This procedure was found to reduce the TOC of the sample from 40 mg/L to 19 mg/L but approximately 90% of the aluminum was carried through in the filtrate. Table 2 shows catechol violet and toxicity responses for the ultrafiltered sample with and without ozone/UV treatment. The increase in catechol violet reactive aluminum after treatment is in good agreement with prediction, but the toxic response was somewhat greater than that expected from aluminum alone, perhaps indicating formation or release of other toxic species during the digestion process (e.g., peroxide). However, the colorimetric data does in its own right provide strong confirmatory evidence for organic binding of the sample aluminum.

Table 2. Colorimetric and toxic responses for colored upland water before and after ozone/UV treatment

	Catechol violet reactivity toxic (30 minute)	Relative response (abs. units)*
Digested	0.039 (130)	1.8
Undigested	0.006 (20)	0.2

* Figures in parenthesis are concentrations in μg Al/L interpolated from response of standard solutions.

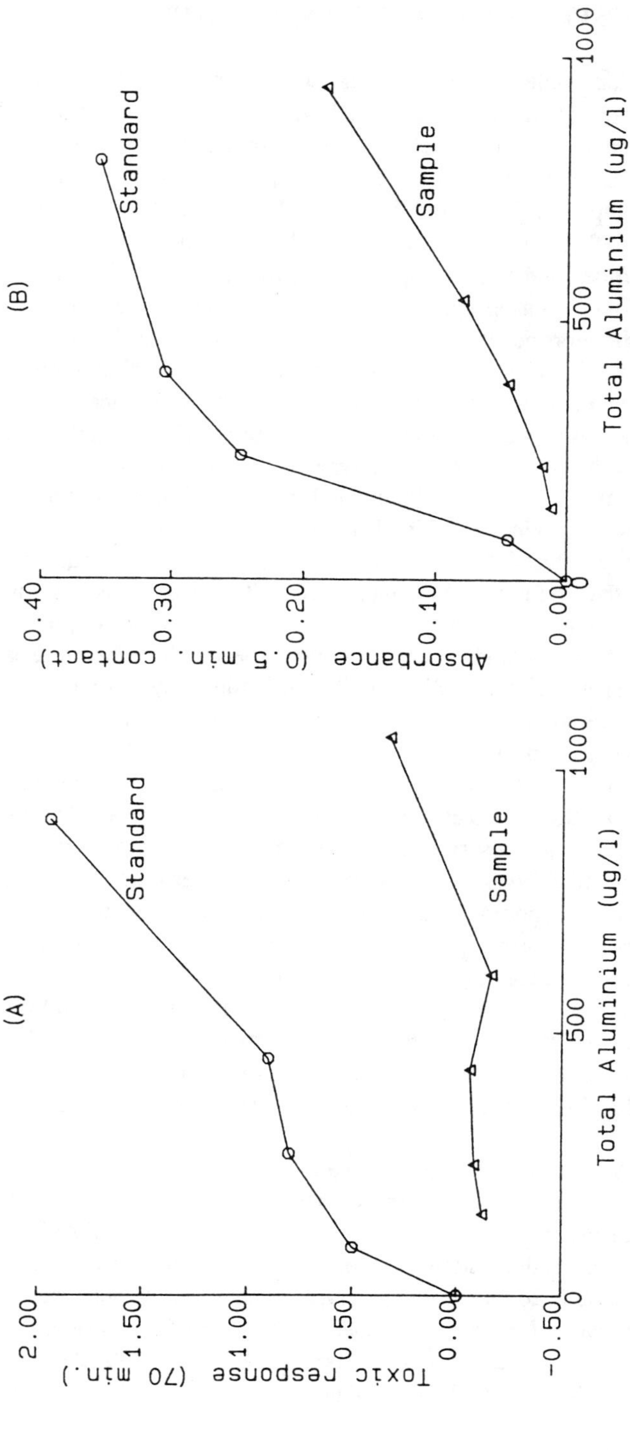

Figure 6. Toxicity (A) and catechol violet reactivity (B) in peat marsh water sample (170 μg Al/L in sample before spiking).

ON-LINE SPECIATION/FISH-STRESS MONITOR

In acid upland waters both total aluminum concentrations and aluminum speciation are known to change rapidly during acid "events" of high stream flow following heavy rain or snowmelt.[13] It is desirable to be able to monitor aluminum speciation throughout such events, as increases in the levels of aluminum in inorganic form are thought to be particularly important in causing fish kills. However, because of the unpredictability and rapidity of their onset, manual sample collection is not always possible. Automatic samplers can be used but this approach suffers from the disadvantage of possible speciation changes during sample storage and transport to the laboratory.

We have therefore developed an on-line discrete sample analyzer shown in schematic form in Figure 7. This was based on an Aztec RC-100 aluminum monitor (Aztec Environmental Control Ltd, Henley-on-Thames, Oxon) designed primarily for use in water treatment works employing alum coagulation. For speciation measurements, a kinetic ion-exchange approach of the type described by Driscoll[14] was adopted. This provides an operational split of "labile" and "non-labile" aluminum species on the basis of their dissociation kinetics. Complexation by organic ligands is thought to markedly reduce the toxic effects of aluminum,[15,16] and this procedure affords a simple means of determining the extent of such complexation. To achieve this, the sample flow rate through the column is arranged to be sufficiently rapid to prevent appreciable dissociation of the monomeric alumino-organic species; these therefore pass through the column in the non-labile fraction, while the labile fraction retained by the column includes the free aquo ion, monomeric hydroxy or fluoro complexes and also cationic hydroxy polymeric species.

In the unmodified RC-100 monitor, total aluminum is determined colorimetrically using the pyrochatechol violet reagent.[17] At preset intervals, samples are pumped from the flow-through main reservoir to one of the two reaction pots where they are acid-digested, then mixed with the catechol reagent. After color development, they are drawn into the glass pump cylinder for absorbance measurement. Modification of the basic instrument to permit speciation measurements was achieved by incorporation of a 15-cm x 0.9-cm id column of 50-100 mesh Dowex 50W-X8 strong cation exchange resin. Sample flow rate through the column was controlled by peristaltic pump. During field testing, collection of air in the resin bed was found to be a problem but this was overcome by introducing a second dummy column upstream of the resin column as an air trap, and a flow restrictor downstream of the column to prevent degassing at its outlet.

The monitor is designed to be capable of unattended operation for periods of up to two weeks but columns are normally changed routinely at weekly intervals. Two calibration ranges are available: 0 - 300 μg Al/L and 0 - 1200 μg Al/L, with automatic switchover. Calibration intervals of 6, 8, 12 or 24 hours can be set. The normal sampling interval is 20 minutes, with a typical measurement sequence being one non-labile determination for every three totals. The limit of detection is 20 μg/L and the precision is typically 2-3% rsd for both total and non-labile measurements at concentrations distant from the limit of detection.

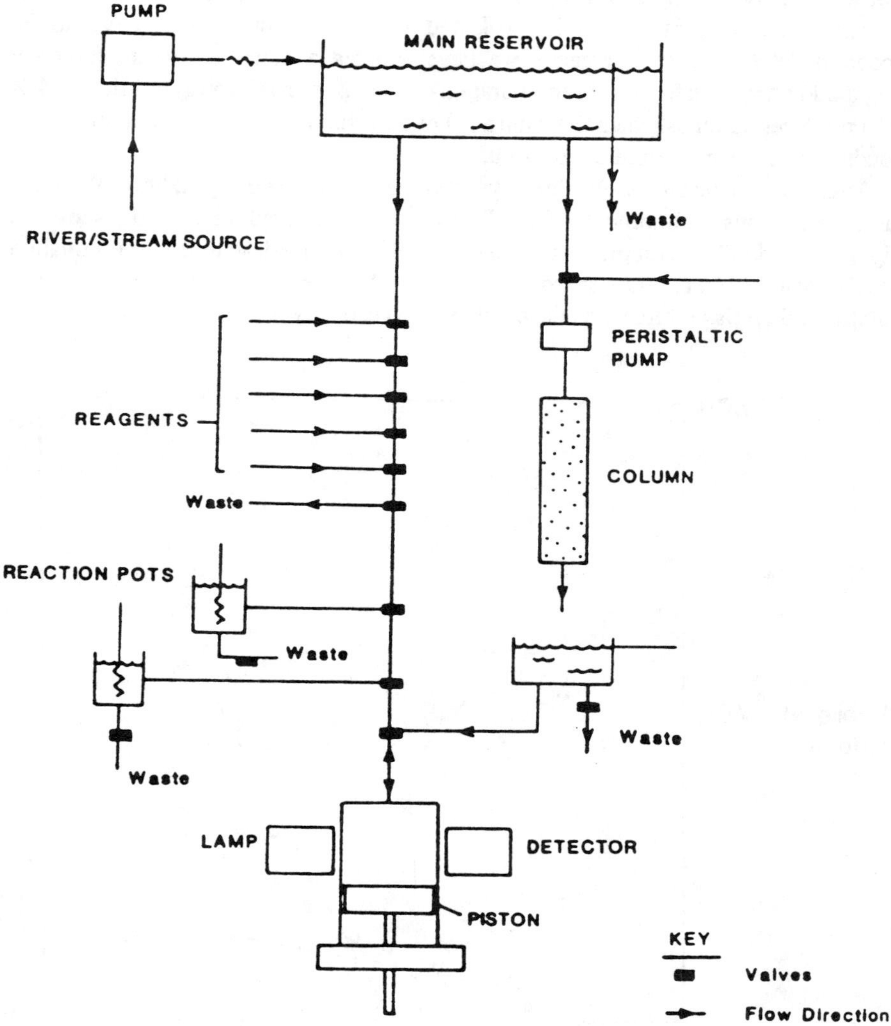

Figure 7. Schematic of Aluminum Speciation Monitor.

In using a column ion-exchange procedure for speciation measurements it is extremely important to ensure that the columns are conditioned such that the sample pH is maintained as near constant as possible on passage through the column. Otherwise the pH changes can dramatically affect the observed labile / non-labile split. For an on-line monitor designed for unattended use this obviously involves some degree of comprise, as samples of varying pH and ionic composition will have to be accommodated. For convenience we have adopted a batch procedure for resin preparation and have found that equilibration of 10 g of resin (H^+ form) in 1 litre of 4 *M* HCl produces columns which maintain a pH of 5 to within ± 0.3 units on samples with conductivities ranging from 25 to 125 μS.

An example of results attainable on synthetic samples prepared by addition of a range of concentrations of aluminum to 5 mg/L of aquatically derived fulvic acid is shown in Figure 8. A familiar complexation capacity plot is observed, with the column-labile fraction being fairly constant at around 10 - 20% of the total for aluminum levels up to 400 μg/L, then rising rapidly as the complexation capacity is exceeded.

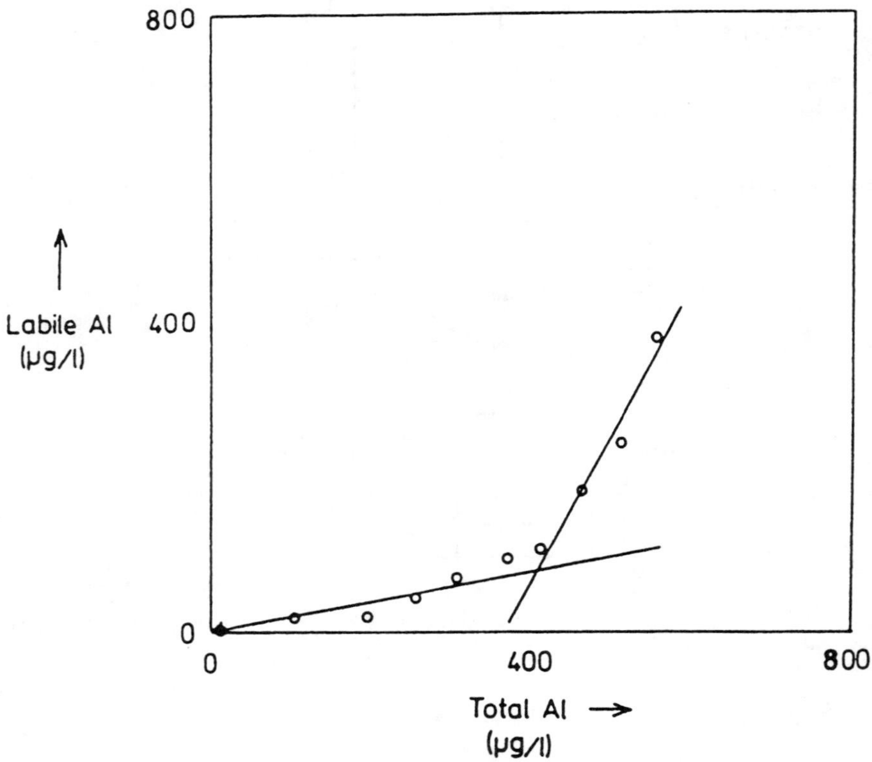

Figure 8. Column labile vs total Al in the presence of 5 mg/L fulvic acid.

The monitor has been incorporated in a Mobile Acid Waters Monitoring System (MAWMS). This has been deployed in the mid-Wales Llynn Brianne catchment in collaboration with Welsh Water, as part of their study of upland water acidification and the influence of land use in the area. A schematic of the MAWMS is shown in Figure 9. In addition to the aluminum monitor, the system includes pH, conductivity and dissolved oxygen instruments, together with a fish stress monitor based on measurement of ventilation frequencies. The fish used in the monitor are obtained from a local hatchery to ensure consistent stock acclimatized to low ionic strength conditions.

Prior to deployment, the performance of the aluminum monitor in the MAWMS was evaluated over a period of five days unattended operation. This test was performed with a pH 5 synthetic solution containing aluminum and fulvic acid, together with NaCl to adjust ionic strength. Non-labile and total aluminum determinations were performed consecutively with a recalibration interval of 6 hours. An indication of the reproducibility of calibration and signal response is given in Table 3. The total amounts of aluminum and natural organic matter passing through the column during this test were in excess of those likely to be encountered in the field between column changes. As the results were very satisfactory, the system was considered ready for field use.

Table 3. Reproducibility of calibration and signal response for a synthetic sample over a five day period of operation

Day	Mean Blank Absorbance	Mean Cal. Constant	Mean Aluminum Conc. (μg/L) Total	Non-labile
1	0.053 ± 0.003	0.417 ± 0.001	270 ± 6	179 ± 4
2	0.053 ± 0.001	0.417 ± 0.003	271 ± 5	188 ± 3
3	0.050 ± 0.003	0.415 ± 0.002	273 ± 7	176 ± 5
4	0.053 ± 0.005	0.413 ± 0.002	273 ± 6	174 ± 4
5	0.056 ± 0.002	0.415 ± 0.001	-	-

Confidence limits are at the $p = 0.05$ level.

Figure 10 shows a plot of MAWMS data obtained for an afforested stream in the Llyn Brianne catchment during a one week period in April 1987. A strong inverse correlation between pH and total aluminum concentration is clearly displayed. Non-labile aluminum concentrations also follow a similar pattern but at much lower levels, indicating a substantial exceedence of the aluminum complexation capacity of the organic matter. Such detailed information on changes in stream chemistry over a short time-span could only be achieved with an on-line system.

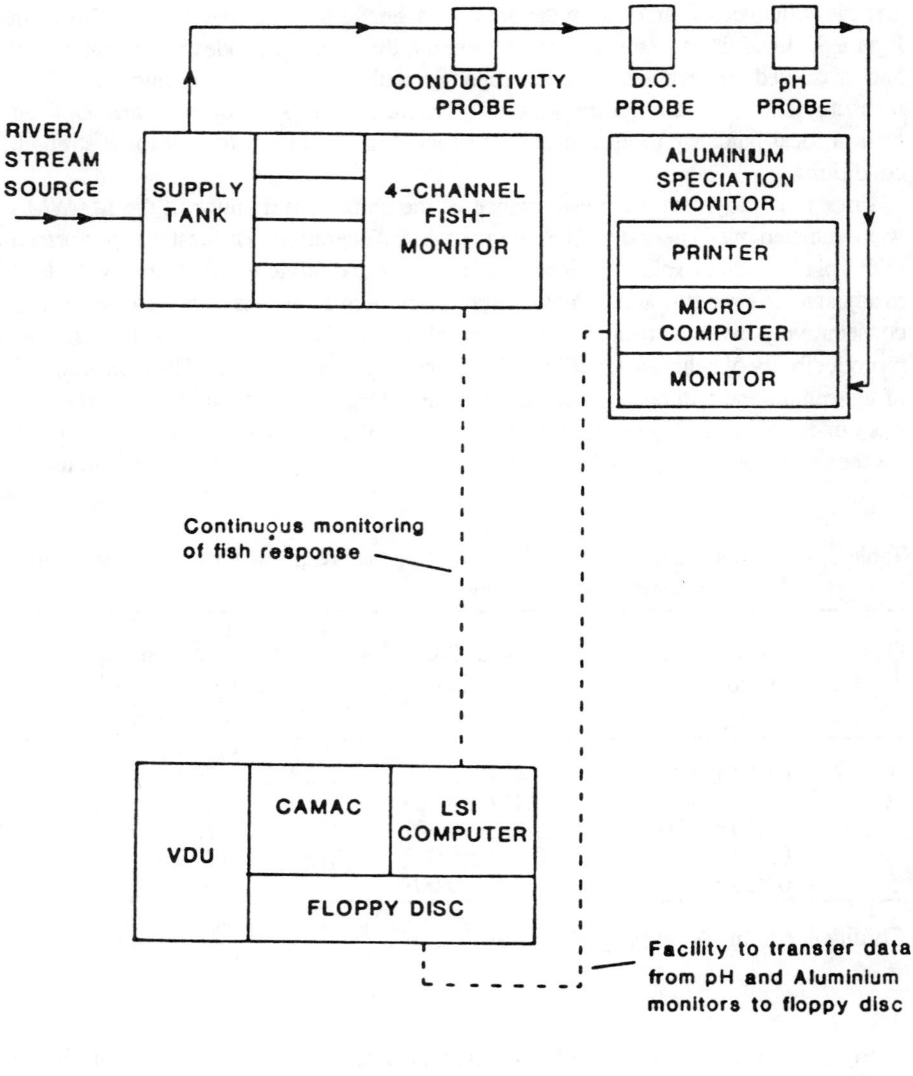

Figure 9. Schematic of Mobile Acid Waters Monitoring System.

The combination of on-line aluminum speciation measurements and fish stress monitoring has also proved to provide an extremely powerful tool for studying fish response during stream dosing experiments designed to simulate the effects of acid episodes. Figure 11 shows MAWMS data for such an experiment carried out by Welsh

Water on a moorland stream in the Llyn Brianne catchment.[18] Here, the monitor has been placed in a zone where the stream has been simultaneously dosed with acid (pH 7 to pH 4.5) and aluminum (20 µg/L to 350 µg/L). A statistically highly significant increase in the magnitude and variability of ventilation frequency occurred within a few hours of the commencement of dosing, while on completion a similarly rapid recovery to normal response was observed.

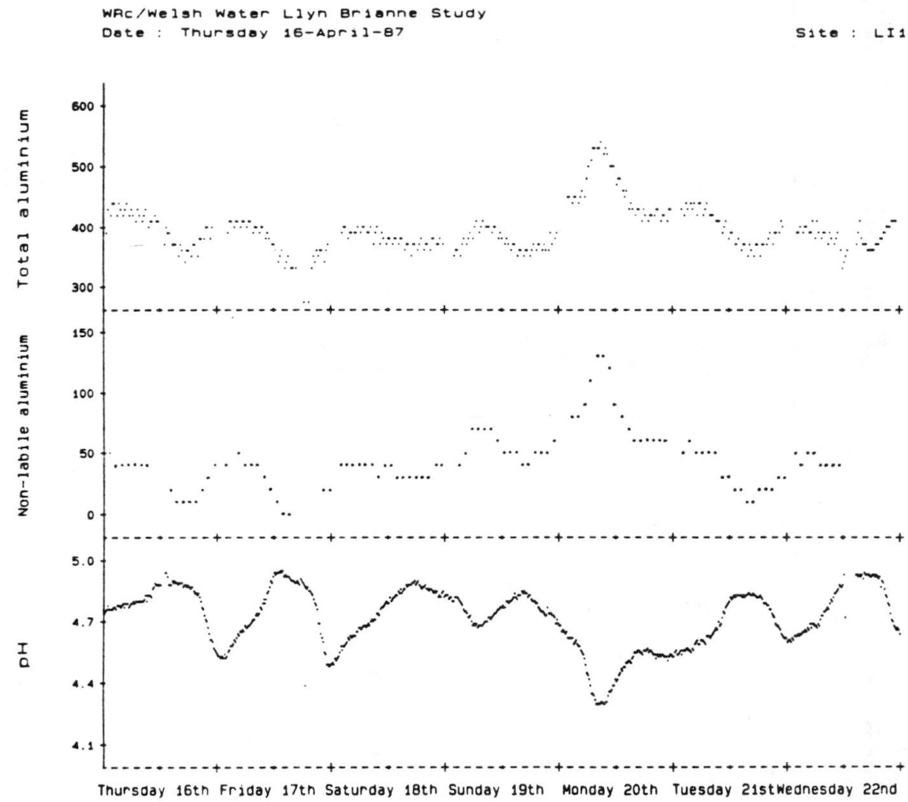

Figure 10. Weekly data output from MAWMS aluminum speciation and pH monitors for afforested stream in Llyn Brianne catchment, Wales.

DISCUSSION AND CONCLUSIONS

The limitations of the thermodynamic modeling approach to metal speciation are well known. Despite a lack of accurate information on the equilibrium constants for reaction of aluminum with natural organics, calculations indicate that such complexes

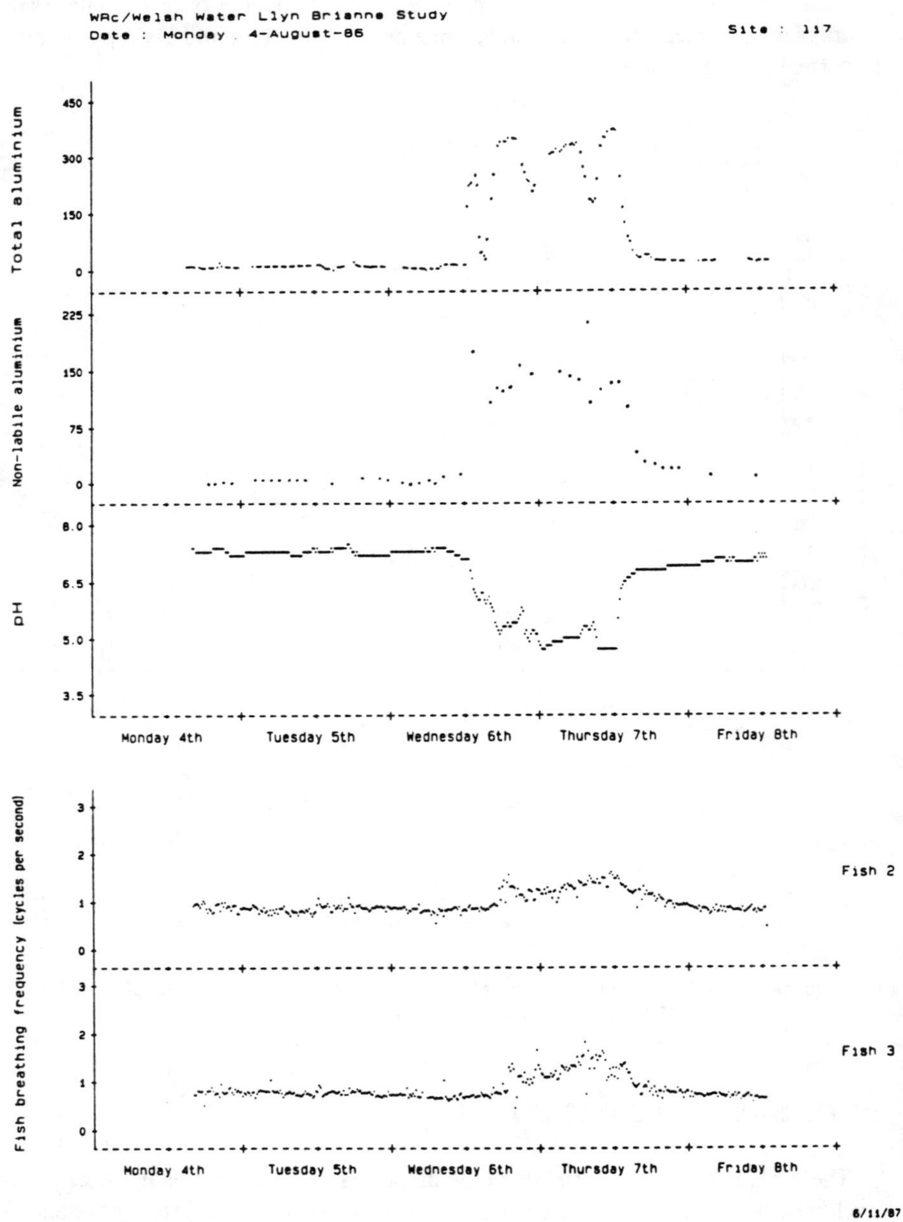

Figure 11. Summary data sheet for MAWMS during stream dosing experiment.

are important in natural waters. Furthermore, the use of thermodynamic models provides information on individual aluminum species that is not available by other means. In this study the use of thermodynamic models was also of value in the design and interpretation of laboratory speciation/toxicity experiments.

The toxicity studies with luminescent bacteria showed clearly the strong dependence of aluminum toxicity on speciation. Aluminum complexes with citric acid, fluoride and organics in natural water samples were shown to be less toxic than uncomplexed aluminum. There was good agreement between the measurement of "labile" aluminum using a kinetic catechol violet procedure and the aluminum toxicity. In the case of fluoride, however, the results may imply some toxicity from the aluminum complexes. If earlier speculation that the toxic forms of inorganic aluminum are $Al(OH)_2^+$ or $AlOH^{2+}$ is correct[7] it would not be unexpected if $AlF(OH)^+$, AlF_2^+ or AlF^{2+} were also toxic. The importance of aluminum complexation in reducing fish toxicity was noted several years ago;[15] more recently, an excellent correlation between the amount of natural organic material and the amelioration of toxicity of aluminum (and copper and zinc) has been established.[16] Results obtained in this study complement those studies to produce a consistent picture of aluminum toxicity in the presence of natural organics.

It is now clear that there is a strong relationship between aluminum speciation and toxicity. Total aluminum determinations alone are therefore of limited use for monitoring or assessing environmental effects. Operational speciation techniques of the type described here can provide useful information but, in our view, it is essential that their results are directly correlated with some measure of biotic response. The combination of an on-line aluminum speciation instrument with a fish-stress monitor, as described here, provides a powerful tool for further study of the role of aluminum in acid episodes which are of critical importance in upland waters.

ACKNOWLEDGMENT

We acknowledge the collaboration of the University of Wales Institute of Science and Technology and Welsh Water, in particular, and cooperation of Dr. S. Ormerod.

REFERENCES

1. Beckman Instruments Inc. *Microtox System Operating Manual.* (Carlsbad, California: Beckman Instruments Inc. 1982).

2. Hunt, D. T. E., and J. D. Creasy. "The calculation of equilibrium trace metal speciation and solubility in aqueous systems by a computer method, with particular reference to lead," *Water Research Centre* TR 151 (1980).

3. French, P. "Thermodynamic modeling of aluminum speciation and solubility in natural waters," *Water Research Centre* ER 1041-M (1985).

4. May, H. M., P. A. Helmke, and M. L. Jackson. "Gibbsite solubility and thermodynamic properties of hydroxy-aluminum ions in aqueous solution at 25°C," *Geochim. Cosmochim. Acta.* 43:861-868 (1979).

5. Baes, C. F., and R. E. Mesmer. *The Hydrolysis of Cations* (New York: Wiley 1976).

6. Dryssen, D. "The solubility and complex formation of aluminum hydroxide," *Vatten* 40:3-9 (1984).

7. Helliwell, S., G. E. Batley, T. M. Florence, and B. G. Lumsden. "Speciation and toxicity of aluminum in a model fresh water," *Env. Tech. Lett.* 4:141-144 (1983).

8. Plankey, B. J., and H. H. Patterson. "Kinetics of Aluminum - fulvic acid complexation in acidic waters," *Env. Sci. Technol.* 21:595-601 (1987).

9. British Drug Houses. *Zwitterionic Biological Buffers* (undated trade literature).

10. Gunn, A. M., D. T. E. Hunt, and D. A. Winnard. "Aluminum speciation and its effect on toxicity." In *Proceedings of International Conference on Chemicals in the Environment*, Lisbon 14 July 1986. J. N. Lester, R. Perry, and R. M. Sterrit, Eds. (London: Selper Ltd. 1986) pp 628-635.

11. Barnes, R. A. "The determination of specific forms of aluminum in surface water," *Chem. Geol.* 15:177-191 (1975).

12. Smith, R. W. "Relations among equilibrium and non-equilibrium aqueous species of aluminum hydroxy complexes," In *Non-equilibrium Systems in Natural Water Chemistr* y. Adv. Chem. Series 106:250-279 (1971).

13. Rogeberg, E. J. S., and A. Henrikesn. "An automatic method for fractionation and determination of aluminum species in fresh-waters," *Vatten* 41:48-53 (1985).

14. Driscoll, C. T. "A procedure for the fractionation of aqueous aluminum in dilute acidic waters," *Int. J. Environ. Anal. Chem.* 16:267-283 (1984).

15. Driscoll, C. T., J. P. Baker, J. J. Bisogni, and C. L. Schofield. "Effect of aluminum speciation on fish in dilute acidified waters," *Nature* 284:161-164 (1980).

16. Hutchinson, N. J., and J. B. Sprague. "Reduced lethality of Al, Zn and Cu Mixtures to american flagfish by complexation with humic substances in acidified soft waters," *Env. Toxicol. Chem.* 6:755-765 (1987).

17. Dougan, W. K., and A. L. Wilson. "The absorptiometric determination of aluminum in water: A comparison of some chromogenic reagents and the development of an improved method," *Analyst* 99:414-430 (1974).

18. Ormerod, S. J., N. S. Weatherley, P. French, S. Blake, W. M. Jones. "The physiological response of brown trout *Salmo trutta* to induced episodes of low pH and elevated aluminum in a Welsh Hill-stream," *Annals Soc. Zool. Belg.* 117(Suppl 1):435-47 (1987).

CHAPTER 13

CHRONIC NO-OBSERVED-EFFECT CONCENTRATIONS OF ALUMINUM FOR BROOK TROUT EXPOSED IN LOW-CALCIUM, DILUTE ACIDIC WATER

Laverne Cleveland, Edward E. Little, Raymond H. Wiedmeyer, Denny R. Buckler

INTRODUCTION

The adverse effects of acid deposition on aquatic resources has been extensively studied during the past decade. Much of the research has been focused on the effects of acidity on the biota in low alkalinity lakes and streams.[1,2,3] Recently, more attention has been directed toward the interactive effects of acidity and trace elements in acidified waters.[4,5,6,7,8] Aluminum and other trace elements from point source discharges, watershed soils, or lake and stream sediments may reach toxic levels in surface waters.

Aluminum, which is ubiquitous in the environment, has been the subject of research for more than five decades. Although it has been intensely studied, no standard methodology has yet been developed for its determination in natural or experimental waters. Kramer[9] described aluminum speciation under various conditions, as well as problems associated with the development of a standard methodology for elucidating the aqueous chemistry of aluminum. Some of the most salient problems associated with understanding aqueous aluminum chemistry result from multiple reactions; slow rates of reactions; amphoteric behavior; pronounced influence of pH on solubility; high adsorbability; and the difficulty of separating solid and liquid phases.[9,10]

Although the chemistry of aluminum in dilute acidic waters is reasonably well understood,[11,12] further research is needed on the various forms of aluminum that exist under certain conditions and the interactive effects of acidity and aluminum on aquatic organisms. The present study was conducted to determine the chronic no-observed-effect concentrations of aluminum for brook trout (*Salvelinus fontinalis*) exposed in low-calcium, dilute acidic water and to examine the speciation of aluminum during the exposures. The test conditions were representative of surface waters with moderate acid neutralizing capacity. The pH and aluminum combinations tested may be applicable to waters being impacted by acid deposition as well as alum-treated or limed waters. The survival and hatching of eyed eggs of brook trout, and survival, growth, behavioral, and biochemical responses of the resultant larvae and juveniles were measured during 60-day exposures.

METHODS

Two partial life-cycle studies were conducted for 60 days in a modified flow-through proportional diluter.[13] In the first exposure (exposure A), eyed eggs of brook trout and the resultant larvae and juveniles were exposed to nominal aluminum concentrations of 38, 75, 150, and 300 μg/L at pH 5.5; controls (no aluminum added) were held at pH 5.5 and 7.2. In the second exposure (exposure B), eyed eggs and the resultant larvae and juveniles were exposed to nominal aluminum concentrations of 50, 100, 200, and 400 μg/L at pH 6.5; controls were held at pH 6.5 and 7.2. Because aluminum is less soluble under less acidic conditions and its toxicity to fish decreases, we used slightly higher aluminum exposure concentrations in exposure B than in A. The exposures were conducted in soft water containing about 3.0 mg/L calcium. Aluminum sulfate hexadecahydrate (98% pure: 8.56% aluminum) was purchased from Fisher Scientific Company, Fairlawn, New Jersey. The eyed eggs were obtained from Beity's Resort, Valley, Washington, and held overnight in control water at pH 7.2 and 12°C before the study was initiated.

In general, we used the methods of Cleveland et al.[8] to prepare the experimental water, to obtain and monitor exposure water quality, and to evaluate the effects of acid and aluminum on brook trout survival, growth, behavior, and whole-body ribonucleic acid (RNA) and deoxyribonucleic acid (DNA). We placed 50 eyed brook trout eggs into each of four 177-mL glass hatching containers, which were suspended in each duplicate exposure tank. The tanks were covered to shield the eggs from light, and aeration provided continuous circulation of the exposure water. Egg mortality, hatch, and incomplete hatch were recorded daily until hatching was complete. Dead eggs were removed and discarded and newly hatched larvae were transferred to duplicate containers within each exposure tank. Eggs yielding larvae that were unable to complete the hatching process and had the chorion attached were considered to be incompletely hatched. The median hatching date for each treatment was considered to be day 0 of the larval exposure. On day 15, two groups of 20 larvae were randomly selected from the hatching containers, photographed for length determination,[14] and placed in the smaller growth chambers within each duplicate tank. The remaining larvae were placed in the larger chamber of each tank for behavioral, biochemical, and aluminum bioconcentration measurements.

Mortality of fish in the growth chambers was recorded daily and analyzed for days 15, 30, 45 and 60 of the exposures. These fish were photographed again on day 30 for length determinations. At day 30, each small chamber was thinned to 10 fish each to avoid the effects of over-crowding on growth. Lengths and weights were determined from direct measurements on days 45 and 60. Behavioral evaluation on 10 fish from each duplicate treatment included measurements of locomotory activity, feeding and swimming capacity, and buoyancy of individual fish at 30 and 60 days of exposures A and B. Frequency of movement was the number of times a fish changed position as it began movement or changed direction during a 2-min interval. Feeding was the number of strikes directed at prey-- *Daphnia magna* or *Artemia salina*-- during a 5-min period. Swimming capacity was measured in a stamina tunnel as fish were subjected to incremental increases in water velocity until fatigue occurred. The buoyancy of fish was assessed by assigning a score of 3 to fish that remained on the aquaria bottoms, 2 to fish that swam in midwater, and 1 to fish that swam at the water surface. Whole-body RNA and DNA were determined[8] at 15 and 30 days of exposure A and B.

Whole-body aluminum residues were determined on a wet-weight basis for 10 fish per duplicate treatment on day 15 and 5 fish per duplicate on days 30, 45, and 60.[15] Tissue residue data were statistically analyzed to derive surface response plots based on whole-body aluminum residues, aluminum exposure concentrations, and days of exposure.

Water samples were collected at the beginning of the exposures and weekly thereafter to determine aluminum concentrations. Our analytical approach involved separating the "fast reacting" or labile aluminum species (inorganic and organic) from the "slow reacting" polymeric forms. A modification of the oxine/methyl isobutyl ketone (MIBK) extraction procedure of Barnes[16] was used to separate the two forms of aluminum. The following four steps were incorporated into the method: (1) determination of total aluminum in acidified unfiltered samples (weekly determinations for each treatment); (2) determination of total aluminum in filtered acidified samples (two determinations per treatment); (3) determination of extractable aluminum in filtered samples (two determinations per treatment); and (4) determination of extractable aluminum in unfiltered samples (3-4 determinations per treatment). The commercial in-line apparatus used to filter water samples contained 0.1-μm mixed cellulose acetate and nitrate filters. The filters were rinsed with 50 mL of 1% nitric acid and then with 50 mL of reverse osmosis (RO) water before use. Aluminum stock solutions were prepared in RO water obtained from a commercial RO water purification system and anionic, cationic, and mixed-bed exchange resins. In addition, the REDEQL-EPAK chemical equilibrium model of Ingle et al.[17] was used to estimate aluminum speciation during the two exposures.

To determine filterability of aluminum at a nominal pH of 5.5, we used filters of several types and pore sizes to filter 100-ml water samples that contained nominal aluminum concentrations of 150 μg/L.

Atomic absorption standards (1000 μg/L: J.T. Baker Chemical Company, Phillipsburg, NJ) were used for the preparation of standard solutions of aluminum. All standards were preserved in 1% nitric acid solutions. Aluminum concentrations in spiked samples and exposure water samples were determined by atomic absorption spectrophotometry. In performing analyses on a Perkin-Elmer Model 5000 spectrophotometer, equipped with an HGA 500 graphite furnace and an AS-40 autosampler, we used standard methods suggested by the manufacturer. Samples of 20 μl were analyzed.

The MIBK extract was transferred to a 100-mL Teflon* separatory funnel, and 15-25 ml of 10% HCl solution were added. The funnel was shaken vigorously for 30 s, and the phases were allowed to separate. The aqueous layer was drained off and the extraction repeated; the two aqueous layers were then combined and diluted to 5% HCl with RO water. Aluminum concentrations in exposure water samples were quantified by referring to a calibration curve obtained from aluminum standards prepared in 1% nitric acid. The linear range of the instrument was 10-100 μg/L, and the detection limit was 2 μg/L.

Samples taken during the first and last 30-day period from each treatment of exposures A and B were analyzed for organic carbon content. The samples were separated into dissolved and particulate organic carbon fractions by filtration according to standard methods.[18] Dissolved organic carbon was measured with a Technicon Autoanalyzer II Industrial System Dissolved Organic Cartridge.[19] Particulate organic carbon was determined with a Coulometric Model 5020 total carbon analyzer, and total organic carbon was calculated as the sum of dissolved and particulate fractions.

One-way analysis of variance was used to compare treatment effects on survival, growth, behavior, and nucleic acid content in both exposures. Percent data were arcsine transformed, and square root transformations were made on strike frequency and water column position data prior to statistical analyses. The Least Significant Difference means comparison test was used to distinguish differences among treatment means ($P \leq 0.05$). Swimming capacity data were also analyzed by simple regression to determine the relation between swimming capacity and aluminum exposure concentration. Whole-body residues of aluminum were analyzed as a completely randomized design in which the treatments were arranged factorially. The linear statistical model contained the effects of aluminum exposure concentrations, days of exposure, and interactions of exposure concentrations and days of exposure. Linear and quadratic polynomial orthogonal contrasts were computed to produce surface response plots for whole-body residues of aluminum. Statistical Analysis Systems[20] programs were used to perform all analyses on the mainframe computer system of the University of Missouri, Columbia, Missouri.

RESULTS

Water quality during exposures A and B (Table 1) simulated the dilute acidic conditions in some surface waters of the northeastern United States as well as conditions that may exist in alum-treated or limed surface waters. Total aluminum measured in unfiltered acidified samples from both exposures were close to the nominal concentrations (Table 2). Total dissolved aluminum measured in filtered acidified samples ranged from 13 to 21 μg/L in exposure A and from 12 to 22 μg/L in exposure B; total extractable aluminum in samples that were extracted immediately without filtration ranged from 39 to 211 μg/L in exposure A and from 27 to 194 μg/L in B; and total dissolved aluminum in filtered extracted samples ranged from 14 to 27 μg/L in

* Registered trademark of E. I. du Pont de Nemours and Company, Inc., Wilmington, Delaware.

exposure A and from 13 to 26 µg/L in B (Table 2). Results from the REDEQL-EPAK modeling suggest that a large portion of the total aluminum present in these studies was in the solid, or undissolved, form (Table 2). This was confirmed by the results obtained from our filtration procedures which showed that concentrations of dissolved aluminum were consistently low (Table 2). Filtration of exposure water with 3.0-µm Teflon, 0.4-µm polycarbonate, 0.2-µm nylon and Fluoropore[R] (Millipore Corporation, Bedford, MA), and 0.1-µm mixed cellulose-acetate and nitrate filters gave similar results. The filtration results, coupled with the REDEQL-EPAK estimates, suggest the presence of particulate or polymeric aluminum species larger than 3.0 µm.

Concentrations of dissolved organic carbon averaged 1.8 mg/L in exposure A and 1.9 mg/L in B; for particulate organic carbon the concentrations were 0.1 mg/L in A and 0.21 mg/L in B. Total organic carbon averaged 2.1 mg/L in both exposures.

Table 1. Mean water quality characteristics ± standard deviation (SD) during exposure of brook trout to aluminum and acidic pH

Nominal exposure condition		Measured water quality chracteristics[a]							
		Calcium (mg/L)		pH		Alkalinity (µeq/L)		Hardness (µeq/L)	
Aluminum (µg/L)	pH	Mean	SD	Mean	SD	Mean	SD	Mean	SD
Exposure A									
0	7.2	3.3	0.2	7.0	0.4	207	11	256	9
0	5.5	-	-	5.7	0.4	10	10	256	9
38	5.5	3.2	0.0	5.7	0.4	14	19	255	8
75	5.5	3.3	0.2	5.7	0.4	13	10	253	8
150	5.5	3.2	0.1	5.6	0.4	9	7	255	8
300	5.5	3.3	0.2	5.6	0.3	9	5	255	7
Exposure B									
0	7.2	3.3	0.2	6.9	0.4	205	13	246	12
0	6.5	-	-	6.5	0.3	73	10	248	13
50	6.5	3.2	0.0	6.5	0.2	68	10	246	14
100	6.5	3.3	0.1	6.5	0.2	67	9	246	13
200	6.5	3.3	0.07	6.6	0.2	67	11	247	11
400	6.5	3.2	0.2	6.6	0.2	69	18	245	14

[a]Sample size (n) was 2 to 6 for calcium, 85 to 90 for pH, and 18 for alkalinity and hardness

Table 2. Mean measured concentrations (standard deviations in parentheses) and estimated concentrations of aluminum (µg/L) complexed with hydroxide (OH) during exposure of brook trout for 60 days

Nominal aluminum concentration (µg/L)	Mean pH	Type of sample: Unfiltered acidified (total aluminum)	Unfiltered extracted (total extractable aluminum)	Estimated aluminum complexed with OH[a]: Solid	Estimated aluminum complexed with OH[a]: Dissolved	Filtered acidified (total dissolved aluminum)	Filtered extracted (total dissolved extractable aluminum)
Exposure A							
0	7.0	5 (3)	NM	0	5	NM	NM
0	5.7	8 (9)	NM	0	7	NM	NM
38	5.7	29 (16)	39 (18)	11	17	13 (4)	14 (9)
75	5.7	68 (26)	52 (19)	50	17	20 (9)	15 (2)
150	5.6	142 (35)	97 (25)	121	20	20 (0.8)	24 (2)
300	5.6	292 (51)	211 (59)	271	20	21 (10)	27 (8)
Exposure B							
0	6.9	5 (6)	NM	0	5	NM	NM
0	6.5	4 (4)	NM	0	4	NM	NM
50	6.5	57 (34)	27 (16)	35	21	22 (18)	13 (5)
100	6.5	88 (27)	38 (23)	64	23	13 (2)	26 (23)
200	6.6	169 (59)	70 (22)	145	24	12 (0)	26 (3)
400	6.6	350 (53)	194 (24)	332	18	14 (2)	22 (2)

Note: NM = not measured.
[a]The µg/L values were derived from the % total aluminum estimated by the REDEQL-EPAK chemical equilibrium model of Ingle et. al.[17]

For clarity of presentation, biological results are reported in reference to total aluminum concentrations. Refer to Table 2 for corresponding concentrations of other aluminum species. Brook trout embryo mortality was not significantly affected during exposure A or B; however, the number of hatching fish that were unable to completely shed the chorion was significantly greater than in control treatments, in both exposures (Table 3). The number of incompletely hatched fish was significantly greater in the pH 5.7 than in the pH 7.0 control treatment of exposure A and significantly greater in the pH 6.5 than in the pH 6.9 control treatment in exposure B (Table 3).

Table 3. Mean percent mortality, hatch, and incomplete hatch (standard deviation in parentheses) of brook trout embryos during exposure to aluminum at acidic pH

Mean aluminum concentrations (µg/L)	Variables			
	Mean pH	Mortality (%)	Hatch (%)	Incomplete Hatch (%)
Exposure A				
5[a]	7.0	5.0 (2.6)	95.3 (5.4)	3.5 (2.4)
8[a]	5.7	7.5 (3.7)	92.3 (3.6)	10.0 (7.6)[b]
29	5.7	6.3 (2.0)	93.8 (2.0)	27.5 (4.7)[bc]
68	5.7	6.0 (3.4)	94.0 (3.4)	35.4 (8.8)[bc]
142	5.6	7.0 (3.5)	94.3 (7.2)	42.4 (7.8)[bc]
292	5.6	7.0 (1.9)	92.8 (2.1)	47.7 (4.2)[bc]
Exposure B				
5[a]	6.9	6.5 (2.6)	93.3 (5.9)	2.5 (2.8)
4[a]	6.5	7.3 (3.1)	93.0 (2.6)	5.9 (2.7)[b]
57	6.5	6.8 (2.8)	93.0 (2.8)	10.0 (3.7)[b]
88	6.5	4.0 (1.5)	96.0 (1.1)	14.6 (4.3)[bc]
169	6.6	5.0 (3.1)	94.8 (2.9)	18.5 (5.9)[bc]
350	6.6	6.0 (3.7)	94.0 (3.7)	28.2 (6.6)[bc]

[a]Background concentrations in controls.
[b]Significantly different from pH 7.0 control (Exposure A) or pH 6.9 control (Exposure B), $P \leq 0.05$ (LSD test).
[c]Significantly different from pH 5.7 control (Exposure A) or pH 6.5 control (Exposure B), $P \leq 0.05$ (LSD test).

Larval mortality after 15 days for exposures A and B ranged from 1 to 4%. Significant effects on mortality at 15 days of exposure B were neither consistent nor dose related (Table 4). After 30, 45, and 60 days, the mortality of juveniles was significantly increased at 292 µg/L aluminum in exposure A and 350 µg/L in B compared with control groups (Table 4).

Table 4. Mean percent mortality and standard deviation (SD) of brook trout during exposure to aluminum at acidic pH

Mean aluminum concentration (μg/L)	Mean pH	Exposure duration: Day 15 Mean	Day 15 SD	Day 30 Mean	Day 30 SD	Day 45 Mean	Day 45 SD	Day 60 Mean	Day 60 SD
Exposure A									
5[a]	7.0	2.1	2.4	4.5	5.0	4.5	5.0	10.8	11.8
8[a]	5.7	1.8	1.3	2.0	0	2.0	0	7.5	6.4
29	5.7	3.0	1.5	3.5	0.6	3.5	0.6	3.5	0.6
68	5.7	0.8	1.5	1.0	1.1	1.0	1.1	6.8	6.7
142	5.6	4.0	2.8	4.0	1.1	6.3	3.9	9.0	6.1
292	5.6	2.3	2.3	27.0[bc]	26.6	57.7[bc]	39.6	70.3[bc]	22.3
Exposure B									
5[a]	6.9	1.0	1.1	1.0	1.1	1.0	1.1	1.0	1.1
4[a]	6.5	2.0	1.1	2.0	2.0	2.0	2.0	4.5	5.0
57	6.5	2.8[b]	1.8	4.3[b]	2.5	4.3	2.5	6.8	4.8
88	6.5	0.8[c]	1.5	1.0	1.2	1.0	1.2	3.5	4.4
169	6.6	2.3	2.5	3.0	2.0	3.0	2.0	3.0	2.0
350	6.6	3.8[b]	2.0	8.5[bc]	6.6	37.5[bc]	17.7	48.5[bc]	21.8

[a]Background concentration in controls.
[b]Significantly different from pH 7.0 control (Exposure A) or 6.9 control (Exposure B), $P \leq 0.05$ (LSD test).
[c]Significantly different from pH 5.7 control (Exposure A) or 6.5 control (Exposure B), $P \leq 0.05$ (LSD test).

At 15 days of exposure A, brook trout exposed to aluminum concentrations of 142 μg/L or greater were significantly larger than those in both control groups, but at 30, 45 and 60 days, fish exposed to 292 μg/L aluminum were consistently smaller than those in both control groups (Table 5). At 30, 45, and 60 days of exposure A, fish exposed to 29 μg/L aluminum or greater were significantly smaller than those in the pH 7.0 control group, though the effects on length were not consistent at 45 days (Table 5). In exposure B, brook trout exposed to 88 μg/L aluminum or greater for 30 days and to 169 μg/L or greater for 45 and 60 days were significantly smaller than the fish in both control groups (Table 5); and fish exposed to 57 μg/L aluminum for 30 days and 88 μg/L for 60 days were smaller than those in the pH 7.0 control group (Table 5). Generally, at 30, 45, and 60 days, fish in all treatment groups were significantly smaller than those in the pH 7.0 control group in exposure A or in the pH 6.9 control group in exposure B (Table 5).

The feeding response, measured as frequency of strikes at prey, was significantly lower for fish in the pH 5.7 control groups compared to those in the pH 7.0 control groups after 30 days in exposure A; however, other behavior patterns were not affected (Table 6). At 30 days of exposure A, aluminum concentrations of 29 µg/L or higher significantly altered the feeding behavior of brook trout. Compared with fish in the pH 5.7 control groups, strike frequency was 61% lower in fish exposed to 142 µg/L aluminum and 97% lower in fish exposed to 292 µg/L. Strike frequencies were significantly higher in brook trout exposed to 29 and 68 µg/L aluminum than in the pH 5.7 controls but significantly lower than in the pH 7.0 controls (Table 6). The frequency of movement was 56% less in fish exposed to 142 µg/L aluminum and 71% less in those exposed to 292 µg/L than movements of the pH 5.7 control fish (Table 6). A significant difference in water column position, which reflects the development of the swim bladder and buoyancy, occurred among brook trout exposed to 142 and 292 µg/L aluminum. Fish of these exposure groups remained at the bottom of the aquaria and rarely moved into the water column, but their swimming capacity was not significantly reduced (Table 6).

At 30 days of exposure B, strike frequency was significantly lower in brook trout of the pH 6.5 control than in those of the pH 6.9 control; however, all other responses were similar among the control groups (Table 6). Strike frequencies of fish exposed to 169 and 350 µg/L aluminum were 52% and 94% lower, respectively, than those of the pH 6.5 controls (Table 6). The frequency of movement of brook trout exposed to 350 µg/L aluminum was 70%-75% less than that of fish in the pH 6.5 and 6.9 control treatments. Development of buoyancy, reflected in water column position, was retarded in brook trout exposed to 350 µg/L aluminum; and swimming capacity was reduced in fish exposed to 169 µg/L aluminum but not in those exposed to 350 µg/L (Table 6).

After 60 days in exposure A, the strike frequencies were 43% and 73% lower in brook trout exposed to 292 µg/L aluminum than in those of the pH 5.7 and 7.0 control treatments, respectively. Strike frequencies were significantly lower in the pH 5.7 controls than the pH 7.0 controls (Table 7). In other treatments, strike frequencies were intermediate and not significantly different from those in the controls. Frequency of movement was significantly lower (54%) in fish exposed to 292 µg/L aluminum than in those of the pH 5.7 control group. The swimming capacity remained significantly lower in fish exposed to 68, 142, and 292 µg/L aluminum than in the control fish (Table 7).

Partial recovery of behavioral function was evident after 60 days in exposure B (Table 7). Swimming capacity and locomotory activity were similar among all treatment groups. The strike frequencies were significantly lower in brook trout exposed to 350 µg/L aluminum than in those of the pH 6.9 but not pH 6.5 controls (Table 7).

Table 5. Mean length (mm) and in parentheses weight (mg) and standard deviation (SD) of brook trout during exposure to aluminum at acidic pH

Mean aluminum concentration (µg/L)	Mean pH	Exposure duration: Day 15 Mean	Day 15 SD	Day 30 Mean	Day 30 SD	Day 45 Mean	Day 45 SD	Day 60 Mean	Day 60 SD
Exposure A									
5[a]	7.0	14.8	2.2	16.4	2.8	28.7	2.0	35.3	3.5
		-	-	-	-	(195	43)	(375	116)
8[a]	5.7	14.5	2.0	14.3[b]	2.3	27.8	1.8	33.0[b]	2.3
		-	-	-	-	(170[b]	34)	(303[b]	90)
29	5.7	14.3	2.5	14.5[b]	2.0	28.0	1.9	32.4[b]	3.0
		-	-	-	-	(176[b]	39)	(279[b]	95)
68	5.7	15.2	2.1	14.9[b]	2.4	28.0	1.8	33.1[b]	2.8
		-	-	-	-	(180	36)	(270[b]	77)
142	5.6	16.4[bc]	1.9	14.0[b]	2.3	27.6[b]	2.3	32.2[b]	2.5
		-	-	-	-	(160[b]	39)	(237[bc]	60)
292	5.6	16.7[bc]	1.8	12.2[bc]	1.9	21.8[bc]	3.0	25.9[bc]	4.5
		-	-	-	-	(77[bc]	35)	(125[bc]	68)
Exposure B									
5[a]	6.9	16.5	1.7	22.1	2.3	29.7	2.4	36.8	3.2
		-	-	-	-	(248	48)	(439	142)
4[a]	6.5	15.7	2.0	20.4[b]	2.5	29.5	1.5	35.3	2.3
		-	-	-	-	(224	37)	(376[b]	105)
57	6.5	16.1	2.0	19.5[b]	2.6	29.2	2.0	35.1	3.1
		-	-	-	-	(224	47)	(372[b]	110)
88	6.5	16.3	2.0	18.2[bc]	2.9	28.9	1.6	34.8[b]	2.9
		-	-	-	-	(210[b]	34)	(361[b]	100)
169	6.6	15.8	2.1	15.9[bc]	2.9	28.0[bc]	1.8	33.4[bc]	2.7
		-	-	-	-	(182[bc]	36)	(286[bc]	83)
350	6.6	15.3	1.9	14.4[bc]	2.5	24.1[bc]	4.2	30.1[bc]	4.6
		-	-	-	-	(112[bc]	60)	(204[bc]	84)

Note: Sample size (n) ranged from 60 to 80 during the first 30 days and from 14 to 42 during the final 30 days of exposure.

[a]Background concentrations in controls.
[b]Significantly different from pH 7.0 control (Exposure A) or pH 6.9 control (Exposure B), $P \leq 0.05$ (LSD test).
[c]Significantly different from pH 5.7 control (Exposure A) or pH 6.5 control (Exposure B), $P \leq 0.05$ (LSD test).

Table 6. Mean behavioral responses and standard error (SE) of brook trout after 30 days of exposure to aluminum at acidic pH

Mean aluminum concentration (μg/L)	Mean pH	Strike frequency Mean	Strike frequency SE	Movements/2 min Mean	Movements/2 min SE	Water column position Mean	Water column position SE	Swimming capacity (cm/s) Mean	Swimming capacity (cm/s) SE
Exposure A									
5[a]	7.0	26.4	5.4	65.9	8.3	2.2	0.1	14.8	0.9
8[a]	5.7	12.9	3.2[b]	74.5	8.0	2.2	0.1	15.0	0.8
29	5.7	16.3	3.7[bc]	63.6	8.6	2.2	0.1	15.9	1.2
68	5.7	20.0	5.5[bc]	73.6	9.6	2.2	0.1	15.0	0.8
142	5.6	5.0	2.2[bc]	32.6	6.4[bc]	2.6	0.1[bc]	12.7	0.8
292	5.6	0.4	0.3[bc]	21.4	3.9[bc]	2.8	0.1[bc]	12.6	0.5
Exposure B									
5[a]	6.9	38.0	6.1	81.1	6.1	2.0	0.1	15.2	0.7
4[a]	6.5	20.1	4.5[b]	67.7	6.9	2.2	0.1	15.5	0.9
57	6.5	14.8	4.6[bc]	59.6	9.3	2.2	0.1	13.4	0.5
88	6.5	21.0	4.7[b]	55.6	6.9	2.1	0.1	15.1	1.1
169	6.6	9.6	2.9[bc]	47.2	7.7	2.2	0.1	12.6	0.6[bc]
350	6.6	1.2	0.9[bc]	20.0	5.3[bc]	2.7	0.1[bc]	15.6	1.2

[a]Background concentrations in controls.
[b]Significantly different from pH 7.0 control (Exposure A) or pH 6.9 control (Exposure B), $P \leq 0.05$ (LSD test).
[c]Significantly different from pH 5.7 control (Exposure A) or pH 6.5 control (Exposure B), $P \leq 0.05$ (LSD test).

During exposure A, the RNA content of brook trout exposed to 292 μg/L aluminum for 30 days averaged 30 μg/mg protein and was lower than that in fish of both control groups which averaged 60 to 71 μg/mg of protein; DNA content for all treatments ranged from 30 to 37 μg/mg protein and was not affected; and the RNA-DNA ratio in fish exposed for 15 days to 29 μg/L aluminum or greater ranged from 1.2 to 1.7 and was significantly lower than that in fish of both control groups which ranged from 2.2 to 2.7. After 30 days, only the RNA-DNA ratio of 1.0 for fish exposed to 292 μg/L aluminum differed significantly from that of fish in the pH 7.0 controls which was 2.1. Generally, there were no consistent effects of aluminum on RNA or DNA content of brook trout in exposure B.

Bioconcentration of aluminum by brook trout during exposures A and B was characterized by an initial increase during the first 15 to 30 days of exposure, followed by a decline in whole-body concentrations during the final 30 days (Figure 1). Brook trout accumulated slightly more aluminum at pH 5.6 to 5.7 (exposure A) than at pH 6.5 to 6.6 (exposure B).

Chronic no-observed-effect concentrations of aluminum for brook trout varied with acidity, length of exposure, and the biological endpoint measured (Table 8). The no-observed-effect concentrations are given as total and dissolved extractable aluminum. Refer to Table 2 for the corresponding total dissolved and total extractable aluminum concentrations.

Table 7. Mean behavioral responses and standard error (SE) of brook trout after 60 days of exposure to aluminum at acidic pH

Mean aluminum concentration (μg/L)	Mean pH	Strike frequency		Movements/2 min		Swimming capacity (cm/s)	
		Mean	SE	Mean	SE	Mean	SE
Exposure A							
5[a]	7.0	32.2	6.2	79.8	8.7	23.0	1.4
8[a]	5.7	15.2	4.9[b]	66.4	9.7	19.8	1.2
29	5.7	19.7	4.2	66.6	10.5	20.3	1.4
68	5.7	23.9	5.6	74.4	9.4	15.4	0.7[bc]
142	5.6	24.4	4.9	70.9	5.6	16.0	1.1[bc]
292[d]	5.6	8.7	6.8[bc]	30.3	16.3[bc]	15.2	2.7[bc]
Exposure B							
5[a]	6.9	25.6	7.0	72.1	8.1	22.2	1.6
4[a]	6.5	12.0	3.2	84.7	11.4	18.6	1.5
57	6.5	31.9	7.0	75.0	8.9	19.6	1.5
88	6.5	12.6	4.0	79.9	10.4	20.6	1.1
169	6.6	11.2	3.2	78.5	9.1	20.0	1.5
350[d]	6.6	2.0	1.1[b]	81.5	22.4	18.2	1.0

[a]Background concentrations in controls.
[b]Significantly different from pH 7.0 control (Exposure A) or pH 6.9 control (Exposure B), $P \leq 0.05$ (LSD test).
[c]Significantly different from pH 5.7 control (Exposure A) or pH 6.5 control (Exposure B), $P \leq 0.05$ (LSD test).
[d]Sample size (n) equals 6 for this treatment.

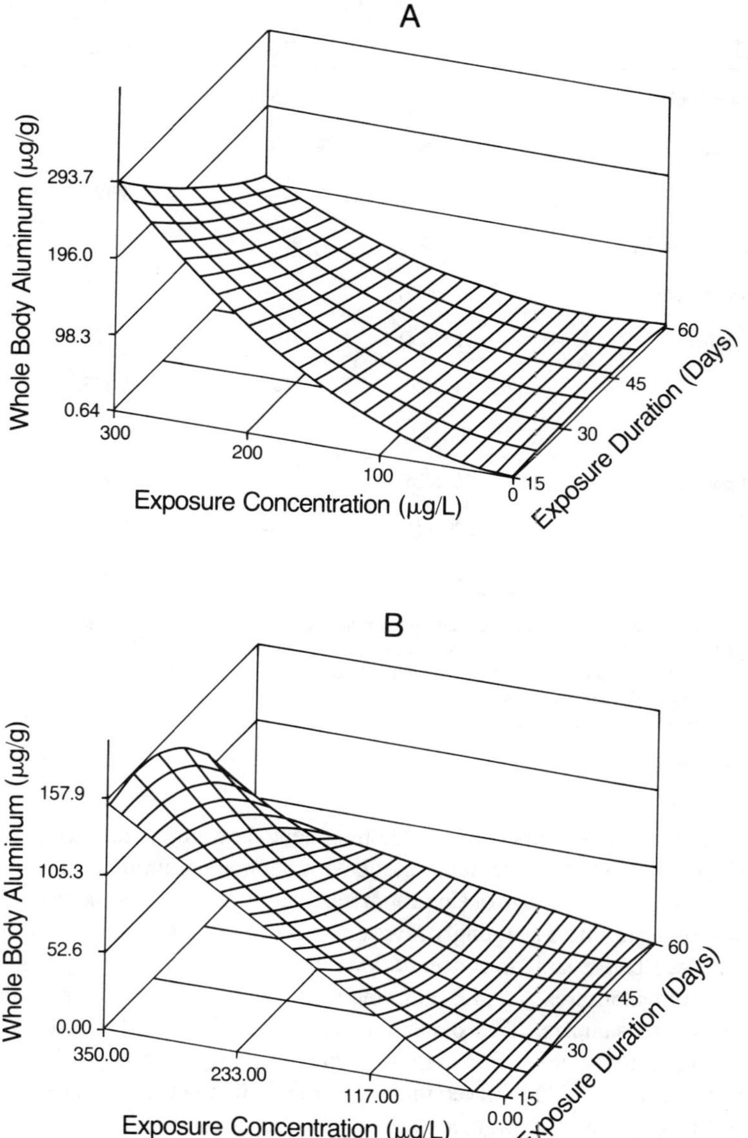

Figure 1. Three dimensional surface response plots of whole-body aluminum residues in brook trout at pH 5.6-5.7 (A) and pH 6.5-6.6 (B).

Table 8. Highest no-observed-effect aluminum concentrations (µg/L) for biological responses of brook trout

	pH 5.6 to 5.7		pH 6.5 to 6.6	
Variable	30 days	60 days	30 days	60 days
Movement/2min	68(15)	142(24)	169(26)	>350(22)
Strike frequency	8(-)	142(24)	4[d](-)	>350(22)
Buoyancy	68(15)	---	169(26)	---
Swimming capacity	>292(27)	29(14)	88(26)	>350(22)
Incomplete hatch[a]	8[b](-)	---	57(13)	---
Fry mortality	142(24)	142(24)	169(26)	169(26)
Length	142(24)	142(24)	57(13)	88(26)
Weight	142(24)[c]	68(15)	88(26)[c]	88(26)
RNA Content	142(24)	---	>350(22)	---
DNA Content	>292(27)	---	>350(22)	---
RNA-DNA ratio	>292(27)	---	57(13)	---

Note: Values represent total aluminum and (in parentheses) total dissolved monomeric aluminum.
[a]Embryo responses were determined after all embryos had hatched or died.
[b]Background aluminum measured in pH 5.7 control treatment.
[c]Determined at 45 days of exposure.
[d]Background aluminum measured in pH 6.5 control treatment.

DISCUSSION

Analyses of water samples from each treatment showed that the concentrations of total dissolved aluminum and total dissolved monomeric aluminum remained nearly constant over the entire range of aluminum treatments. Filtration of the samples through 0.1-µm-pore-size filters seemingly removed most of the insoluble aluminum species that were present at similar concentrations in the two exposures. In the extracted samples from exposures A and B, aluminum concentrations were higher in unfiltered than in filtered samples. The oxine extraction procedure used measures dissolved monomeric and low molecular weight polymeric (inorganic and organic) aluminum species.[16] Driscoll et al.[12] observed that high organic non-labile monomeric aluminum was correlated with total organic carbon levels in natural water samples from the Adirondack region of New York; however, the dilute exposure water used in our study contained low levels of total organic carbon, and probably also little extractable organically complexed aluminum. Furthermore, Okura et al.[21] demonstrated that the oxine extraction procedure measures only monomeric aluminum if the reaction time between samples and 8-hydroxyquinoline is limited to 10 s, and Smith and Hem[22] observed that, during polymerization, aluminum hydroxide polymers showed metastable ionic behavior until they contained 100 to 400 atoms. The higher concentrations of aluminum in unfiltered samples in the present study probably consisted of (1) small amounts of organically complexed aluminum and (2) metastable aluminum hydroxide polymers that formed in both exposures and depolymerized to the more reactive monomeric species during the extraction procedure. The presence of less extractable aluminum in unfiltered samples from exposure B than in exposure A was consistent with the formation of more stable polymers in the higher pH range of exposure B.

Our study and those of others have indicated that both dissolved monomeric and polymeric hydroxy complexes of aluminum adversely affects fish. Inasmuch as the total aluminum concentration increased from 29 to 292 µg/L in exposure A and from 57 to 350 µg/L in exposure B whereas the total dissolved aluminum and total dissolved monomeric aluminum remained constant, the dose-response observed in no-effect aluminum concentrations for the various biological endpoints can be attributed to either differential responses of the endpoints to dissolved aluminum or to the influence of polynuclear or particulate aluminum species. Driscoll et al.,[12] who demonstrated that excess fluoride or citrate decreased the toxicity of aluminum to brook trout and white suckers (*Catastomus commersoni*) at the pH range of 4.4 to 5.2 concluded that inorganic forms of aluminum were more toxic than organic forms. Baker and Schofield[23] showed that larvae and postlarvae of brook trout and white suckers, and eggs of white suckers, held at pH 5.2 to 5.4 were most susceptible to oversaturated solutions of aluminum; they suggested that the mechanisms of toxicity involved precipitation and coagulation of aluminum hydroxide on gills or other surfaces or the adsorption and nucleation of aluminum polymers at surface interfaces. Furthermore, Schofield and Trojnar[24] observed increased aluminum toxicity to brook trout with increasing pH (4.4 to 5.2) and suggested that the toxicity of the aluminum cation was increased by changes in speciation involving hydroxy complexing. Consequently, the measurement of total dissolved (filtered), and readily reactive (extractable) monomeric and polymeric species of aluminum would yield a good estimate of the amount of toxic aluminum present in natural waters. Since organic carbon mediates aluminum toxicity to fish,[12] measurement of total aluminum (unfiltered), which includes organically complexed species, may result in an overestimate of the amount of toxic aluminum present.

The results of whole-body aluminum residue analyses imply that as brook trout advanced from larvae to juveniles they began either to decrease their rate of aluminum uptake or to eliminate aluminum at a rate that exceeded uptake. A decrease in whole-body aluminum residues observed in the older juveniles may have been related to differential tissue accumulation of aluminum. Brumbaugh and Kane[15] observed in smallmouth bass (*Micropterus dolomieui*) gill tissue that the concentrations of aluminum were higher and more variable than in carcass, liver, kidney or gut tissue, and that this difference between tissues probably influenced the variability among whole-body samples. Buergel and Soltero[25] found differential accumulation of aluminum in tissues of rainbow trout from an alum-treated lake, an untreated lake, and a hatchery; aluminum concentrations in fish from all three sources were highest in gill tissue and lowest in muscle. The decline in whole-body aluminum residues in juvenile brook trout in the present study may be related to growth and dilution by muscle tissue that accumulated less aluminum than did the other tissues. However, the possibility that larval and juvenile fish have different modes and rates of aluminum uptake or excretion cannot be ignored.

The present study showed that acidity, length of exposure, and the biological endpoint measured are critical factors in determining chronic no-effect concentrations of aluminum for fish inhabiting dilute acidic waters. The ability of brook trout to hatch was the earliest and most sensitive index of aluminum toxicity in both exposures A (pH 5.6 - 5.7) and B (pH 6.5 - 6.6). During the first 30 days in exposures A and B, feeding behavior of brook trout was the most sensitive index of aluminum toxicity whereas growth and survival had intermediate sensitivity. Although RNA-DNA ratios at 15 days of exposure A were similar in sensitivity to feeding behavior, RNA and DNA were the least sensitive parameters measured at 30 days. After 60 days, in contrast, survival and growth were the most sensitive indices in both exposures. The variable sensitivities of these endpoints over time were also observed among eyed eggs, larvae, and juveniles exposed to pH 5.5 alone and in the presence of 318 μg/L aluminum.[8]

Inasmuch as behavior reflects the integrated whole-animal response to contamination, we measured swimming and feeding behaviors since these responses are essential to survival. Previous studies revealed changes in swimming and feeding behaviors among 30 day old brook trout exposed to 300-318 μg/L aluminum at pH 5.5.[8,26] In the present study, feeding behavior was highly sensitive; the highest no-observed-effect concentration of aluminum was less than 29 μg/L at pH 5.6 and less than 57 μg/L at pH 6.6 after 30 days of exposure. Such deficiency in an important behavioral function would compromise the adaptability of early life stages, particularly at the onset of exogenous feeding[27] and would diminish long-term survival.

The adverse effects of acidity alone on brook trout were also apparent in the present study. Significant differences in successful hatch, strike frequencies, and growth (length and weight) of fish exposed to the pH 5.7 and 7.0 control treatments (exposure A) and 6.5 and 6.9 control treatments (exposure B) were consistent with observations reported by Kwain and Rose[28] and Beamish and Harvey[29] of effects of high acidity on fish.

Aluminum toxicity to aquatic organisms under laboratory and natural conditions is a function of water quality. In waters, aluminum can exist as inorganic monomers, inorganic polymers and also form complexes with many organic and inorganic compounds. Further, it adsorbs onto inorganic, organic, and biotic surfaces.[9] Water quality (e.g., pH, total organic carbon) dictates the types of reactions that aluminum undergoes, and ultimately the quantity and magnitude of toxic responses that it elicits in aquatic organisms.[8,11,12,23] Because aluminum is a ubiquitous, naturally occurring element, resource managers should be aware of environmental variables (e.g., pH) that mobilize or exacerbate the toxic effects of aluminum. Assessments of site-specific problems of aluminum contamination must consider that water quality can greatly influence aluminum toxicity.

ACKNOWLEDGMENTS

We thank R.D. Archeski, S.B. Bevier, P.R. Heine, G.W. Holroyd, D.A. Remington, L.C. Sappington, and M. Walther for technical assistance, and Dr. M. Ellersieck for advice on statistical analyses.

REFERENCES

1. Trojnar, J.R. "Egg hatchability and tolerance of brook trout *Salvelinus fontinalis* fry at low pH," *J. Fish. Res. Board Can.* 34:574-579 (1977).

2. Jagoe, C.J., and T.A. Haines. "Alterations in gill epithelial morphology of yearling Sunapee trout exposed to acute stress," *Trans. Am. Fish. Soc.* 112:689-695 (1983).

3. Rask, M. "The effect of low pH on perch, *Perca fluviatilis L.* III. The perch population in a small, acidic, extremely humic forest lake," *Ann. Zool. Fenn.* 21:15-22 (1984).

4. Hutchinson, T.C., and F.W. Collins. "Effect of H^+ ion activity and Ca^{2+} on the toxicity of metals in the environment," *Environ. Health Perspect.* 25:47-52 (1978).

5. Brown, D.J.A. "Effect of calcium and aluminum concentrations on the survival of brown trout (*Salmo trutta*) at low pH," *Bull. Environ. Contam. Toxicol.* 30:582-587 (1983).

6. Hall, R.J., C.T. Driscoll, G.E. Likens, and J.M. Pratt. "Physical, chemical, and biological consequences of episodic aluminum additions to a stream," *Limnol. Oceanogr.* 30:212-220 (1985).

7. Robinson, J.W., and P.M. Deano. "The synergistic effects of acidity and aluminum on fish (Golden shiners) in Louisiana," *J. Environ. Sci. Health* A20:193-204 (1985).

8. Cleveland, L., E.E. Little, S.J. Hamilton, D.R. Buckler, and J.B. Hunn. "Interactive toxicity of aluminum and increased acidity to early life stages of brook trout," *Trans. Am. Fish. Soc.* 115:610-620 (1986).

9. Kramer, J.R. "Aluminum speciation and analysis," *Environ. Geochem. Rep. Number 198 3/3* (1983). McMaster University, Hamilton, Ontario, Canada.

10. Campbell, P.B.C., M. Bisson, R. Bougie, A. Tessier, and J.P. Villeneuve. "Speciation of aluminum in acidic freshwaters," *Anal. Chem.* 55:2246-2252 (1983).

11. Burrows, W.D. "Aquatic aluminum: chemistry, toxicology, and environmental prevalence," *CRC Crit. Rev. Environ. Control* 7:167-216 (1977).

12. Driscoll, C.T., Jr., J.P. Baker, J.J. Bisogni, Jr., and C.L. Schofield. "Effect of aluminum speciation on fish in dilute acidified waters," *Nature (Lond.)* 284:161-164 (1980).

13. Mount, D.I., and W.A. Brungs. "A simplified dosing apparatus for fish toxicology studies," *Water Res.* 1:21-29 (1967).

14. Martin, J.W. "A method of measuring lengths of juvenile salmon from photographs," *Pr og. Fish-Cult.* 29:238-240 (1967).

15. Brumbaugh, W.G., and D.A. Kane. "Variability of aluminum in organs and whole bodies of smallmouth bass *Micropterus dolomieui*," *Environ. Sci. Technol.* 19:828-831 (1985).

16. Barnes, R.B. "The determination of specific forms of aluminum in natural waters," *Chem. Geol.* 15:177-191 (1975).

17. Ingle, S.E., J.A. Keniston, and D.W. Schultz. "Aqueous chemical equilibrium computer program," Unites States Environmental Protection Agency. EPA-600/3-80-049 (May 1980).

18. *Standard Methods for the Examination of Water and Wastewater*, 14th ed., (Washington, DC: American Public Health Association, 1975).

19. Technicon Industrial Systems. *Dissolved Organic Carbon in Water and Wastewater*, Technicon Industrial Method No. 451-76 W, (1976), Tarrytown, New York.

20. Statistical Analysis System (SAS). "SAS Users Guide: Statistics, A.A. Ray, ed.," Cary, North Carolina, SAS Institute Inc (1982).

21. Okura, T., K. Goto, and T. Yotsuyinagi. "Forms of aluminum determined by an 8-quinolinolate extraction method," *Anal. Chem.* 34:581-582 (1962).

22. Smith, R.W., and J.D. Hem. "Effects of aging on aluminum hydroxide complexes in dilute aqueous solutions," Geological survey water supply paper 1827-D. United States Government Printing Office (1972). Washington, DC.

23. Baker, J.P., and C.L. Schofield. "Aluminum toxicity to fish in acidic waters," *Water Air Soil Pollut.* 18:289-309 (1982).

24. Schofield, C.L., and J.R. Trojnar. "Aluminum toxicity to brook trout (*Salvelinus fontinal is*) in acidified waters," *Environ. Sci. Res.* 17:341-366 (1980).

25. Buergel, P.M., and R.A. Soltero. "The distribution and accumulation of aluminum in rainbow trout following a whole-lake alum treatment," *J. Freshwater Ecol.* 2:37-44 (1983).

26. Hunn, J.B., L. Cleveland, and E.E. Little. "Influence of pH and aluminum on developing brook trout in low calcium water," *Environ Pollut.* 43:63-73 (1987). £6

27. LaCroix, G.L., D.J. Gordon, and D.J. Johnston. "Effects of low environmental pH on the survival, growth, and ionic composition of postemergent Atlantic salmon (*Salmo salar*)," *Can. J. Fish. Aquat. Sci.* 43:768-775 (1985).

28. Kwain, W., and G.A. Rose. "Growth of brook trout (*Salvelinus fontinalis*) subject to sudden reductions of pH during their early life history," *Trans. Am. Fish. Soc.* 114:564-570 (1985).

29. Beamish, R.J., and H.H. Harvey. "Acidification of the La Cloche Mountain Lakes, Ontario, and resulting fish mortalities," *J. Fish. Res. Board Can.* 29:1131-1143 (1972).

CHAPTER 14

COMPOSITION AND CONSEQUENCES OF ALUMINUM IN WATER, BEVERAGES AND OTHER INGESTIBLES

Roy U. Schenk, Johan Bjorksten, Luther Yeager

INTRODUCTION

Aluminum is one of the most common metals in the environment, constituting 8.4% of the earth's crust. Yet aluminum is not a part of any known animal metabolic process. In contrast, a great many less common metals are involved in enzymes and other metabolic processes. This raises questions as to the probability of potential harmful effects of this element.

Although Harvey Wiley as early as 1929 questioned the use of alum in foods,[1] it was not until 1962 that aluminum was found to be present in increased quantities in aging aorta.[2] This was the first clear clue that aluminum may be harmful to animals and humans. Aluminum is used as a cross-linking agent in the tanning of leather; and this process of cross-linking was implicated by Bjorksten[3] as a primary process whereby aluminum causes damage in living tissue.

A substantial number of beverages, foods and water sources have been analyzed for aluminum content over the years.[4-7] However, little urgency has attached to this effort because, with a few exceptions,[1,3,8] aluminum was generally considered by most workers to be relatively innocuous and non-toxic. This attitude changed rapidly after Alfrey reported that aluminum was implicated in dialysis encephalopathy.[9] When this was confirmed by other workers, a voluminous literature rapidly developed on aluminum toxicity.

In the recent, thorough review by Ganrot,[10] values reported in the literature for aluminum in plasma were shown to have declined by a factor of 100 since the early analyses. It is Ganrot's belief that this is due to the inadequacy of the earlier methods of analysis and to aluminum contamination of the testing system at the parts-per-billion level encountered in these analyses.

Animal and human toxicity to aluminum is generally due to a slow accumulation of aluminum. However, when renal dialysis bypassed the gastrointestinal barrier to aluminum, rapid absorption of aluminum by the transferrin in the blood occurred. As a result poisoning by aluminum has occurred in considerably less than a year to some kidney dialysis patients.[11] For persons not exposed to this source of aluminum, it has nonetheless been predicted that aluminum toxicity will become a major impediment to extended longevity when cancer and heart disease have been removed as major sources of mortality.[3]

The higher values reported for aluminum in plasma by earlier workers also raises questions as to the reliability of the earlier analyses for aluminum in food, water, beverages, body tissues, etc. However, for many of the values, particularly for food, values of parts-per-million were more common, and so may be less susceptible to error by contamination. In addition, contamination by soil, which regularly contains 5% or more aluminum, can be anticipated with many foods. Foods ingested while containing the contaminating soil carry significant amounts of aluminum into the digestive system. However, the amount of this aluminum which is capable of being solubilized is not clear, but is probably quite low.

The evidence of aluminum toxicity due to renal dialysis supports the belief that increased uptake of aluminum into the blood induces a greater uptake into the body organs. Furthermore, higher blood concentrations give a higher percentage of ultrafilterable aluminum.[12] It appears probable that absorption into the organs is related to the ultrafilterable aluminum. If this is indeed so, then surges of aluminum into the blood would be more serious than a more steady, lower amount of aluminum. This has significant implications regarding the sources of aluminum in the diet.

Rapid excretion of ingested aluminum into the urine[10] suggests that most aluminum absorption may take place in the stomach. This would not be surprising since aluminum is quite substantially more soluble in the acidic conditions of the stomach than in the modestly alkaline environment of the intestines.

Foods cooked in aluminum cooking containers regularly contained an increased aluminum content.[4] This was especially true with longer cooking and with greater food acidity. Although the results reported have been quite variable, the levels of aluminum reported are often quite high. This aluminum, particularly in acid foods, must be considered to be soluble and so substantially more available for absorption. The increased absorption of aluminum from antacids in the presence of citrate tends to confirm this.[13,14]

Because of the variable results reported in the past and uncertainty concerning reliability, the substantial usage of alum in the treatment of water, the fact that many new manufactured food products are on the market, plus the increasing usage of aluminum as a packaging material, it seemed desirable to obtain data on a wide range of water sources, beverages, foods and other substances that may be ingested, utilizing current methodology and using great care to reduce and to test for extraneous aluminum contamination. The ultimate objective of this testing is to develop an understanding of the most common sources of significant aluminum ingestion, absorption, and retention.

Towards this goal, water samples from several sources in over 20 states and 3 foreign countries were analyzed along with many common beverages and a moderate number of foods, primarily dry solids. Analysis of foods poses a problem. Currently, digestion in nitric acid is utilized, but this produces a value approaching that for total aluminum rather than aluminum solubilizable in the digestive system. While it is true that food initially encounters the high acidity of the stomach, which favors solubilization of aluminum, nonetheless some aluminum compounds are simply too insoluble to be completely or even largely solubilized in this environment. As a result, alternative methods for determining potentially absorbable aluminum in food are needed.

METHODS

Water samples were collected at various locations throughout the U.S., primarily in the western two-thirds of the country, and in three foreign countries. When feasible, three or more samples were collected at each city at different drinking water outlets rather than from the water treatment plant. This was done so as to determine aluminum content of water as it is actually consumed. The samples were collected into 4-mL borosilicate glass vials with screw caps. These had been pre-rinsed with an aqueous solution containing 6 g of disodium EDTA per liter and then twice with distilled water prior to being drained and dried in an environment protected from dust. The vials were rinsed at least two times with the water sample to be collected and were filled and sealed tightly. A vial containing distilled water was included in each container of 40 vials to serve as a control. Aluminum contents of the control samples increased by 1 to 3 ppb during transit. This value was subtracted as a correction to the values obtained in the analyses. The samples were analyzed within one to two months of the time of collection. All vials were serially numbered and supplied with correspondingly numbered log sheets for descriptive data to be computer recorded and compiled for rapid retrieval.

Samples of beverages were collected in the same vials using the same methods. Samples of solid edible products were collected in pre-rinsed 10-mL plastic containers with attached, friction fitted plastic lids. The solid, semi-solid and solid-containing liquid samples were liquified by adding 0.5 mL of nitric acid (Mallinckrodt AR Select) to 0.25-g samples in tared, pre-rinsed 20-mL borosilicate culture tubes. The tubes were covered with a glass sphere and digested overnight in a 45°C water bath. The samples were then diluted with distilled water, and the solutions were analyzed by the same method as the liquid samples. Corrections were made from blank values of culture tubes containing acid only which were digested under the same conditions.

The liquid-filled vials were checked for cap tightness, immersed in the sodium EDTA solution briefly, rinsed in distilled water, and dried before opening. Aluminum determinations were made in triplicate on a Perkin-Elmer Zeeman 3030 Atomic Absorption Spectrophotometer with the following conditions:

Wavelength	309.9 nm
Slit	0.7 nm
Pyrocoated graphite furnace with L'vov platform.	
Drying temperature	120°C
Char Temperature	1000°C
Atomization temperature	2500°C

Result deviations averaged 10% for higher aluminum contents and 15% for low values.

The AS-60 Autosampler cups were pre-rinsed in the sodium EDTA solution and distilled water before use. The operator wore vinyl gloves (similarly rinsed) to transfer samples from the vials to AS-60 cups, and the autosampler tray cover was immediately replaced after loading to obviate pickup of airborne dust. Data were stored in an IBM Personal Computer XT with WordPerfect software for selective retrieval.

RESULTS AND DISCUSSION

Water

Samples of water were collected primarily in the western and central parts of the U.S., together with a few samples from eastern states. In addition, several samples from three foreign countries were analyzed. The samples were collected from outlets from which water was consumed rather than from the original treatment plant in order to determine what aluminum levels were actually being consumed. These results are shown in Table 1. When only one sample, or when three or more samples were obtained, all but the high and low samples are combined in the center column.

When more than one sample was taken from various locations in a municipal water system, substantial variations in aluminum content were often found. The reasons for this substantial variation are not certain, but they could be simply a result of differences in the plumbing at each site, since the total aluminum levels are minute.

In general, most drinking water was found to have less than 100 parts per billion (micrograms per liter) of aluminum, though often one of the samples tested higher. Several water sources in the west coast states were found to have undetectably low levels of aluminum. On the other hand, water samples from New York City, New York, Kansas City, Kansas, Southgate, Kentucky, Freemont, Ohio, and most of the cities in Finland were found to have aluminum contents exceeding 100 ppb.

Table 1. Aluminum Content of Drinking Water

	Al (ppb)		
	Low	Other	High
CALIFORNIA			
Berkeley	31	56	74
Eureka	52	97	274
Gold Run		0	
Hopland		152	
Laytonville		98	
Manhattan Beach	38		128
Orick	11		25
Palo Alto		118	
Redcrest		109	
Sacramento	27	34	37
San Francisco	15	59	86
San Raphael	11	14	142
COLORADO			
Denver	48	90	116
Estes Park	42		48
Granby	47	67	102
Loveland	44	59	78
Rocky Mt. Nat'l Park	95		166
HAWAII			
Hana, Maui		124	
Kailua, Kona	35	59	88
Lehaina, Maui		43	
Paia, Maui		12	
IDAHO			
Coeur d'Alene	44	46	61
Kellog	28		31
Wallace	33	47	63

(Continued)

Table 1. (Continued)

	Al (ppb)		
	Low	Other	High
ILLINOIS			
Bellwood		76	
Bloomington	32	42	43
Chicago	43	61	75
Chillicothe		107	
Creve Coeur		34	
Henry		13	
Hillside	3	72	130
New Salem	32		35
Peoria	28	60	467
Pekin	72		88
Peru		18	
Petersburg	8		29
Pittsfield		11	
Pontiac	4	42	47
Rockford	20	42	221
Springfield	24	55	61
INDIANA			
Bedford	44	77	117
Bloomington	26	62	106
Dale	1	3	7
Evansville		6	
Huntingburg		39	
Indianapolis	13	31	38
Jasper		75	
Kentland		1	
Lincoln City		11	
Loogootee		17	
Terre Haute	7	26	137
Vincennes	5	6	15

(Continued)

Table 1. (Continued)

	Al (ppb)		
	Low	Other	High
KANSAS			
Goodland	16	42	69
Hayes	12	42	61
Kansas City	109	121	245
Lawrence	35	66	193
Topeka	18	19	23
KENTUCKY			
Covington	9	72	92
Southgate	333	383	400
LOUISIANA			
Baton Rouge		210	
Kenner		15	
Lafayette	12		21
Lake Charles		15	
MICHIGAN			
Bay City	12		16
Detroit	11	27	123
Lansing		16	
Midland		7	
Mount Pleasant		6	
MINNESOTA			
Austin	44	45	47
Rochester	24	36	93

(Continued)

Table 1. (Continued)

	Al (ppb)		
	Low	Other	High
MISSOURI			
Boonville	44	59	204
Columbia	29	61	63
Florida		2	
Hannibal	36	38	92
Howard County	51		62
Kansas City	4	72	368
Monroe City	38	42	61
Odessa		95	
Paris	14	45	47
St. Louis	88	207	233
MONTANA			
Billings	21	38	53
Bozeman	11	14	34
Butte	59	82	98
Cooke City		59	
Custer Nat'l Park		70	
Missoula	19		86
Red Lodge		39	
Saltese		56	
NEW YORK			
New York	254	276	299
NEVADA			
Reno	55	79	126
Wells	7	14	40
Winnemucca	5	13	45

(Continued)

Table 1. (Continued)

	Al (ppb) Low	Other	High
OHIO			
Ashtabula	2	30	76
Freemont	128	200	245
Madison	11	28	42
Tipp City	72	133	199
Toledo	90	106	177
Trotwood	35	85	96
ORGEON			
Bandon	14	26	45
Brookings		21	
Bullard's Beach St. Park		15	
Coos Bay	12	22	483
Corvallis		29	
Florence		145	
Newport	0	19	20
North Bend	2	6	46
Packwood		0	
Philomath	0		31
Portland	4	32	70
Waldport	3	49	63
Woodland		0	
PENNSYLVANIA			
Philadelphia		688	
SOUTH CAROLINA			
Charleston	5	249	1029
Sullivan's Island	2	75	223

(Continued)

Table 1. (Continued)

	Al (ppb)		
	Low	Other	High
SOUTH DAKOTA			
Badlands Nat'l Park		43	
Bryant	2		32
Deadwood	40		43
Keystone	21	30	53
Mitchell	45	47	104
Mt. Rushmore Nat'l Park		334	
Rapid City	32	36	60
Sioux Falls	39		66
Wall	96	126	128
Watertown	30	95	502
Willow Lake	131		371
TENNESSEE			
Memphis		45	
TEXAS			
Alvin		21	
Bryan		20	
Clear Lake City		13	
Dallas		15	
Galveston		6	
Houston	5	57	356
Kingwood		23	
League City		4	
Sealy		10	
Sugarland	11		251
Texas City		20	
Waco	1	51	520
West University Place		30	
Woodway	12		51

(Continued)

Table 1. (Continued)

	Al (ppb)		
	Low	Other	High
UTAH			
Bonneville Flats		45	
Ogden	19	23	37
Roy	21	33	51
Salt Lake City	21	41	45
WASHINGTON			
Auborn	27	30	50
Bellevue	12	41	118
Bellingham		29	
Edgewick		75	
Ellensberg	36	37	49
Enumclaw	0	20	22
Kent	35		44
Moses Lake	26	57	75
Mount Ranier		0	
Puget Sound	26		55
Randle		0	
Seattle	21	49	85
Spokane	32	39	72
Sunshine		0	
Winslow	2	12	52
Woodland	0		12
WISCONSIN			
Janesville	23	36	58
La Crosse	46		112
Madison	12	24	40
Waunakee	12	18	118

(Continued)

Table 1. (Continued)

	Al (ppb) Low	Other	High
WYOMING			
Cheyenne	75	91	136
Fort Bridger	31		39
Rawlin		96	
Rock Springs	44	54	119
Sheridan	50	63	97
Sundance	16		61
Yellowstone Nat'l Park	539		608
Yellowstone Park, well water		33	
CANADA			
Brentwood Bay		40	
Tsawwassen		100	
Vancouver	7	39	152
Victoria	24	54	95
FINLAND			
Fantsnas	45		147
Helsinki	20	162	702
Koski	169	437	445
Lappeenrauta	161	197	237
Punkaharju	117	251	413
Tommolansalmi	137		142
Valko		59	
ISRAEL			
River Jordan		66	
VENEZUELA			
Mara Cay	36	118	341

Table 2. Aluminum Content of Fruit Juices

Juice	Aluminum Content micrograms/Liter
Orange, Donald Duck	322
Orange, Minute Maid	311
Orange, Valio (Finland)	429
Orange, Sunkist, fresh squeezed	43
Lemon-Lime Diet Drink, Minute Maid	272
Mandarine Orange Syrup, Food Club	299
Grapefruit, hand pressed	298
Grapefruit, Texas Red, fresh squeezed	44
Grapefruit, Texsun	314
Grapefruit, Texsun Pink	87
Lemon, Sunkist, fresh squeezed	48
Realemon, Reconstituted, Borden	3700
Peach Nectar, Heart's Delight	4130
Tomato, Valio (Finland)	148

If one calculates total intake of aluminum from water based on an estimated consumption of 2 liters of water per day, in most instances the intake would be less than 0.2 mg per day from water. Since a typical ingestion of aluminum would be about20 to 40 mg per day,[6] this suggests that water is not normally a major source for aluminum ingestion.

Beverages

Beverages tend to have a substantially higher aluminum content than does water. Exceptions to this include some fresh squeezed fruit juices, which had less than 50 ppb of aluminum.

Fruit juices tested (Table 2), excluding two samples with very high aluminum, had an average content of 220 ppb of aluminum. Except for persons consuming massive quantities of these beverages, this seems unlikely to provide substantial quantities of aluminum. The one caveat that must be noted here is that lemon juice, orange juice, and citric acid do enhance absorption of aluminum from antacids.[13,14] This makes it probable that absorption of the aluminum present in these juices is also enhanced. Likewise it suggests that these juices may increase aluminum absorption from other substances intermixed in the digestive tract. However, since this aluminum is presumably absorbed as the strongly bound citrate chelate,[15] it may also be excreted more readily. Thus, substantially more study needs to be done to assess the impact of the enhanced aluminum absorption by such juices.

Carbonated, non-alcoholic beverages have substantially higher but quite variable aluminum contents (Table 3). The samples tested varied from 107 to 2084 ppb of aluminum, with an average of 830 ppb. Since the aluminum in these beverages is soluble, and they are acidic and often contain citrate, it is likely that enhanced aluminum absorption from these products will occur. With the substantial quantities of these beverages consumed by many persons, they may well constitute a significant source of absorbed aluminum.

Carbonated, alcoholic beverages also have substantial and variable aluminum contents. Surprisingly, beer in aluminum cans did not generally contain more aluminum than did other of these beverages. Evidently, the internal coating adequately protects the contents. Beer tended to be lower in aluminum, averaging 300 ppb, than did the carbonated wines and wine coolers, which averaged 970 ppb of aluminum. Again the aluminum contents varied substantially. This suggests that processing methods can be modified to attain substantially lower average levels of aluminum in these beverages.

Teas, which are reported to accumulate aluminum in the plant's tissues,[16] were indeed found to produce beverages with a high aluminum content (Table 4). Some instant teas were found to have lower levels of aluminum than freshly brewed teas. This may be due to filtration of fine colloidal aluminum-containing substances. There is some evidence that aluminum in the tea beverage may not be as readily absorbable as is aluminum from other beverages due to chelation with the high molecular weight polyphenols present in teas in large quantities.[16] Tea bags are also a significant source of aluminum, adding on the average an additional 342 ppb of aluminum to the 1% tea extract. This is due to treatment of the paper with aluminum compounds during the processing of the paper. This is a source of aluminum which could be readily and substantially reduced.

Herbal teas tended to be substantially lower in aluminum than were the regular teas. With the exception of one chamomile tea, 1% extracts of these products had average aluminum contents (420 ppb) somewhat lower than those of the 3% coffee extracts (640 ppb). These tea and coffee concentrations were chosen to approximate those recommended for beverage purposes.

Wines generally had higher aluminum concentrations than did distilled liquors (Table 5). The aluminum in wines is also associated with substantial fruit acids, including citric acid. As a result, the high aluminum concentrations in some of these wines may well be a source of substantial absorbable aluminum. The lower level of consumption of the distilled liquors together with the lower aluminum concentrations indicate that these are at most a minor source of aluminum in the diet.

Table 3. Aluminum Content of Carbonated Beverages

Beverage	Aluminum Content micrograms/Liter
Non-Alcoholic	
Ginger Ale, Schweppes	231 - 581
Ginger Ale, Canada Dry	934
Coca Cola	208 - 407
Diet Coke	2064
Dr. Pepper	186 - 698
Diet Pepsi	586 - 2084
Royal Crown Cola	103
Seven Up	1312
Diet Seven Up	387 - 929
Slice	1681
Alcoholic	
Budweiser (Al can)	178 - 288
Huber (glass)	350 - 578
Pabst Blue Ribbon (Al can)	360
Old Milwaukee (Al can)	271 - 293
Old Style (Al can)	67
Sun Country Cooler	967 - 1131
Seagram's Wine Cooler	616 - 918
Reunite Natural Peach	213 - 1372
Reunite Lambrusco	688 - 1727
Andre Champagne	1095

Table 4. Aluminum Content of Hot Water Extracted Beverages

Substance Tested	Aluminum Content micrograms/Liter
Teas - 1% Extract	
Jasmine Tea, bulk, Fujitan Tea Branch	1107
Earl Grey, bulk	1268
China Green, bulk	1632
Keemun, bulk	719
Kenya	1095
Apricot flavored, bulk	1154
Mango flavored, bulk	1097
Strawberry flavored, bulk	766
Orange Blossom Oolong, bulk	1559
Instant Tea, 100% tea, Lipton	2445
Ice Tea Mix with Nutrasweet, Lipton	1370
Instant Tea with Nutrasweet, Tetley's	378
Instant Nestea with Nutrasweet	462
Instant Nestea with sugar	603
Hot tea from restaurant	2230
Iced tea from restaurant	3550
Tea Bags*	
Jasmine, Voyager	1173
Jasmine, Quonk Yick	1154
Constant Comment, Bigelow	1026
Ceylon Breakfast	1157
English Breakfast,Twinnings	770
English Tea Time, Bigelow	848
Darjeeling, Twinnings	878
Darjeeling, Jacksons	1508
Earl Grey, Twinnings	618
Black Currant Flavored, Twinnings	1189
Black, L O'Rient	2931
Black, Ten Ren	1928
Regular, Lipton	2514
Oolong, Pagoda	1008
Oolong, Pagoda, Second brewing of above	424

*Tea bags add an average of 342 ppb of aluminum

(Continued)

Table 4. (Continued)

Substance Tested	Aluminum Content micrograms/Liter
Herbal Teas - 1% Extract	
Almond Pleasure, Lipton	527
Gentle Orange, Lipton	140
Lemon Soother, Lipton	417
Quietly Chamomile, Lipton	398
Chamomile, bulk	1065
Chamomile, Celestrial Seasonings	429
Country Apple, Celestial Seasonings	475
Emperor's Choice, Celestial Seasonings	208
Lemon Mist, Celestial Seasonings	224
Mandarine Orange Spice, Celestial Seasonings	329
Red Zinger, Celestial Seasonings	378
Roastaroma, Celestial Seasonings	511
Sleepy Time, Celestial Seasonings	497
Lemon Iced Delight, Celestial Seasonings	312
Feeling Free, Bigelow	378
Looking Good	649
Sweet Orange Spice, Magic Mountain	220
Instant Coffee - 1% solution	
Nescafe, Brava	20
Taster's Choice, Decaf	72 - 581
Taster's Choice, Regular	110
Maxwell House, Decaf	125
Whole Coffee - 3% Extract	
Maryland Club	1163
Gevalia, Traditional Whole Bean (ground)	838
Sanka, Auto-Drip	574
Folgers, Auto-Drip	375
Maxwell House, Private Coll'n, Kenya AA	235

Table 5. Aluminum Content of Wines and Spirits

Beverage	Aluminum Content micrograms/Liter
Wines	
Chablis, Carlo Rossi	1246
Chablis, Blanc, Gallo	1496
Chablis, Inglenook Navelle	1068
Chablis, Taylor	3200
Chablis, Los Hermanos	905
Schloss Garten, Zeller Schwarze Katz	1327
Mosel, Alko (Finland)	427
Mateus Rose	886
Dry Vermouth, Ambassador	1044
Wild Irish Rose	440
Cherry, Manischevitz	2840
Port, Gallo	2650
Zinfandel, Paul Masson	515
Red Rose, Gallo	834
Burgundy, Inglenook Navelle	1773
Rice Wine, Kikomen Manjo	555
Cabernet Sauvignon, Firestone	724
Cherry, Home-made	616
Wild Grape, Home-made	388
Spirits	
Brandy, E & J	274
Brandy, J. Bonet	192
Gin, Store Brand	148
Rum, light, Ron Rico	597
Tequila, Jose Cuervo	635
Vodka, Gilbey's	499
Vodka, Hallers	420
Whiskey, Bourbon, Kentucky Tavern	168
Whiskey, Scotch, Cutty Sark	474
Whiskey, Scotch, Usher's	183
Whiskey, Hiram Walker 10	410

Food products show extremely wide variations in aluminum content (Table 6). Even within specific types of foods, there are often wide variations. For example, canned soups varied from 32 to 3600 ppb of aluminum, and cereal products varied from 40 to 29,330 ppb of aluminum. A lot of this variation appears to be associated with aluminum introduced in various food additives. One of the common sources of high levels of aluminum is the anti-caking agent, sodium aluminum silicate. This is present in salt, non-dairy creamers, and a multitude of other powdered materials. It appears that it may be carried into other manufactured food products when salt is added to these products. Alum is another significant source of aluminum. Since it is soluble in water of moderate acidity, it may be a more significant source of absorbable aluminum. Aluminum phosphates are the most common food additive. Their lower solubility suggests they may not be as serious a source of absorbable aluminum.[17]

Unprocessed foods showed a substantial variation in aluminum content, varying from 107 ppb in eggs to 7160 ppb in lettuce. Many processed foods also showed low aluminum contents. This suggests that processing methods are available to produce products with low aluminum contents.

Analyses of other substances which are ingestible showed that many of these products had very high aluminum contents (Table 7, note that results are reported in mg/Kg). Many of the antacids tested have aluminum hydroxide as a major active ingredient, and these do test quite high in aluminum. However, other products such as Tums and various vitamin and mineral products also contain substantial amounts of aluminum. It is likely that in these latter products the aluminum is contained in the compressing agent of the tablets, e.g., calcium phosphate, which is not particularly soluble in the digestive system.

In future studies, it is particularly important to establish whether all sources of aluminum are harmful. Aluminum which is not absorbed from the intestinal system into the blood is presumably of little concern from a health standpoint. Therefore, it is imperative to determine to what extent various food additives and other sources of aluminum are absorbed into the blood, and what foods and digestive conditions influence the total aluminum absorption. Since aluminum in food is rapidly absorbed and the vast bulk is rapidly excreted into the urine,[10] analysis of the urine before and after ingestion of products containing substantial amounts of aluminum may well provide an effective estimate of the extent to which various aluminum products are absorbed.

Table 6. Aluminum Content of Food Products

Substance Tested	Aluminum Content micrograms/Kg
Iodized Salt, Morton's	31,300
Salt, table, Diamond Crystal	36,600
Apple Cider Vinegar, Heinz	2,420
Tarragon Flavored Vinegar, Heinz	3,020
Sweet Pickle Relish, liquid, Heinz	5,400
Sweet Pickles, liquid, Vlasik	126
Baby Dill Pickles, liquid, Vlasik	277
Sweet Pickle Slices, liquid, Roundy's	177
Sweet Pickles, Kohl's	9,970
Sweet Pickles, liquid, Kohl's	9,340
Tomato Ketchup, Heinz	4,300
Mustard, two sources	3,770
Peanut Butter, Kroger, natural	6,290
Gelatin, Cherry, Jell-O	25,200
Worchestershire Sauce, Lea & Perkins	373
Teriyaki Marinade & Sauce, Kikkoman	1,540
Soy Sauce, La Choy	80
Soy Sauce, Shoyu	317
Non-dairy Creamer, store brand	94,300
Non-dairy Creamer, Coffeemate	25,700
Same as above, single serve packet	57,700
Skim Milk, several sources	1,409
Fresh Milk, whole, Gerlands, Houston	102
Fresh Milk, Kotisare (Finland)	2,300
Powdered Milk, Valio (Finland)	7,900
Powdered Milk, Carnation	28
Buttermilk, Flav-o-rich	668
Sour Cream, Pure Milk Co.	311
Cream, Half & Half	242
Cheese, Stilton	6,010
Cheese, Rondeau	4,620
Cheese, Blue	3,830
Cheese, American, Cheez Whiz	9,540
Cheese, Swiss, Borden	5,250
Cheese, Cheddar, Kraft, Cracker Brl	14,100
Ice Cream, Flav-o-Rich	10,730
Chocolate Powder, Nestle's Quik	6,210

(Continued)

Table 6. (Continued)

Substance Tested	Aluminum Content micrograms/Kg
Chocolate Candy, Fazer	148
Chocolate Cookie, Oreo	12,700
Clear Meat Bouillon	180
Soup, Vegetable, Campbell's	1,078
Soup, Vegetable & Bean, Campbell's	835
Soup, Chunky Beef, Campbell's	3,600
Soup, Chicken with rice, Campbell's	32
Soup, Split Pea, low Na, Campbell's	102
Soup, Onion, Restaurant	2,780
Soup, Cream of Mushroom, Restaurant	3,285
Margarita Mix, Mr & Mrs "T"	2,900
Egg, Chicken	107
Eggs, Scrambled	2,865
Omelette	3,140
Fish, Salmon	5,440
Fish, Herring	127
Oyster, raw	8,760
Anchovies	2,270
Pork Sausage	6,580
Bacon	6,500
Sugar, raw	5,060
Salad Dressing, Kraft Miracle Whip	3,700
Salad Dressing, Weight Watchers	2,500
Sunflower Seed	2,360
Almond Slices, Sun Giant	3,000
Millet Seed	5,200
Cashews, Planters	3,040
Pecans, Texas Papershell	204
Lettuce	7,160
Papaya	1,460
Pineapple	1,300
Peas, frozen, Pict Sweet	1,640
Potato, Red	3,630
Potato, Sweet	1,010
Squash, yellow, Green Giant	20,400
Grits	5,580
Noodles, Stouffers	3,500

(Continued)

Table 6. (Continued)

Substance Tested	Aluminum Content micrograms/Kg
Spaghetti, Creamette	7,160
Rice, yellow, Rice-a-Roni	1,970
Chow Mein Rice, L'Orient	3,380
Oats, Quaker, 3 minute	2,210
Oats, Quaker, Masa Hari	4,180
Rice Crispies, Kellog's	7,110
Raisin Bran, Post's	29,330
Raisin Bran, Kellog's	4,110
Wheat cereal, Malt-o-Meal	790
Wheat cereal, Nutri-Grain	40
Pancake Mix, Pillsbury Hungry Jack	27,300
Bread, white	351
Bread, pumpernickel	13,200
Wheat Germ, Kretschmer	2,200
Yeast, Brewers, Golden Harvest	1,670
Baby Formula, Mlocl's	16,800

Table 7. Aluminum Content of Other Ingestible Substances

Substance	Aluminum Content milligrams/Kg
Baking Soda, Arm & Hammer	4
Alka-mints	15
Antacid, Gaviscon	240
Gelusil	81
Mylanta	562
Remegel	449
Rolaids	314
Tums	51
Kaopectate	17
Peptobismol	10
Ascorbic Acid, 500 mg tablets, Walgreens	54
Choline, 500 mg tablets, Rugby	26
Calcium, 500 mg tablets, Rugby	33
Zinc, 50 mg chelated tablets, Rugby	36
Carotene Powder, 2.4%	40

REFERENCES

1. Wiley, H. W. *History of a Crime Against the Food Law*, 1976 reprinting of 1929 edition. (New York: Arno Press). pp. 400-2

2. Zinsser, H., J. Bjorksten, et al. "The freezing pool: A unified sequence of the aging process," In: *Medical and Clinical Aspects of Aging*, Blumenthal, Ed., (New York: Columbia Univ. Press, 1962), pp 460-483.

3. Bjorksten, J. *Longevity: A Search,* R. Schenk, Ed. Bjorksten Res. Fdn., Madison, WI, pp 214-222 (1981).

4. Campbell, I. R., et al. "Aluminum in the environment of man," *AMA Archives of Indust. Health* 15:359-448 (1957).

5. Lione, A. "The prophylactic reduction of aluminum intake," *Food Chem Toxic.* 21:103-109 (1983).

6. Greger, J. L. "Aluminum content of the American diet," *Food Technol.* 39(5):73-80 (1985).

7. Miller, R. G., et al. "The occurrence of aluminum in drinking water," *J. Am. Water Wks Assoc.* 76(1):86-91 (1984).

8. Spira, L. *Clinical Aspect of Chronic Poisoning by Aluminum and its Alloys.* J. Bale & Sons U Curno, London, England (1933).

9. Alfrey, A. C., et al. "The dialysis encephalopathy syndrome," *N. Eng. J. Med.* 294:184-188 (1976).

10. Ganrot, P. O. "Metabolism and possible health effects of aluminum," *Envir. Health Persp.* 65:363-441 (1986).

11. Alfrey, A. C., et al. "Metabolism and toxicity of aluminum in renal failure," *Am J. Clin. Nutr.* 33:1509-16 (1980).

12. Elliott, H. L., et al. *Plasmapheresis, Aluminum, and Dialysis Dementia.* Lancet ii: 1255 (1978).

13. Slanina, P., et al. "Dietary citric acid enhances absorption of aluminum in antacids," *Clin Chem.* 32:539-541 (1986).

14. Weberg, R., and A. Bèrstad. "Gastrointestinal absorption of aluminum from single doses of aluminum containing antacids in man," *Eur. J. Clin Invest.* 16:428-432 (1986).

15. Martin, R. B., et al. "Transferrin binding of Al^{3+} and Fe^{3+}," *Clin. Chem.* 33:405-407 (1987).

16. Stagg, G. V., and D. J. Millin. "The nutritional and therapeutic value of tea - A review," *J. Sci Food Agric.* 26:1439-1459 (1975).

17. Kaehny, W. D., et al. "Gastrointestinal absorption of aluminum from aluminum-containing antacids," *New Eng. J. Med.* 296:1398-90 (1977).

CHAPTER 15

THE CHEMISTRY OF NEUROLOGICALLY ACTIVE, NEUTRAL, AND WATER SOLUBLE ALUMINUM COMPLEXES

William O. Nelson, Tammy G. Lutz, Chris Orvig

INTRODUCTION

Aluminum has generally been regarded as a benign element for many years; however, in the last decade or two a considerable body of evidence has accumulated implicating Al in a variety of dysfunctions, both neurological (e.g., Alzheimer's disease, dialysis encephalopathy, amyotrophic lateral sclerosis) and osteological (osteoporosis, osteomalacia). In Alzheimer's disease, for instance, intracellular sites of aluminum accumulation are nuclei of neurons, which bear neurofibrillary tangles,[1] and senile plaques, where aluminosilicates are found.[2] The role played by Al in each of these conditions is poorly understood, and there remains controversy over whether it is involved at all. Over the last twenty years, a significant number of interactions of Al with various body fluid constituents have also been discovered,[3] suggesting that the chemistry of aluminum is ripe for development from the vantage point of possible biological applications. These applications might involve ligands for the removal of *in vivo* Al from Al overload victims, or ligands for the study of *in vivo* Al transport. The work reported here was undertaken with both of these possibilities in mind.

Of direct relevance to the increasing number of putative detrimental health effects associated with aluminum is its ubiquity in the food treatment and processing, pharmaceutical, and packaging industries.[4-7] Some sources of ingestible aluminum are listed in Table 1. There is general agreement that the average North American diet contains from 5 to 50 milligrams a day of Al (which is poorly absorbed), but this can be far exceeded, for example, by individuals who daily consume significant quantities of Al-based antacids. Recent results from Sweden suggest that the coadministration of citric acid and $Al(OH)_3$ can cause increased levels of Al in the brain and bone of rats[8] as well as in the blood of humans.[9] There is, therefore, already a pressing need for rational examination of the chemistry of the element with ligands which are present *in vivo* or in complexes with properties which may shed light on the absorption problem.

Table 1. Some sources of ingestible aluminum

$Al(OH)_3$	- antiperspirants, antacids, buffered aspirins. - gels to lower urinary phosphate and oxalate.
Alums	- M $Al(SO_4)_2$ M = K, Na, NH_4 - pickling, astringents, antiseptics, emetics - baking powder - purification of drinking water
Phosphates	- baking products - emulsifying agent in cheese
Basic Carboxylates	- $Al(OH)_x(RCO_2^-)_{3-x}$ - antacids, deodorants

The development of aluminum coordination chemistry has been hindered by the complicated and extensive hydrolysis undergone by Al^{3+} in aqueous solution as a function of pH.[10] Except for certain amino- and hydroxy-carboxylate ligands, most solution studies of Al-ligand interactions have been performed in acid at pHs of 3 or less in order to suppress the formation of polymeric oxo-hydroxoaluminum species.[11,12] Consequently, these studies only reveal part of the aluminum chelation story. The appropriate coordination chemistry has recently become of concern to inorganic chemists,[13] and this is often manifested in the study of the various Al species formed in aqueous solution with a ligand of some environmental, agricultural, or nutritional importance (e.g. citrate).[13-16]

COMPLEXES OF 3-HYDROXY-4-PYRONES

In formulating compounds which might be relevant to the *in vivo* Al transport problem, our approach has been to prepare Al-ligand complexes which are (1) of overall neutral charge (providing some degree of lipophilicity), (2) of sufficient water solubility to allow a reasonable *in vivo* concentration to be attained, and (3) of adequate stability to hydrolysis to allow the complex to persist intact in aqueous solution at neutral pH (7) or physiological pH (7.4). These requirements are rather stringent for a metal-binding ligand system, but not unachievable. The first candidate compounds which were prepared to test for these properties are shown in Figure 1. The pyrones were selected because they are known to form complexes with a variety of divalent[11,17-23] and trivalent metal ions,[11,19,24-34] and there are formation constants in the literature for Al^{3+} complexes of Kojic acid[11,19] and maltol[11,27,29] which suggest that thermodynamically stable complexes are formed. The pyrones are 3-hydroxy-4H-pyran-4-one (pyromeconic acid-Hpa), 3-hydroxy-2-methyl-4H-pyran-4-one (maltol-Hma), 5-hydroxy-2-(hydroxymethyl)-4H-pyran-4-one (Kojic acid - Hka), and 5-hydroxy-2-(chloromethyl)-4H-pyran-4-one (chlorokojic acid-Hck).

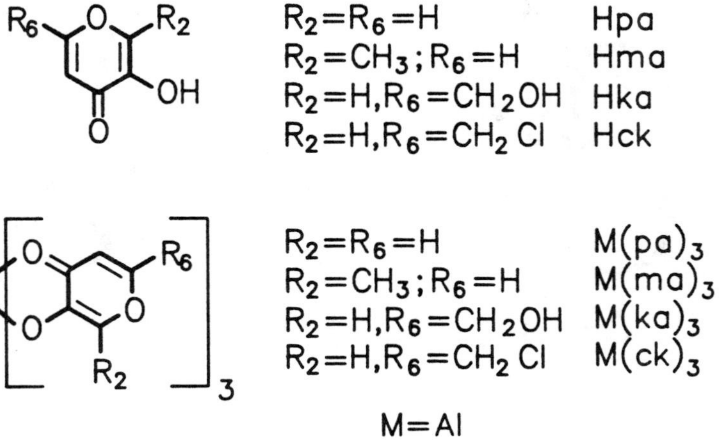

Figure 1. The pyrones and their Al complexes.

The complexes were easily prepared in, and isolated from, aqueous solution.[35,36] Most are water soluble to some degree (from less than 0.1 mM ($Al(ck)_3$) to greater than 60 mM ($Al(ma)_3$)). Elemental analyses were consistent with the formulation of simple *tris*(ligand)aluminum(III) species as shown in Figure 1; a number of the compounds are quite hygroscopic. Infrared, mass, and proton NMR spectra also supported the formulation. Solution conductivity (resistance) measurements showed that all the complexes which have water solubility of at least 1 mM - $Al(pa)_3$, $Al(ma)_3$, and $Al(ka)_3$ - are of neutral charge at neutral pH. This unusual (for Al) combination of water solubility and neutral charge suggested that octanol/water partition coefficients (p) would further elucidate the possible utility of these complexes in the study of *in vivo* Al transport. Partition coefficients were determined by monitoring the lower energy band in the ultraviolet spectrum of each complex in aqueous solution before and after extracting with water-saturated 1-octanol. The most water soluble complex ($Al(ma)_3$)(60mM) also has the highest p (0.67), the numbers for the other complexes being less than 0.10.

After the data confirmed that the molecular formulae were indeed correct, the disposition of the three ligands around the aluminum center required elucidation. Shown in Figure 2 are the four possible arrangements of three unsymmetric bidentate ligands (represented as A-B). It is well known[37] that Al is a relatively labile metal ion and, therefore, it is impossible to differentiate between the two possible stereoisomers--the left-handed Λ and the right-handed Δ--because they are interconverting so quickly in solution at room temperature (ca. 20°C). It should be possible, however, to differentiate between the meridional (mer) and the facial (fac) geometric isomers. The ^1H nuclear magnetic resonance (NMR) spectra of the complexes suggested that the mer and fac isomers were rapidly interconverting even at a temperature as low as -70°C in

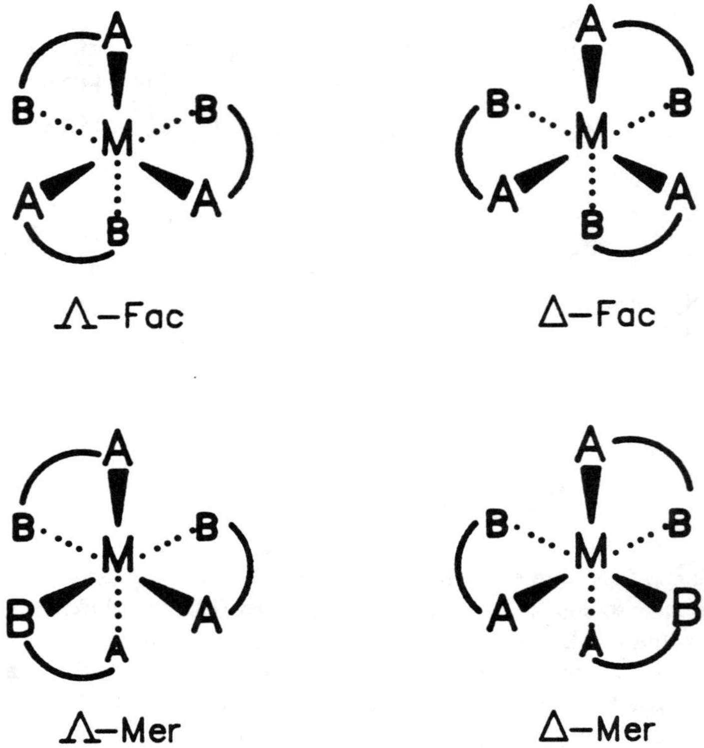

Figure 2. Geometrical and optical isomers of a simple complex of the form $M(A\text{-}B)_3$.

deuterated methanol solution. At this temperature some broadening of the NMR peaks was observed, but not enough to indicate that the resolution of two geometric isomers was achieved.

The structure in the solid state was ascertained unambiguously by X-ray diffraction measurements on single crystals.[35] Crystals of $Al(ma)_3$ were grown by solution diffusion in diethyl ether/methanol.[38] The molecular geometry was found to be mer in the solid state as shown in Figure 3. Because the ketone oxygens on two of the ligands coordinate Al trans to one another, the third ligand was not constrained to one orientation and this is illustrated in Figure 3. Indeed, the result found in the X-ray diffraction experiments was 58% of the molecules in orientation A, and 42% of them in orientation B, giving the averaged view represented by orientation C.

Both $Al(ma)_3$ and $Al(ka)_3$ are water soluble and of neutral charge in aqueous solution at physiological pH, however, their stability to hydrolysis was undetermined. ^{27}Al NMR was chosen for this study because, under favorable conditions, it allows the observation of one peak per mononuclear Al-containing complex, thereby acting as a direct probe

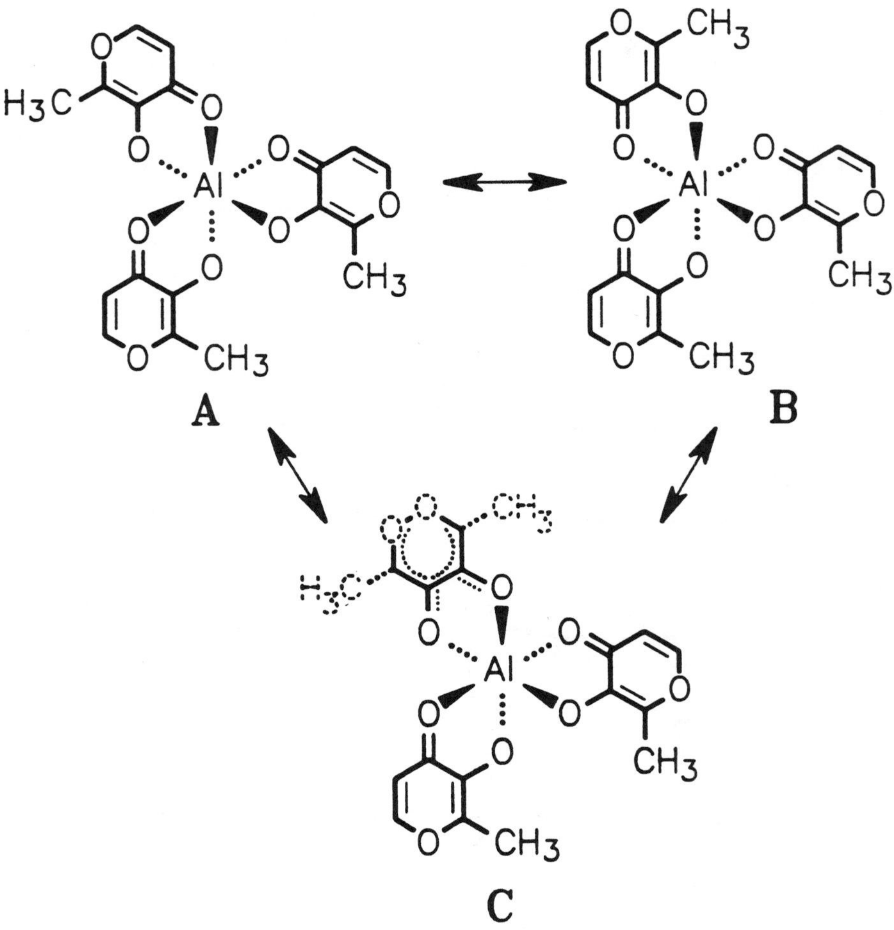

Figure 3. Disorder in the crystal structure of $Al(ma)_3$.

of the reactivity of the complex in solution. Although aluminum is a quadrupolar nucleus, ^{27}Al is monoisotopic and 20% as receptive as hydrogen so that even though the spectra may contain lines of considerable breadth, they are easy to collect and they contain considerable information if the molecular species are of sufficient symmetry.[39]

The variable pH ^{27}Al NMR evinced the hydrolysis steps involved in the decomposition of these complexes in water over a wide pH range. Figure 4 shows this experiment for a 0.05 M solution of $Al(ma)_3$. Clearly observable in acidic solution are several partially hydrolyzed species formed from $Al(ma)_3$, whose NMR peak is at 39 ppm ($W_{1/2}$ = 900 Hz). These species resulted from the sequential protonation of the maltol ligands with decreasing pH and are, in order of appearance: $[Al(ma)_2(H_2O)_2]^+$ at 26 ppm; $[Al(ma)(H_2O)_4]^{2+}$ at 13 ppm; and the fully hydrolyzed $[Al(H_2O)_6]^{3+}$ at 0 ppm

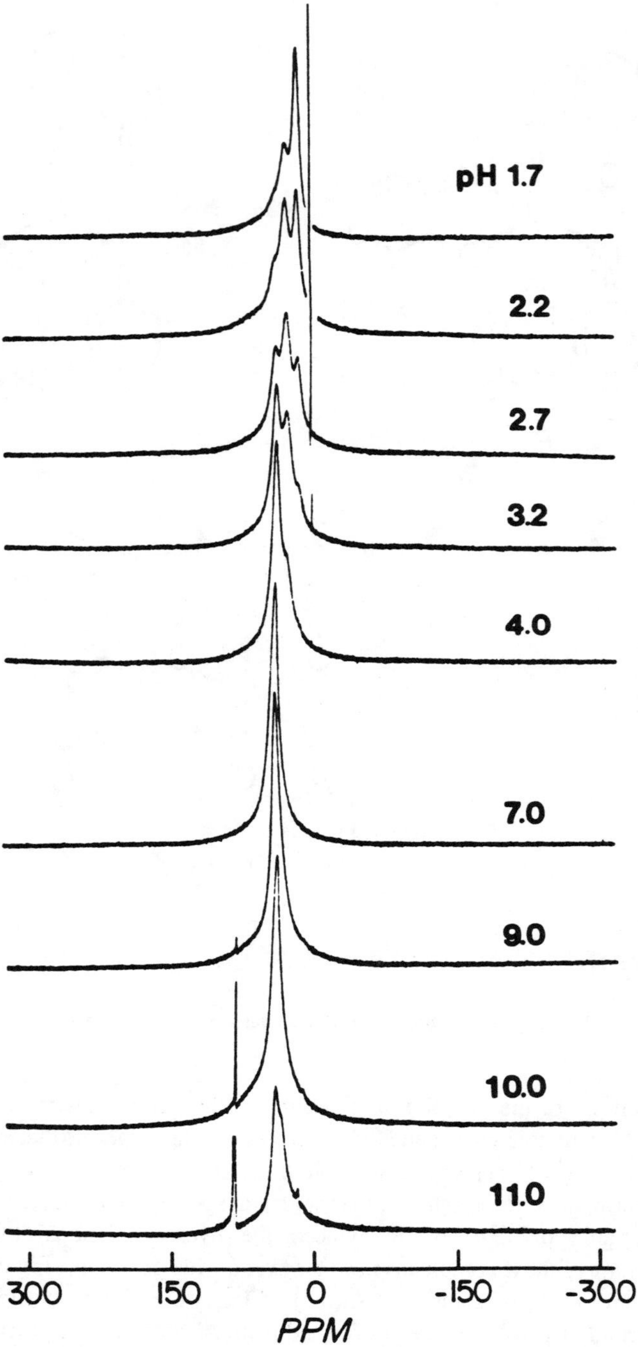

Figure 4. Variable pH ^{27}Al NMR of $Al(ma)_3$. Reprinted with permission from *Inorg. Chem.* 26:2171 (1987). Copyright 1987 American Chemical Society.

(the ^{27}Al NMR standard). The assignments are supported by the relative intensities of the peaks as the pH is reduced. When the pH is raised, the basic hydrolysis proceeds as maltol ligands are replaced sequentially by hydroxide to form, ultimately, the aluminate anion $[Al(OH)_4]^-$ which has a sharp signal at 80 ppm. Similar results were obtained for $Al(ka)_3$, although at lower concentration. There is, therefore, a window of stability to hydrolysis for $Al(ma)_3$ at this concentration of about pH 4 to 9. At lower concentration (0.004 M), both $Al(ma)_3$ and $Al(ka)_3$ were quite similar displaying stability to hydrolysis from pH 4.5 to 8. This experiment, when considered with the solubility and partition coefficient data, suggests that $Al(ma)_3$ would be a good candidate for *in vivo* studies because it is (1) significantly water soluble; (2) significantly lipophilic; and (3) stable to hydrolysis over a pH range of about 4 to 9. An anticipated result in light of these properties might be considerable *in vivo* mobility.

COMPLEXES OF 3-HYDROXY-4-PYRIDINONES

In an effort to broaden the base of Al complexes which might be of biological interest, we synthesized a series of 3-hydroxy-4-pyridinone ligands for complexation with aluminum; these are illustrated in Figure 5.[40] (The 3-hydroxy-4-pyridinones form in the acid hydrolysis of heated milk powders.) The ligands are 3-hydroxy-2-methyl-4(1H)-pyridinone (Hmpp), 3-hydroxy-1,2-dimethyl-4-pyridinone (Hdpp), 3-hydroxy-2-methyl-1-(2'-aminoethyl)-4-pyridinone(Hmapp·2HCl), and 3-hydroxy-2-methyl-l-hexyl-4-pyridinone (Hmhpp). The ring N-R in the pyridinones allows the preparation of a greater variety of pyridinones; the pyrones were limited in their variety and availability because the oxygen is only divalent.

There has been considerable research recently on several hydroxypyridinones as ligands for a variety of biologically important metal ions. Contributions from Kontoghiorghes on iron and plutonium,[41] and from Raymond who has been actively studying the related l-hydroxy-2-pyridinones[42-44] and 3-hydroxy-2-pyridinones[44] with the actinides and iron, have shown these species to be good ligands for various metal ions. Thus far we have prepared Al complexes of three of the pyridinones as shown in Figure 5.[45,46]

As with the pyrone complexes, the elemental analyses, and the infrared, proton NMR, and mass spectral data were all consistent with bidentate, monobasic ligation of Al^{3+} in *tris*(ligand) complexes of neutral charge. The variable pH ^{27}Al NMR data for $Al(mpp)_3$ show a wider window of stability to hydrolysis (pH 3.5 to 10) than that for $Al(ma)_3$ even though $Al(mpp)_3$ is less soluble than $Al(ma)_3$ and the spectra were recorded at lower concentration (0.035 M vs. 0.05 M in Figure 4). Analogous hydrolysis products ($[Al(mpp)_2(H_2O)_2]^+$, $[Al(mpp)(H_2O)_4]^{2+}$, $[Al(H_2O)_6]^{3+}$) appeared in the same order as a function of decreasing pH as they did in $Al(ma)_3$. The solubilities of the three compounds suggest that they might have interesting *in vivo* properties; the N-protonated ($Al(mpp)_3$) and N-methylated ($Al(dpp)_3$) complexes are both somewhat water soluble and the N-hexylated ($Al(mhpp)_3$) complex is quite lipophilic.

In $Al(dpp)_3$, the exchange between the meridional and facial arrangements of ligands can be studied by proton NMR because the exchange is less facile than in the pyrone

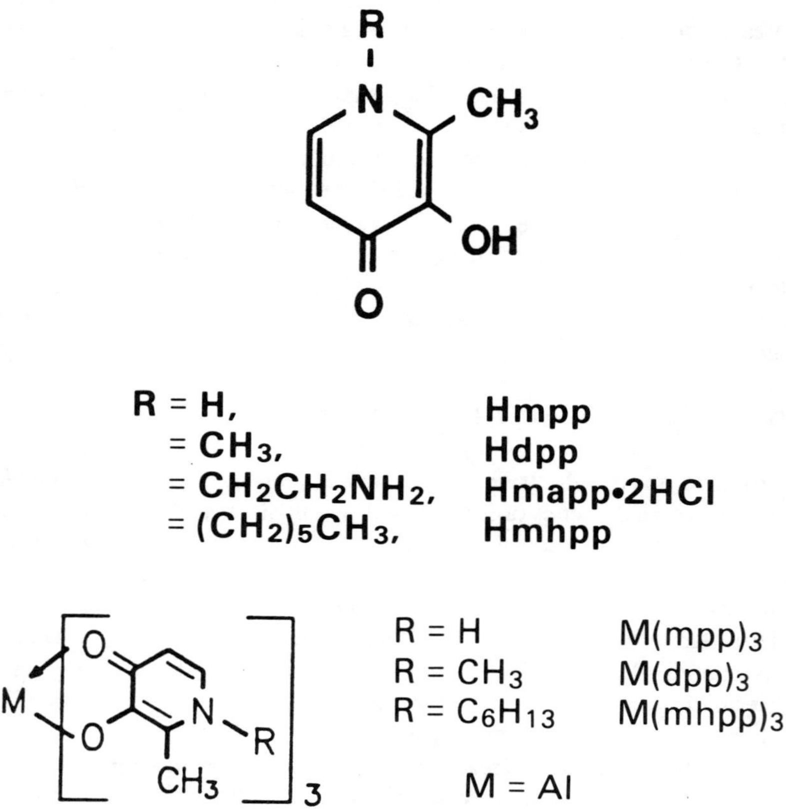

Figure 5. The hydroxypyridinone ligands and their complexes.

complexes. By cooling the sample, the exchange can be slowed to a measurable rate because its frequency is slowed relative to the chemical shift difference between the exchanging signals. The result is that rapidly averaging signals become resolved as can be seen in Figure 6. The exchange is monitored using the singlets in the NMR spectrum for the methyl group attached to the ring C atom (2.4 ppm in Figure 6).

When $Al(dpp)_3$ is crystallized from water, the facial isomer preferentially forms as the dodecahydrate $Al(dpp)_3 \cdot 12H_2O$. The contrast between this highly ordered compound and the disordered $Al(ma)_3$ is vivid. Figure 7 shows the results of an X-ray diffraction experiment with $Al(dpp)_3 \cdot 12H_2O$ (it is a stereoview which should be viewed with a pair of standard stereoglasses). The contents of the unit cell are shown in the view which is two cells deep. The structure consists of discrete facially oriented $Al(dpp)_3$ units (2 per unit cell), and water molecules (24 per unit cell), half of which form hexagonal channels in the corners of the unit cell and half of which form bridges between the channels and adjacent $Al(dpp)_3$ units. The impressive symmetry of this complex is obvious from Figure 7: the water channels are of hexagonal symmetry ($\bar{3}$

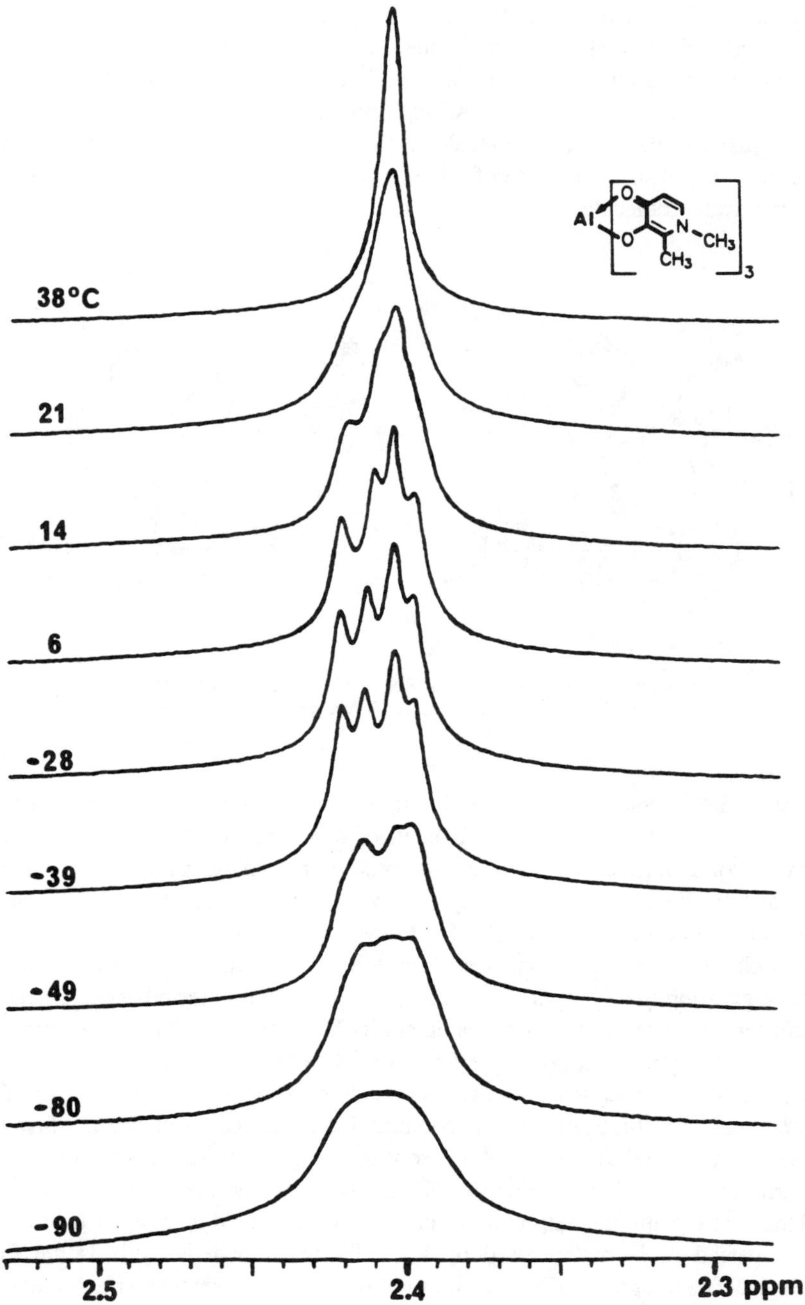

Figure 6. Variable temperature ^1H NMR $Al(dpp)_3$. Reprinted with permission from *Inorg. Chem.* 27:1045 (1988). Copyright 1988 American Chemical Society.

or S_6) while the $Al(dpp)_3$ units are of threefold symmetry (3 or C_3). These Al molecules are connected to the hexagonal water channels by a chain of two waters. One of these waters bridges from a ketone oxygen of a ligand on one $Al(dpp)_3$ to a hydroxyl oxygen of a ligand rotated by 120° on an $Al(dpp)_3$ one unit cell below. Each of these intermolecularly bridging water molecules is itself bridged to the hexagonal channels by another water. Thus, every one of the O atoms chelating an Al atom is hydrogen bonded to a water molecule.

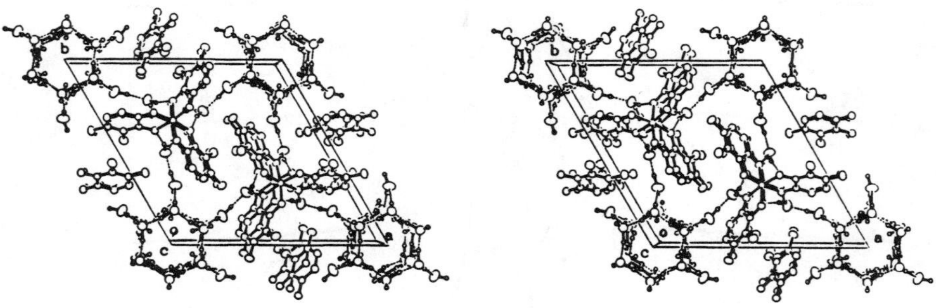

Figure 7. A stereoview of the exoclathrate. Reprinted with permission from *J. Am. Chem. Soc.* 109:4121 (1987). Copyright 1987 American Chemical Society.

The structure is unique not just in aluminum chemistry but also in the study of hydrogen bonding. These hexagonal channels of water are closely related to the ice I structure of solid water except that they are proton ordered herein and ice is not.[45-47] Within each six membered ring, the hydrogen bonding is circular and the circle is homodromic[48] because all the O-H···O hydrogen bonds run in the same counterclockwise direction when viewed down the *c* axis in Figure 7. This is consistent with the suggestion that homodromic circles should be the preferred arrangement in water clusters because of the lower resultant dipole moment.[48] This arrangement is crystallographically imposed by the perfect threefold axis coincident with the channel axis (*c*). Staggered columns of water molecules have been previously reported;[49] however, this is the first structure to contain single hexagonal channels of water molecules so closely related to the solid state structure of water. As a result of the rigid lattice structure in which the $Al(dpp)_3$ units are held outside the channels, we have termed this an exoclathrate, a structure wherein the guest is held outside the host water structure instead of being enclosed by it.[50] To our knowledge this is the first exoclathrate. The bridging water molecules hold everything together so rigidly that all the hydrogen atoms were located during the solution of the structure.[46] The analogous gallium complex $Ga(dpp)_3{\cdot}12H_2O$ also forms an exoclathrate when crystallized from aqueous solution.[46]

The importance of hydrogen bonding to the properties of these chemical compounds has manifested itself several times in our studies. The crystal structures of two of the 3-hydroxy-4-pyridinone ligands have been solved,[40] and both Hmpp and Hdpp form dimers in the solid state as a result of the formation of two O-H···O=C hydrogen bonds in each dimer unit. In addition, in Hmpp there is an available N-H which is also used for hydrogen bonding (N-H···O=C) to form a polymer with the result that all possible hydrogen bonding sites are involved in such interactions. The X-ray diffraction experiments on $Al(ma)_3$, $Al(dpp)_3$, Hmpp, and Hdpp show that hydrogen bonding is an extremely important (if not the determining) factor in the solid state structures of these species. Solution ^{27}Al NMR studies suggest that there may be a correlation between solid state structure, solution structure, and ^{27}Al NMR line-width, although this aspect of the work is still underway.

The overall formation constant (β_3) for the $Al(dpp)_3$ complex has been determined to be 10^{32}:[51]

$$Al^{3+} + 3L^- \xrightarrow{\beta_3} AlL_3$$

The complex is extremely stable, so much so that it is completely formed at a pH of 6 despite the fact that the pK for the deprotonation of Hdpp is 9.8. This high stability combined with the other properties listed previously suggests that this ligand, or variations on it, may have application in the treatment of Al-burdened patients. It has recently been found to be non-toxic in studies of its iron decorporation properties.[52]

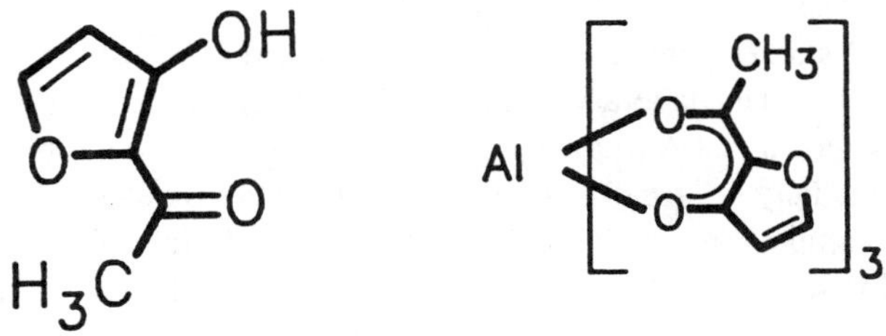

Figure 8. Isomaltol and its Al complex.

ISOMALTOL

We have recently prepared the *tris*(isomaltolato)aluminum(III) complex (Figure 8) as part of a larger project to examine the complexes of post transition metal ions with food additives, and possible metabolites of food additives.[53] Isomaltol (1-(3-hydroxy-2-furanyl)ethanone) is a structural isomer of maltol (Hma) and is a trace constituent of bread obtained by the reaction of starch with enzymes. It is prepared in a two-step procedure which involves the dehydration of the glucose moiety of lactose to form *O*-galactosylisomaltol which is then hydrolyzed to isomaltol.[54] The Al complex is somewhat water soluble although not nearly as soluble as $Al(ma)_3$. All the data are consistent with the formulation of the compound as another neutral *tris*(ligand)aluminum complex.

The ^{27}Al NMR spectrum (3 ppm; $W_{1/2} = 180$ Hz) is the only available datum relevant to the fac vs. mer question. When compared with the data available for our other AlL_3 complexes, it suggests that the complex is fac in solution because it has a lower $W_{1/2}$ than any other complex in our series. It presents a binding center more like a delocalized acetylacetonate than the localized hydroxyketones of which we have cited several examples. This additional factor contributes to the very low $W_{1/2}$. The delocalized symmetric center is likely to have a lower $W_{1/2}$ than the localized less symmetric center as a result of a lower electric field gradient at the Al nucleus.[55]

The most important feature of isomaltol as a ligand is the presence of the functionalizable ketone group adjacent to the metal ion which should provide a mechanism by which two or three of the molecules can be incorporated into a tetradentate or hexadentate chelating agent for possible Al removal *in vivo*. This is currently underway in our laboratories.

IN VIVO TOXICOLOGY

The unusual combination of water solubility, lipid solubility, and hydrolytic stability suggested that those complexes with all these properties should be examined for *in vivo* toxicity. In particular the *n*-octanol/water partition coefficients are a rough model of lipid solubility and brain capillary permeability when the molecular weight is low enough (below an upper limit of 400 to 657 dalton).[56]

A number of studies with several of these complexes are currently underway. A toxicological study in the laboratory of Dr. D.R. Crapper-McLachlan at the University of Toronto has shown that $Al(ma)_3$ and $Al(ka)_3$ are potent neurotoxins in rabbits when injected intracranially (13 μmol), and that $Al(ma)_3$ is 20 times more neurotoxic than the Al-lactate complex (only 1/20 of the dose must be administered to induce a lethal encephalopathy and to result in the same final concentration of Al in the brain.)[57] $Al(ka)_3$ is less neurotoxic than $Al(ma)_3$. Considering the hydrolytic stability (pH 4-9), water solubility (0.06 M), and partition coefficient (0.67) of $Al(ma)_3$, we concluded that the neutral water soluble complexes cross brain cell walls more easily than the charged Al-lactate complex, and the $Al(ma)_3$ is more effective than the less lipid extractable $Al(ka)_3$. Neutral molecules of low enough molecular weight can cross cell membranes via a passive diffusion mechanism whereas a carrier-mediated mechanism is necessary for molecules which are charged. It has also been determined that $Al(ma)_3$ is very active in tissue Al uptake studies with different tissues from a number of different animals.[57]

Professor Dan Perl has found $Al(ma)_3$ to be very mobile in the olfactory system of rabbits when administered intranasally.[58] He has found that the complex produces Al lesions in many parts of the olfactory system.

The chemistry described in this chapter is contributing to new areas of aluminum research as a result of the *in vivo* mobility imparted to Al in complexes of the type which are described, particularly $Al(ma)_3$. A direct result is that $Al(ma)_3$ is now becoming widely used in the study of aluminum neurotoxicity and aluminum mobility.[59] Good chelating agents for *in vivo* Al is the other aim of the work, and clearly the 3-hydroxy-4-pyridinones, with their high stabilities and variable functionalities on the ring nitrogen atom, will provide a useful building block.

Maltol is a natural product, easily obtained by the alkaline hydrolysis of streptomycin,[60] and it is a commonly used additive in breads, cakes, malted beverages, and chocolate milk. We note this because of the Swedish results previously cited wherein the coadministration of Al salts and citric acid led to *in vivo* Al accumulation.[8,9] The coadministration of toxic metals and good ligands for them may be a situation to eschew.

ACKNOWLEDGMENTS

We are extremely grateful to Drs. Donald McLachlan and Steven Rettig for fruitful collaborations on the toxicology and the X-ray crystallography, respectively. We also thank Dr. David Clevette, Martin Finnegan, Tim Karpishin, and Alexis Smith for experimental assistance, plus the Natural Sciences and Engineering Research Council (Canada) and the University of British Columbia for financial support.

REFERENCES

1. Perl, D.P. and A.R. Brody. "Alzheimer's Disease: X-ray spectrometric evidence of aluminum accumulation in neurofibrillary tangle-bearing neurons," *Science* 208:297-299 (1980).

2. Candy, J.M., J. Klinowski, R.H. Perry, E.K. Perry, A. Fairbairn, A.E. Oakley, T.A. Carpenter, J.R. Atack, G. Blessed, and J.A. Edwardson. "Aluminosilicates and senile plaque formation in Alzheimer's Disease," *The Lancet* i:354-357 (1986).

3. Crapper-McLachlan, D.R. and B.J. Farnell. "Aluminum and neuronal degradation," in *Metal Ions in Neurology and Psychiatry*, S. Gabay, J. Harris, and B.T. Ho, Eds. (New York: Alan R. Liss, Inc., 1985), pp. 69-87.

4. Sorensen, J.R.J., I.R. Campbell, L.B. Tepper, and R.D. Lingg. "Aluminum in the environment and human health," *Environ. Health Perspect.* 8:3-95 (1974).

5. Ganrot, P.O. "Metabolism and possible health effects of aluminum," *Environ. Health Perspect.* 65:363-441 (1986).

6. Lione, A. "The prophylactic reduction of aluminum intake," *Food Chem. Tox.* 21:103-109 (1983).

7. Lione, A. "Aluminum toxicology and the aluminum-containing medications," *Pharmac. Ther.* 29:255-285 (1985).

8. Slanina, P., Y. Falkeborn, W. Frech, and A. Cedergren. "Aluminum concentrations in the brain and bone of rats fed citric acid, aluminum citrate or aluminum hydroxide," *Food Chem. Tox.* 22:391-397 (1984).

9. Slanina, P., W. Frech, L.-G. Ekstrom, L. Loof, S. Slorach, and A. Cedergren. "Dietary citric acid enhances absorption of aluminum in antacids," *Clin. Chem.* 32:539-541 (1986).

10. Baes, C.F. Jr., and R.E. Mesmer. *The Hydrolysis of Cations*, (New York: John Wiley and Sons, Inc., 1976), pp. 112-123.

11. Martell, A.E., and R.M. Smith. *Critical Stability Constants, Volumes 1-5.* (New York: Plenum Press, 1974-1982).

12. Motekaitis, R.J., and A.E. Martell. "Complexes of aluminum (III) with hydroxy carboxylic acids," *Inorg. Chem.* 23:18-23 (1984).

13. Martin, R.B. "The chemistry of aluminum as related to biology and medicine," *Clin. Chem.* 32:1797-1806 (1986).

14. Martin, R.B. "Citrate binding of Al^{3+} and Fe^{3+}," *J. Inorg. Biochem.* 28:181-187 (1986) and references contained therein.

15. Karlik, S.J., E. Tarien, G.A. Elgavish, and G.L. Eichhorn. "Aluminum-27 nuclear magnetic resonance study of aluminum(III) interactions with carboxylate ligands," *Inorg. Chem.* 22:525-529 (1983).

16. Öhman, L.-O., and S. Sjöberg. "Equilibrium and structural studies of silicon (IV) and aluminum(III) in aqueous solution. Part 9. A potentiometric study of mono- and polynuclear aluminum(III) citrates," *J. Chem. Soc. Dalton Trans.* 2513-2517 (1983).

17. Wiley, J.W., G.N. Tyson Jr., and J.S. Steller. "The configuration of complex kojates formed with some transition elements as determined by magnetic susceptibility measurements," *J. Am. Chem. Soc.* 64:963-964 (1942).

18. Bryant, B.E. and W.C. Fernelius. "Some metal complexes of kojic acid," *J. Am. Chem. Soc.* 76:5351-5352 (1954).

19. Okac, A. and Z. Kolarik. "Potentiometrische untersuchung von komplexsalzen der kojisaure in wassrigen losungen," *Collect. Czech. Chem. Commun.* 24:266-272 (1959).

20. Morita, H., S. Shimomura, and S. Kawaguchi. "The syntheses and properties of bis(3-hydroxy-2-methyl-4-pyronato) complexes of bivalent metal ions," *Bull. Chem. Soc. Japa n* 49:2461-2464 (1976).

21. Gerard, C. and R.P. Hugel. "Thermodynamic stabilities of maltol and kojic acid complexes with divalent transition metals: Manganese, cobalt, nickel, and zinc," *J. Chem. Res. (S)* 404-405 (1978).

22. Gerard, C. "Etudes des complexes neutres de l'acide kojique et du maltol avec les cations divalents: Mn, Co, Ni, Cu, et Zn," *Bull. Chem. Soc. Fr.* I:451-456 (1979).

23. Gerard, C. and R.P. Hugel. "Complexes du cuivre(II) avec le maltol (hydroxy-3 methyl-2 pyrone-4)," *Compt. Rendu Acad. Sci. Paris Serie II* 295:175-177 (1982).

24. Musante, C. "Sui composti dell'acido kojico con alcuni metalli," *Gazz. Chim. Ital.* 79:679-683 (1949).

25. Sommer, L. and A. Losmanova. "Eisen(III)-chelate der kojisaure und mekonsaure," *Collect. Czech. Chem. Commun.* 26:2781-2805 (1961).

26. Yoneda, H., G.R. Choppin, J.L. Bear, and J.V. Quagliano. "Complexes of the lanthanide elements with kojate and acetylacetonate ions," *Inorg. Chem.* 3:1642-1643 (1964).

27. Chiacchierini, E. and M. Bartusek. "Complexes of maltol with uranyl and aluminum," *Collect. Czech. Chem. Commun.* 34:530-536 (1969).

28. Tuck, D.G. and M.K. Yang. "Coordination compounds of indium. Part XI. The stepwise one-electron reduction of some neutral indium (III) complexes," *J. Chem. Soc. (A)* 3100-3102 (1971).

29. Foye, W.O. and J.-R. Lo. "Metal-binding abilities of antibacterial heterocyclic thiones," *J. Pharm. Sci.* 61:1209-1212 (1972).

30. Agarwal, R.C., S.P. Gupta, and D.K. Rastogi. "Complexing behaviour of kojic acid - studies on its rare earth metal complexes," *J. Inorg. Nucl. Chem.* 36:208-211 (1974).

31. Dutt, N.K., and U.U.M. Sarma. "Chemistry of lanthanons-XLI. Isolation and characterisation of tris chelates of lanthanides with maltol, kojic acid and chloro-kojic acid," *J. Inorg. Nucl. Chem.* 37:1801-1802 (1975).

32. Morita, H., Y. Hayashi, S. Shimomura, and S. Kawaguchi. "Synthesis and properties of tris-(3-hydroxy-2-methyl-4-pyronato) complexes of chromium(III), manganese(III), and iron(III)," *Chem. Lett.* 339-342 (1975).

33. Gerard, C. and R.P. Hugel. "Iron (III) complexes of maltol (3-Hydroxy-2-methyl-4-pyrone), including hydroxo-complexes, in an acidic medium," *J. Chem. Res. (S)* 314 (1980).

34. Anderson, B.F., D.A. Buckingham, G.B. Robertson, and J. Webb. "Tris (3hydroxy-4H-pyran-4-onato) iron(III), $C_{15}N_9FeO_9$," *Acta Cryst.* C39:723-725 (1983).

35. Finnegan, M.M., S.J. Rettig, and C. Orvig. "A neutral water-soluble aluminum complex of neurological interest," *J. Am. Chem. Soc.* 108:50335035 (1986).

36. Finnegan, M.M., T.G. Lutz, W.O. Nelson, A. Smith, and C. Orvig. "Neutral water-soluble post-transition-metal chelate complexes of medical interest: aluminum and gallium tris (3-Hydroxy-4-Pyronates)," *Inorg. Chem.* 26:21712176 (1987).

37. Taube, H. "Rates and mechanisms of substitution in inorganic complexes in solution," *Chem. Rev.* 50:69-126 (1952).

38. Orvig, C. "A simple method to perform a liquid diffusion crystallization," *J. Chem. Ed.* 62:84 (1985).

39. Delpuech, J.J. "Aluminum-27," in *NMR of Newly Accessible Nuclei, Volume 2, Chemically and Biochemically Important Elements*, P. Laszlo, Ed. (New York: Academic Press, Inc., 1983), pp. 153-195.

40. Nelson, W.O., T.B. Karpishin, S.J. Rettig, and C. Orvig. "Physical and structural studies of N-substituted-3-hydroxy-2-methyl-4(1H)-pyridinones," *Can. J. Chem.* 66:123-131 (1988).

41. Kontoghiorghes, G.J. "Structure/iron binding activity of 1-hydroxypyrid-2-one chelators intended for clinical use," *Inorg. Chem. Acta* 135:145-150 (1987) and references contained therein.

42. Riley, P.E., K. Abu-Dari, and K.N. Raymond. "Specific sequestering agents for the actinides. 9. Synthesis of metal complexes of 1-hydroxy-2-pyridinone and the crystal structure of tetrakis (1-oxy-2-pyridonato)aquothorium(IV) dihydrate," *Inorg. Chem.* 22:3940-3944 (1983).

43. Scarrow, R.C., D.L. White, and K.N. Raymond. "Ferric ion sequestering agents. 14. 1-hydroxy-2(1*H*)-pyridinone complexes: Properties and structure of a novel Fe-Fe dimer," *J. Am. Chem. Soc.* 107:6540-6546 (1985).

44. Scarrow, R.C., P.E. Riley, K. Abu-Dari, D.L. White, and K.N. Raymond. "Ferric ion sequestering agents. 13. Synthesis, structures, and thermodynamics of complexation of cobalt(III) and iron(III) tris complexes of several chelating hydroxypyridinones," *Inorg. Chem.* 24:954-967 (1985).

45. Nelson, W.O., S.J. Rettig, and C. Orvig. "The exoclathrate $Al(C_7H_8NO_2)_3 \cdot 12H_2O$. A facial geometry imposed by extensive hydrogen bonding with the ice I structure," *J. Am. Chem. Soc.* 109:4121-4123 (1987).

46. Nelson, W.O, T.B. Karpishin, S.J. Rettig, and C. Orvig. "Aluminum and gallium compounds of 3-hydroxy-4-pyridinones: Synthesis, characterization, and crystallography of biologically active complexes with unusual hydrogen bonding," *Inorg. Chem.* 27:1045-1051 (1988).

47. Peterson, D.W. and H.A. Levy. "A single-crystal neutron diffraction study of heavy ice," *Acta Crystallogr.* 10:70-76 (1957).

48. Saenger, W. "Circular hydrogen bonds," *Nature* 279:343-344 (1979): *Nature* 280:848 (1980).

49. Mak, T.C.W. "Hexamethylenetetramine hexahydrate: A new type of clathrate hydrate," *J. Chem. Phys.* 43:2799-2805 (1965).

50. Jeffrey, G.A. and R.K.W. McMullan. "The clathrate hydrates," *Prog. Inorg. Chem.* 8:43-108 (1967).

51. Clevette, D. J., W. O. Nelson, A. Nordin, C. Orvig, and S. Sjöberg, submitted for publication.

52. Kontoghiorghes, G.J., M.A. Aldouri, A.V. Hoffbrand, J. Barr, B. Wonke, T. Kourouclaris, L. Sheppard. "Effective chelation of iron in β-thalassemia with the oral chelator 1,2-dimethyl-3-hydroxypyrid-4-one," *Brit. Med. J.* 295:1509-1512 (1987).

53. Lutz, T.G., D.J. Clevette, S. J. Rettig, and C. Orvig. *Inorg. Chem.* in press (1989).

54. Hodge, J.E. and E.C. Nelson. "Preparation and properties of galactosylisomaltol and isomaltol," *Cereal Chem.* 38:207-221 (1961).

55. Akitt, J.W. and W.S. McDonald. "Arrangements of ligands giving low electric field gradients," *J. Magn. Reson.* 58:401-412 (1984).

56. Levin, V.A. "Relationship of octanol/water partition coefficient and molecular weight to rat brain capillary permeability," *J. Med. Chem.* 23:682-684 (1980).

57. McLachlan, D.R., and C. Orvig. Manuscript in preparation (1988).

58. Perl, D.P. Personal communication (1987).

59. Savory, J. Personal communication (1986). Roberts, E. Personal communication (1986).

60. Schenck, J.R. and M.A. Spielman. "The formation of maltol by the degradation of streptomycin," *J. Am. Chem. Soc.* 67:2276-2277 (1945).

CHAPTER 16

NEUROTOXICITY OF ALUMINUM

Judith K. Marquis

INTRODUCTION

Concern for both the biological and environmental effects of aluminum has increased in recent years, partly due to its putative association with dementing disorders and partly due to evidence of mobilization of the metal by acid precipitation. The following list presents a summary of the major sources of aluminum that may contribute to total human exposure to the element.

- Major sources of aluminum for human exposure
  1. Drinking water
  2. Residues in foods
  3. Cooking utensils
  4. Food and beverage packaging
  5. Antacid formulations
  6. Antiperspirant formulations
  7. Acidic leaching into groundwater

Aluminum (Al) has been implicated as a neurotoxic agent in a number of laboratory and epidemiological studies. Elevated Al concentrations have been related to impaired motor function and to a number of cognitive deficits in both humans and experimental animals. Neurofibrillary degeneration has been observed in brain regions that exhibit elevated Al levels. In addition, dementia has been correlated to elevated brain Al levels in Alzheimer's disease, as well as in patients who are undergoing chronic dialysis therapy and who ingest significant amounts of Al hydroxide to balance phosphate metabolism. Aluminum has also been implicated as a causative agent in the parkinsonian dementia of Guam. (For a recent, diversified review of Al toxicity, particularly with regard to a possible etiological association with Alzheimer's disease, the reader is encouraged to consult the December 1986 issue of the journal

Neurobiology of Aging.[1] Some of the pathological conditions that have been associated with Al exposure in humans are listed below.

- Pathological conditions potentially associated with aluminum exposure in humans
  1. Dialysis encephalopathy
  2. Parkinsonian dementia of Guam
  3. Alzheimer's disease
  4. Osteomalacia

The apparent association of Al exposure with neurotoxicity requires careful scrutiny in order to determine whether Al is indeed a priority for greater level of concern by regulatory agencies, as well as a priority for further biological studies and, ultimately, more precise assessment of the risks that may be associated with exposure to Al in the environment. Aluminum is the third most abundant element of the earth and has been "generally regarded as safe" in the regulation of chemicals. The ubiquity of Al and the lack of evidence for a specific molecular mechanism of toxicity further confound the problem. In the following paragraphs, I will review evidence that the nervous system may be a major "target organ" for Al toxicity, and I will consider as well evidence that selected binding sites in nerve cells are sensitive to extremely low amounts of Al in their immediate environment.

TRANSPORT OF ALUMINUM ACROSS BIOLOGICAL MEMBRANES

One of the principal questions that must be resolved in order to understand the environmental toxicity of Al is certainly the "bioavailability" of the metal in its many different chemical forms. Naturally occurring Al exists in chemical forms that are mostly insoluble. At all but very acidic conditions, in aqueous media, Al forms complex insoluble polymers that may exhibit very low bioavailability. The possibility that some of the Al polymers are lipoid and can diffuse across cell membranes sets the scene for uptake and storage of Al complexes in the intracellular matrix, a phenomenon that may be accompanied by a myriad of biochemical effects on cellular function. It also suggests that particular subgroups of sensitive or susceptible individuals may be identified in a randomly exposed population. These include individuals whose diet may enhance the formation of fat-soluble Al complexes.[1]

It has been shown that Al administered intraperitoneally can increase the permeability of the blood-brain barrier of rats to small peptides.[2] Thus, it is possible that Al transport into the CNS is not only cumulative but increases with increasing exposure to the metal. Nagyvary and Bradbury[3] demonstrated hypocholesterolemic effects of Al complexes in rats fed dietary Al alginate and suggested the formation of Al-lipid complexes. Slanina et al.[4] described a number of dietary factors, including sources of citric acid, that can produce a significant increase of Al uptake into many different organs in rats and increase Al retention in both brain and bone of exposed rats. These and other factors that increase the bioavailability of Al require more careful study.

As with most xenobiotics, the major sites of action for adverse effects of Al on neuronal function include the synthesis, uptake, transport, and metabolism of neurotransmitter molecules. Studies of aluminum effects on neurochemical markers include *in vivo* exposures in laboratory animals, *in vitro* biochemical assays, and *in vitro* cell culture assays.

- Factors that influence the bioavailability of aluminum
  1. pH - solubility
  2. Counter ions and buffers
  3. Dietary factors, e.g., citrate
  4. Formation of Al-lipid complexes
  5. Damage to membrane barriers
  6. Multiple chemical states of Al salts
  7. Route of exposure and vehicle

IN VIVO STUDIES OF ALUMINUM NEUROTOXICITY

Studies of Al effects on neurotransmitters and their associated enzymes in exposed animals often produce a wide spectrum of results. Many factors enter into this variability, not the least of which are the multiple chemical states that can be assumed by hydrated Al salts.[5,6] In addition, in reviewing available data from animal studies, dose, duration of exposure, route of exposure, method of presentation (e.g., gavage, dry feed, or drinking water), buffering conditions, pH, counter ions, and many other parameters must be compared.

Despite such experimental limitations and variables, there is considerable evidence for *in vivo* alterations by Al ions of the synthetic and metabolic enzymes of central neurotransmission. Most investigators have approached their studies from the perspective of associating Al exposure with the etiology of senile dementia of the Alzheimer type (SDAT), a neuropathological disorder that is often accompanied by evidence of Al deposits in postmortem samples of brain tissue. Although SDAT is a disorder involving multiple neurochemical pathways, the presence of dementia and the densities of senile plaques and neurofibrillary tangles are best correlated with deficiencies of cholinergic function. In addition, both human and animal studies have implicated cholinergic dysfunction in memory deficits. Thus, studies of Al effects on neurotransmitters have focused largely on Al and cholinergic function.

In 1980, Yates and co-workers[7] reported decreased levels of both choline acetyltransferase (ChAT) and acetylcholinesterase (AChE) activity in brain tissue from rabbits with neurofibrillary degeneration induced by intracisternal injection of Al chloride. In brain areas devoid of pathology, no change was found in cholinergic enzyme activity. More recently, a similar study in rabbits, again utilizing intracisternal injection of Al chloride, exhibited no significant neurochemical alterations in the grey matter of the spinal cord.[8] The latter results were also consistent with earlier findings reported by Hetnarski et al.[9] On the other hand, Kosik and co-workers[10] injected Al chloride directly into the spinal cord of rabbits and observed a 39% decrease in ChAT activity. Bates and Boegman[11] also reported a decrease in spinal cord ventral root and sciatic nerve ChAT activity following intracranial injection of Al chloride.

- *In Vivo* data on aluminum neurotoxicity

1. Intraventricular injections:

   - Altered neurotransmitter uptake in isolated synaptosomes

   - Decreased AChE and ChAT activity in CNS (brain and spinal cord) correlated with areas of neurofibrillary degeneration

2. Rat feeding studies

   - Decreased dopamine levels in brain

   - Altered dopamine-β-hydroxylase and phenyl-N-methyl-transferase activity in brain

   - Decreased AChE in brain

   - Effects are dose-related and vary with trace metal content of diet

Other central neurotransmitters are affected by Al exposure. In a rat feeding study, Wenk and Stemmer[12] examined the effects of dietary Al on central catecholamine levels and found that the Al effects depend in a complex fashion on the dietary intake of other trace metal ions, including Cu, Zn, Fe, and Mg. In rats fed 0.1% Al acetate in the dry feed, it was shown that Al decreased the dopamine content of the cortex in rats on a low Cu-low Zn diet. It was also found that Al could restore normal levels of dopamine β-hydroxylase and phenyl-N-methyl transferase activity in the cortex, hippocampus, and cerebellum of Cu- and Zn-deficient rats.

Our own work on Al neurotoxicity began with a limited dietary study in rats utilizing Al chlorohydrate in citrate-buffered drinking water, pH 6.0, and 28 days exposure in adult rats.[13] This study demonstrated an apparent increase in AChE activity in whole brain homogenates from rats drinking 0.12% Al chlorohydrate, while no change in the baseline AChE activity was measured in the rats drinking 0.06% Al chlorohydrate.

By contrast, rat pups exposed for 8, 15, or 22 days post-weaning, under conditions similar to the adult study, exhibited decreased brain AChE activity. This effect was correlated with dose of Al in the 8-day rats but could not be correlated with measured levels of Al in the brains. The relationship between activation and inhibition of AChE activity following *in vivo* exposure to Al remains to be defined, and the possibility that a biochemical tolerance may develop with continuous dietary intake from birth to senescence should be investigated.

IN VITRO STUDIES OF ALUMINUM NEUROTOXICITY

Many of the effects observed in studies of Al toxicity *in vitro* are seen only at exceptionally high levels of Al and may be considered irrelevant in the context of expected whole body exposures. The findings of these studies are nonetheless important to achieve an understanding of the toxicology of Al, as observed adverse effects can then be followed up in studies with decreasing concentrations to establish a physiologically relevant dose-response relationship.

Numerous kinetic studies have demonstrated effects of Al on the uptake of neurotransmitters in isolated cell preparations. Choline uptake into rat brain synaptosomes was shown by Lai and coworkers[14] to be inhibited by Al chloride although at relatively high concentrations (about 0.1-m*M*). These same investigators demonstrated that synaptosomal dopamine uptake is stimulated by 0.1-m*M* Al chloride and inhibited by a four-fold increase in Al concentration. They also found that the synaptosomal uptake of norepinephrine, serotonin, GABA, and glutamate, but not of phenylalanine, was inhibited by Al chloride. It is important to note, however, that all of these studies were carried out in 1-m*M* Na_2PO_4 buffer under conditions that are likely to produce complexation and polymerization of Al in the assay medium.

Sturman et al.[15] correlated a reduction in high affinity GABA and glycine uptake in synaptosomes from rabbits with Al-induced encephalopathy. They also measured an association between compromised amino acid uptake and the number of neurons exhibiting neurofibrillary changes.

Other effects of Al on neurotransmitter function include direct effects on the synapse and the neuronal membrane. For example, Wong and Lim[16] found that Al can stimulate phospholipid methylation in rat brain synaptosomes, but these effects were observed only at millimolar concentrations of Al in the assay medium. Johnson and Jope[17] have recently carried out several elegant studies of Al effects on rat brain synaptosomes. They demonstrated that Al inhibits high affinity choline transport with an IC50 of about 0.5-m*M*. Such relatively high concentrations of Al chloride were also shown to depress glucose utilization in the synaptosomes and to inhibit carbachol-stimulated hydrolysis of phospholipids in rat brain slices. More recently, Koenig and Jope[18] demonstrated that Al inhibits calcium influx into synaptosomes, and they suggested a specific mechanism for Al neurotoxicity as well as Al transport via interaction with voltage-dependent calcium channels in the neuronal membrane.

Direct effects of Al on the physical properties of membrane lipids have also been demonstrated by Vierstra and Haug.[19] In experiments with soil microorganisms, they found that Al produced a dramatic decrease in membrane lipid fluidity and lowered the lipid phase transition temperature of the organisms. Aluminum also alters the physical state of phosphatidylserine-containing membranes and interacts directly with phospholipid membrane constituents.[20]

Studies on cholinergic markers in cultured neurons indicate a high level of sensitivity to the effects of Al. Miller and Levine[21] treated neuroblastoma cells in culture with varying concentrations of Al chloride and demonstrated that the AChE activity of these cells was inhibited by 10-50 μM concentrations of the salt.

- *In Vitro* Data on Aluminum Neurotoxicity

1. Rat brain synaptosomes:
   - Altered dopamine uptake
   - Decreased uptake of choline, 5-HT, GABA
   - Decreased glucose utilization
   - Decrease Ca^{2+} influx; interaction with Ca^{2+} channels
2. Isolated membrane structure:
   - Altered membrane fluidity
   - Decreased lipid phase transition temperature
   - Altered phospholipid constituents in membranes
3. Neurons in culture:
   - Decreased AChE activity in neuroblastoma
4. Acetycholinesterase kinetics:
   - Noncompetitive inhibition (> 10 μM Al)
   - Displaces Ca^{2+} at high concentrations
   - Ca^{2+} independent enzyme activation at 1-5 μM Al

ALUMINUM EFFECTS ON ACETYLCHOLINESTERASE ACTIVITY

In my own laboratory, we have characterized *in vitro* the kinetics of Al interaction with purified AChE. While these studies have provided mixed results, some important concepts have surfaced. Acetylcholinesterase is well understood as the enzyme that functions to terminate the transmitter action of ACh. Recent biochemical studies indicate that AChE is a heterogeneous protein that exists in multiple molecular forms.

These forms are distributed selectively in the presynaptic and postsynaptic compartments and are selectively inhibited in SDAT.[22,23]

Acetylcholinesterase also contains a complex array of regulatory binding sites spatially removed from the enzyme active site. These are sometimes called peripheral or "extracatalytic" sites. Marquis and Black[24] demonstrated a biphasic effect of Al on AChE purified from bovine caudate nucleus, i.e., activation at 1-5-m*M* Al chlorohydrate and noncompetitive inhibition of enzyme activity at higher concentrations. We had already reported such effects in eel electric cell enzyme[25] and, much earlier, Patocka[26] reported *in vitro* activation of erythrocyte AChE by Al chloride.

Clearly, there are at least two populations of peripheral binding sites for Al on this enzyme, each with a different binding affinity. This is not an unusual phenomenon associated with metal ion-protein interaction. In fact, calcium also exhibits a biphasic, concentration-dependent effect on AChE activity.[27] Calcium effects are only seen at concentrations three orders of magnitude greater than the range of effective Al concentrations, and it has been suggested that calcium binds to different peripheral anionic sites than Al.

The modification of neuronal enzyme activity by Al is not entirely a nonspecific phenomenon. We have not been able to demonstrate any effects of micromolar concentrations of Al chlorohydrate on either choline acetyltransferase or Na, K-ATPase activity in rat brain homogenates. Furthermore, we have recently demonstrated that pretreatment of purified bovine caudate AChE by carboxyl-group modifying carbodiimide compounds totally abolishes the "peripheral" activation and inhibition of enzyme activity by Al chlorohydrate with little or no effect on the functional integrity of the active site.[28]

Inhibition of AChE by Al is an experimental finding that does not readily reconcile the theory that Al may be a causative agent in SDAT with the highly consistent neurochemical analyses that have led to the cholinergic theory of SDAT. In fact, one of the more promising approaches for restoring the observed loss of brain ACh in SDAT has been the use of powerful AChE inhibitors such as physostigmine[29] and tetrahydroaminoacridine.[30] The activation of AChE by very low Al concentrations may be more meaningful in this particular context.

Finally, it is important to consider the possible pharmacodynamic implications of noncompetitive alterations in AChE kinetics. Clearly, noncompetitive inhibition of AChE by Al may result in prolonged stimulation of the postsynaptic cell and the likelihood of receptor desensitization. Such a phenomenon has never been investigated. In fact, very few of the proposed effects of Al on neurotransmitter function have been studied in sufficient detail to provide an adequate working biochemical model for Al neurotoxicity.

ACKNOWLEDGMENTS

This work was supported in part by grants from the Alzheimer's Disease and Related Disorders Association, Inc., and the Center for Brain Sciences and Metabolism Charitable Trust, Cambridge, Massachusetts.

REFERENCES

1. Crapper McLachlan, D.R. "Aluminum and Alzheimer's Disease," *Neurobiol. Aging* 7:525-532 (1986).

2. Banks, W.A. and A.J. Kastin. "Aluminium increases permeability of the blood-brain barrier to labelled DSIP and β-endorphin: Possible implications for senile and dialysis dementia," *The Lancet* ii:1227-1229.

3. Nagyvary, J. and E.L., Bradbury. "Hypocholesterolemic effects of Al^{3+} complexes," *Biochem. Biophys. Res. Commun.* 77:592-598 (1977).

4. Slanina, P., W. Frech, A. Bernhardson, A. Cedergren, and P. Mattsson. "Influence of dietary factors on aluminium absorption and retention in the brain and bone of rats," *Acta Pharmacol. Toxicol.* 56:331-336 (1985).

5. Burrows, W.D. "Aquatic aluminum: Chemistry, toxicology, and environmental prevalence," *CRC Crit Rev. Environ. Cont.* 7:167-216 (1977).

6. Driscoll, C.T., J.P. Baker, J.J. Bisogni and C.L. Schofield. "Effect of aluminium speciation on fish in dilute acidified waters," *Nature* 284:161-164 (1980).

7. Yates, C.M., J. Simpson, D. Russell and A. Gordon. "Cholinergic enzymes in neurofibrillary degeneration produced by aluminium," *Brain Res.* 197:269-274 (1980).

8. Simpson, J., C.M. Yates, D.K. Whyler, H. Wilson, A.J. Dewar, and A. Gordon. "Biochemical studies on rabbits with aluminium induced neurofilament accumulations," *Neurochem. Res.* 10:229-238 (1985).

9. Hetnarski, B., H.M. Wisniewski, K. Iqbal, J.D. Dziedzic and A. Lajtha. "Central cholinergic activity in aluminium-induced neurofibrillary degeneration," *Annals of Neurol.* 7:489-490.

10. Kosik, K.S., W.G. Bradley, P.F. Wood, C.G. Rasool and D.J. Selkoe. "Cholinergic function in lumbar aluminum myelopathy," *J. Neuropathol. Exp. Neurol.* 42:365-375 (1983).

11. Bates, L.A. and R.J. Boegman. "The functional implications of experimentally produced neurofibrillary tangles on axonal transport," *Can. Fed. Biol. Soc.* 26:114 (1983).

12. Wenk, G.L. and K.L. Stemmer. "The influence of ingested aluminum upon norepinephrine and dopamine levels in the rat brain," *Neurotoxicology* 2:347-353 (1981).

13. Marquis, J.K. "Aluminum neurotoxicity: An experimental perspective," *Bull. Environm. Contam. Toxicol.* 29:43-49 (1982).

14. Lai, J.C.K., L. Lim and A. N. Davison. "Effects of Cd^{2+}, Mn^{2+}, and Al^{3+} on rat brain synaptosomal uptake of noradrenaline and serotonin," *J. Inorg. Biochem.* 17:215-225 (1982).

15. Sturman, J.A., H. M. Wisniewski and J. W. Shek. "High affinity uptake of GABA and glycine by rabbits with aluminum-induced neurofibrillary changes," *Neurochem. Res.* 8:1097-1109 (1983).

16. Wong, P.C.L. and L. Lim. "The effects of aluminum, manganese and cadmium chloride on the methylation of phospholipids in the rat liver synaptosomal membrane," *Biochem. Pharmacol.* 30:1704-1705 (1981).

17. Johnson, G.V.W. and R.S. Jope. "Aluminum impairs glucose utilization and cholinergic activity in rat brain *in vitro." Toxicology* 40:93-102 (1986).

18. Koenig, M.L. and R.S. Jope. "Aluminum inhibits the fast phase of voltage dependent calcium influx into synaptosomes." *J. Neurochem.* 49:316-320 (1987).

19. Vierstra, R. and A. Haug. "The effect of Al^{3+} on the physical properties of membrane lipids in *Thermoplasma acidophilus," Biochem. Biophys. Res. Commun.* 84:138-143 (1978).

20. DeLeers, M., J.-P. Servais and E. Wulfert. "Micromolar concentrations of Al^{3+} induce phase separation, aggregation and dye release in phosphatidylserine-containing lipid vesicles." *Biochim. Biophys. Acta* 813:195-200 (1985).

21. Miller, C.A. and E.M. Levine. "Effects of aluminum salts on cultured neuroblastoma cells," *J. Neurochem.* 22:751-758 (1976).

22. Marquis, J.K. "Presynaptic acetylcholinesterase," *Trends Pharmacol. Sci.* 6:387-388 (1985).

23. Fishman E.B., G.C. Siek, R.D. MacCallum, E.D. Bird, L. Volicer and J.K. Marquis. "Distribution of the molecular forms of acetylcholinesterase in human brain: Alterations in dementia of the Alzheimer type," *Ann. Neurol.* 19:246-252 (1986).

24. Marquis, J.K. and E.E. Black. "Aluminum activation and inactivation of bovine caudate acetylcholinesterase," *Bull. Environm. Contam. Toxicol.* 32:704-710 (1984).

25. Marquis, J.K. and A.J. Lerrick. "Noncompetitive inhibition by aluminum, scandium and yttrium of acetylcholinesterase from electrophorus electricus," *Biochem. Pharmacol.* 31:1437-1440 (1982).

26. Patocka, J. "The influence of aluminum on cholinesterase and acetylcholinesterase activity," *Acta Biol. Med. Germ.* 26:845-846 (1971).

27. Marquis, J.K. and G.D. Webb. "Interactions of calcium with purified and intact cell acetylcholinesterase of *Electrophorus electricus," Biochem. Pharmacol.* 23:3459-3465 (1974).

28. Marquis, J.K. and E.E. Black. "Activation and inactivation of bovine caudate acetylcholinesterase by trivalent cations," *Biochem. Pharmacol.* 34:533-538 (1985).

29. Thal, L.J., P.A. Fuld, D.M. Masur and N.S. Sharpless. "Oral physostigmine and lecithin improve memory in Alzheimer disease." *Ann. Neurol.* 13:491-496 (1983).

30. Summers, W.K., L.V. Majovski, G.M. Marsh, K. Tachiki and A. Kling. "Oral tetrahydroaminoacridine in long-term treatment of Senile Dementia, Alzheimer Type," *New Eng. J. Med.* 315:1241-1245 (1986).

CHAPTER 17

ALUMINUM NEUROTOXICITY: CRITERIA FOR ASSIGNING A ROLE IN ALZHEIMER'S DISEASE

D.R. Crapper McLachlan

INTRODUCTION

Alzheimer's disease is the most common cause of senile dementia and a major world health problem. The disease is relentlessly progressive, and no known intervention will prevent or arrest the disease. The etiological event of sporadic Alzheimer's disease remains unknown, but familial Alzheimer's disease in at least some families with early onset disease is linked to a mutation which lies on chromosome 21 between the gene for superoxide dismutase and the centromere, the 21q11.2-21q22.2 region.[1] The physiological role of the mutant gene remains to be defined as does the possible involvement of the familial Alzheimer gene in the much more common sporadic form of the disease.

There is no evidence at present that aluminum causes Alzheimer's disease. Rather, the evidence is consistent with the hypothesis that aluminum is involved in the pathogenesis of this disease, that is, aluminum acts as a secondary factor in contributing to neuronal dysfunction.

Alzheimer's disease is associated with several histopathological brain changes including neuron loss, neurons with neurofibrillary degeneration, senile plaques and amyloid deposits in blood vessel walls and the cores of senile plaques. The same pathological changes are seen with high frequency in the brains of aging but neurologically normal individuals, although the number and distribution of lesions is much less.[2] This observation has led some workers to consider Alzheimer's disease as accelerated brain aging. However, the observation that Down's syndrome, a condition associated with an extra chromosome 21, is a risk factor for Alzheimer's disease[3] strongly suggest that the increased expression of a normal gene, i.e., the extra gene dosage on chromosome 21, is sufficient to release the Alzheimer process. Both the familial Alzheimer cases and the sporadic cases are perfectly normal in neurological and cognitive function until later life, and this suggests that the central disorder in

Alzheimer's disease is an age-related alteration in the regulation of gene expression. There probably is no single "cause" for Alzheimer's disease: many factors endogenous and exogenous to the brain may alter gene expression and release the Alzheimer changes. Furthermore, aluminum may be only one factor involved in the pathogenesis of the disease.

CRITERIA FOR ASSIGNING A ROLE IN ALZHEIMER'S DISEASE

A role for aluminum in the pathogenes of Alzheimer's disease has been based on circumstantial evidence alone. The results of a cause and effect experiment which clearly demonstrates a toxic role in Alzheimer's disease have not yet been reported. A number of recent reviews examine aluminum neurotoxicity and the possible sites of aluminum involvement in Alzheimer's disease.[4,5,6] There is now no doubt that aluminum is present in abnormal amounts in the brain of individuals affected with Alzheimer's disease; the problem is to measure the precise role this common environmental agent plays in the sequence of cellular processes resulting in the disease. Alzheimer's disease occurs only in humans; no animal or *in vitro* cellular system satisfactorily models this common neurodegenerative process. Therefore, cause and effect experiments between any factor considered to be important in either the etiology or pathogenesis of the disease will be difficult to perform in the foreseeable future. The question then arises as to what criteria may be applied to evaluate whether aluminum has a significant involvement in Alzheimer's disease. We submit that at least three independent lines of experimental evidence must be obtained to satisfactorily evaluate the role of aluminum in Alzheimer's disease:

a) Identification in Alzheimer affected tissue of important biochemical events disturbed by aluminum, such as critical steps in energy metabolism, calcium homeostasis or gene expression, the demonstration that aluminum is at the target site in amounts sufficient to exert a toxic action and documentation of an identical biochemical effect in an animal or cellular model.

b) Epidemiological studies relating bioavailable aluminum to Alzheimer's disease.

c) Clinical evidence that the quality of life and survival of Alzheimer patients is improved by either removing aluminum from the environment or removing aluminum from the brain of affected individuals by chemotherapeutic means.

TOXIC BIOLOGICAL EFFECTS OF ALUMINUM

A very large number of toxic biological effects of aluminum have been documented. A partial list is given in Table 1. Each of these toxic manifestations will have to be examined in detail in order to evaluate the precise role of aluminum in Alzheimer's disease. For the purposes of this chapter, the effect of aluminum upon gene expression will be addressed at some length. The intent, however, is not to imply that other neurotoxic effects of aluminum are no less important.

Table 1. Summary of structural and biochemical effects of aluminum in biological systems

Effects	Reference
(A) NUCLEAR EFFECTS	
1. Binds to DNA phosphate and bases	7,8
2. Increases histone-DNA binding	9
3. Decreases RNA in neuroblastoma	10
4. Blocks initiation sites for RNA polymerase	11
5. Blocks RNA polymerase activity *in vitro*	12
6. Blocks ADP ribosylation	13
7. Alters sister chromatid exchange	14
8. Alters steroid-induced chromosome puffing	15
9. Inhibits corticosterone receptor binding to DNA	16
10. Alters poly(A) RNA content rabbit fore brain	17
11. Reduces mRNA coding for neurofilaments in rabbit anterior horn cells	18
12. Decreases cell division and alters DNA synthesis	19,20
(B) CYTOPLASMIC EFFECTS	
1. Induces conformational changes in calmodulin Blocks calmodulin-dependent Ca/Mg-ATPase which is important in the extrusion of Ca^{2+} from cells	21,22
2. Increases intracellular calcium content	23,24

Table 1. (Continued...)

	Effects	Reference
3.	Reduces sugar phosphorylation	25
4.	Decreases respiration	26
5.	Inhibits hexokinase; stabilizes terminal phosphoryl group on ATP	27,28
	Forms long lived complex with ATP	29
	Competitive inhibitor of ATP	30,31
	Inhibits brain glycolysis, depression of yeast, and rat brain cytosolic and mito chondral hexokinase activity	32
6.	Stimulates brain pyruvate kinase	32
7.	Enhances adenylate cyclase stimulation by fluoride but inhibits activation by serotonin and guanine nucleotides in *Fasciola hepatico*; a requirement for activation of the regulatory component of adenylate cyclase by fluoride	33,34
8.	Increases number of lysomes, reduces thiamine pyrophosphatase and nucleoside diphosphatase in the Golgi apparatus	35
9.	Inhibits synthesis of tetrahydrobiopterin	36
10.	Elevates cAMP and cGMP levels	37
11.	Increases ubiquitin response in neurites of cultured nervous tissue	38
12.	Binds to ferritin and is partially sequestered by this mechanism; may alter iron storage	39
(C) CYTOSKELETAL EFFECTS		
1.	Induces neurofibrillary degeneration composed of 10 nm fibres identical in composition to normal neurofilament	40,41, 42,43,44
2.	Alters slow axonal transport, although disputed; No effect upon anterograde transport	45,46,47

Table 1. (Continued.....)

	Effects	Reference
3.	Alters phosphorylation of cytoskeletal proteins, MAP2, and NF200KD Postulated to be secondary to effect on cAMP dependent protein kinase	48
4.	Promotes assembly of microtubules which are more slowly depolymerized than Mg assembled tubules	49,50
(D)	EFFECTS ON MEMBRANES AND MEMBRANE BOUND ENZYMES	
1.	Alters physical properties of membrane lipids	51
2.	Binds to both positive and negative charged sites in membranes, *in vitro*	52
3.	Alters membrane structure	53,54
4.	Alters adenylcyclase activity required for activation of regulatory component of adenylate cyclase *in vitro* by fluoride	33,34
5.	Inhibits at relatively high concentrations Na-K ATPase	55
6.	Enhances brain-specific lipid peroxidation Accelerates peroxidation of membrane lipids stimulated by iron salts Decreases activity of superoxide dismutase in rat brain	56,57
7.	Alters blood brain barrier (BBB) Non competitive inhibition of saturatable transport system for N-tyrosinated peptides and encephalin from brain Increases permeability of BBB to neuropeptides	58 59,60,61
(E)	SYNAPTIC AND NEUROTRANSMITTER EFFECTS	
1.	Alters dendritic shape and synaptic density in chronic preparations	62,63
2.	Blocks high affinity uptake of GABA and glutamate from synaptosomes	64,65
3.	Blocks synaptosome uptake of neurotransmitter amines, choline, dopamine, and noradrenaline	55,66
4.	Inhibits actylcholinesterase	10,67,68
5.	Blocks uptake of calcium and acetylcholine binding	69

Table 1. (Continued....)

	Effects	Reference
6.	Reduces glucose uptake by synaptosomes extracted from rat cortex	70
7.	Depresses norepinephrine and dopamine levels in cortex and activity of enzymes dopamine-β-hydroxylase and phenylethan-olamine-N-methyltransferase when fed to rats on copper, zinc and iron deficient diets	71,72
8.	Controversial reduction in choline acetyltransferase in rabbit hypoglossal nucleus and spinal cord grey matter	73,74,75
9.	Inhibits fast phase of voltage-dependent calcium influx into synaptosomes	76
(F) BLOOD		
1.	Alters activity of cholinesterase	77,78
2.	Interacts with transferrin	79

ALUMINUM AND GENE EXPRESSION

In experimental aluminum encephalopathies, aluminum accumulates rapidly upon DNA containing structures within the nucleus. The exact ligands for intranuclear aluminum are unknown, however, aluminum distribution in the nucleus is related to the state of compaction of chromatin: highly condensed heterochromatin binds 15-18 times more aluminum per unit of DNA than does light euchromatin.[80] Since light euchromatin is deficient in the linker histone H1 family of proteins, aluminum binding may, in part, involve histone and DNA ligands. Indeed, chemical evidence indicates that aluminum alters the binding of linker histones to DNA.

Increasing concentrations of sodium cations employed to electrostatically displace chromosomal proteins from DNA resulted in characteristic dissociation patterns of chromatin proteins when analyzed on polyacrylamide gels.[9] The protein salt elution profiles exhibited by Alzheimer-affected nuclei were reproduced in age-matched control human cerebral cortical nuclei, *in vitro*, when concentrations of aluminum sufficient to yield aluminum to DNA ratios in nuclei of 30 aluminum atoms per 200 base pairs were achieved. Age matched control human cerebral nuclei contain about 9 aluminum atoms per 200 base pairs in the heterochromatin fraction, whereas 30 aluminum atoms per 200 base pairs lies within 1 standard deviation of the mean for Alzheimer heterochromatin as measured by atomic absorption spectroscopy.[80] Thus, sufficient aluminum is present in the diseased state to exert a potentially important alteration in protein-DNA interactions.

An important mechanism controlling gene expression is the maintenance of a transcriptionally competent chromatin structure. An open structure is required to permit RNA polymerase access to coding regions of DNA. Active genes occur on open euchromatin regions of DNA which are depleted in H1 linker histones. The H1 family of histones are involved in condensing chromatin into a highly compact state. In Alzheimer's disease, neocortical nuclei exhibit an increased proportion of the total DNA in the heterochromatin fraction compared to age-matched control nuclei.[81] The altered chromatin structure may contribute to the reduced amount of messenger RNA reported in Alzheimer affected cerebral cortex compared to controls.[82,83]

The kinetics of nucleosome release indicates that Alzheimer chromatin is less accessible to digestion by the enzyme micrococcal nuclease.[84] Importantly, the dinucleosomes released contained approximately twice the content of linker histone $H1^{o}$ than dinucleosomes released from control cerebral cortex, although the absolute amount of $H1^{o}$ in bulk chromatin was not increased. Taken together with the observed increased binding of linker histones to DNA, as measured by increasing salt concentration, the data suggest that aluminum might bind between the linker histones and DNA. For example, coordination at amino acid asp-98 and glu-99 of human $H1^{o}$ and DNA phosphates via an aluminum bridge is a possible neurotoxic site. Direct proof of a cause and effect relation between aluminum in altered gene expression, however, will require identification of Alzheimer specific alterations in gene expression, the demonstration of the presence of aluminum at the appropriate site and the documentation of an aluminum effect on gene expression which reproduces that found in the disease. While such evidence is not yet available, the experimental techniques appear to be near at hand.

In addition to alterations in linker histone binding to DNA, aluminum has a number of *in vitro* effects on nuclear processes including a 50% reduction in DNA template accessibility of chromatin spreads to *eschirichia coli* RNA polymerase,[11] an inhibition of polytene chromosome puffing in response to steroid hormone, ecdysterone,[15] and a 25% decrease in ^{3}H-corticosterone binding to rabbit hippocampal nuclei extracted during an aluminum induced encephalopathy.[16] Aluminum also induces a 34% decrease in sister chromatin exchange in cultured lymphocytes and an increase in unscheduled DNA synthesis.[14] Recently, the *in vivo* effects of aluminum have been assessed by comparing control and aluminum treated rabbit brain total and poly(a)RNA yields in various stages of the aluminum encephalopathy.[17] Employing a quantitative dot blot method on total RNA extracted from aluminum and control treated rabbit forebrains, the mRNA levels for four cytoskeletal proteins: calmodulin, α-tubulin, β-actin and neurofilament 68,000 Mr (NF68) were measured at various stages of the encephalopathy. During the middle stage of the encephalopathy (3, 7 and 10 days post-injection) calmodulin mRNA/g of tissue was significantly reduced by 74%, Table 2(A), while the messenger content for α-tubulin, NF68 and β-actin either remained unaltered or rose somewhat. In the terminal stages of the encephalopathy (12 days post-aluminum injection), all four mRNA species were significantly reduced: calmodulin

Table 2. Calmodulin messenger RNA yield and amount of calmodulin protein.

A) Relative yield calmodulin messenger RNA in Alzheimer and control human brain and rabbit aluminum treated and control brain				B) Amount of calmodulin protein measured by radioimmune assay for Alzheimer and control human brains and rabbit aluminum treated and control brain		
Brain Source	Relative	mRNA Yield	Alz/Cont. %		Protein µg/g soluble protein	
Human	Alz(n=10)	Cont.(n=12)		Alz.(n=18)	Cont.(n=17)	Alz/Cont %
	5.5	8.1	68% $p<.04$	446	635	70% $p<.001$
Rabbit	Al(n=3)	Cont.(n=3)		Al(n=6)	Cont.(n=6)	
(late enceph.)	1719	2304	74% $p<.005$	302	308	N.S.

Note: Numbers in brackets are the number of brains employed in each assay.

76%, α-tubulin 63%, NF 68 81%, and β-actin 76% of control. These results suggest that aluminum has specific, early, effects on some messenger RNA levels, such as calmodulin, which could be related either to homeostatic response to aluminum toxicity or direct effect of aluminum on the transcription of the gene.

The relative yields of seven messenger RNAs extracted from Alzheimer and control neocortex were examined by northern and quantitative dot blot analysis, Table 2 in McLachlan et al.[85] Messenger yields of seven structural proteins were examined and revealed that the messenger yield for neurofibrillary acidic protein, β-actin, somatostatin and ALU expressed as a percent of total RNA was not altered in Alzheimer's disease compared to control. However, α-tubulin and calmodulin were reduced to 65% and 68% of control, respectively and NF 68 was reduced to 27% of the control. Relative calmodulin mRNA yields are shown in Table 2A.

The calcium-binding proteins calmodulin and calbindin $D_{28}K$ were measured by radioimmunoassay in cerebral cortex from brains with degenerative diseases associated with dementia.[86] Samples were obtained from 17 normal control brains, 18 Alzheimer affected brains and 15 brains of patients who died of a dementia process other than Alzheimer's disease. Calmodulin content in the non-Alzheimer affected group was not significantly different from control. However, Alzheimer affected frontal, temporal, and parietal cortex and subjacent white matter was significantly reduced to about 70% of that found in the other two groups Table 2B. Calmodulin extracted from temporal cortex also demonstrated reduced efficacy as an activator of the 3', 5', cyclic nucleotide phosphodiesterase.[86] Thus calmodulin messenger and peptide are both reduced in Alzheimer's disease by about the same magnitude as that found in the aluminum induced encephalopathy in rabbit. While these observations partially fulfill the first criterion enunciated above--that aluminum suppresses messenger RNA production and the amount of peptide in the cerebral cortex of an important second messenger system, calmodulin--direct measures of aluminum at the site of the promotor region of the calmodulin gene had not yet been achieved.

EPIDEMIOLOGY: BIOAVAILABLE ALUMINUM AND ALZHEIMER'S DISEASE

A recent epidemiological study by Vogt[87] tested the hypothesis that a relation (co-variance) exists between the concentration of aluminum in the drinking water and the frequency of dementia of the Alzheimer type. This important study indicated a statistically significant increased mortality rate from this disease with increased aluminum concentration. Mortality rates from presenile and senile dementia during the period 1969-1983 were evaluated for a standard population of over 3,250,000 people. The data was divided into two groups: (a) The record on death certificates of presenile or senile dementia as the primary cause of death; this accounted for 1504 cases; (b) Presenile or senile dementia mentioned as one of four associated conditions at death; this occurred in 16,560 certificates. Since Alzheimer's disease is the most common cause of senile dementia (50-75% of cases), other causes of senile dementia were also evaluated. The second most common cause of senile dementia (10-20%) is multi-infarct dementia, which is secondary to vascular disease of the brain. Therefore, the death certificates were examined for evidence of disease of the circulatory system including: ischemic heart disease, cerebral vascular disease, hypertensive disease, and diabetes mellitus, each of which predisposes to vascular dementia. Death associated with paralysis agitans was also recorded. The population of Norway was divided into five zones that were based on the aluminum content in ground and drinking waters. A survey conducted between 1982 and 1983 revealed that Norway has 384 public water works which supply 71% of the population: 349 of these public works employ surface waters, and 35 employ groundwaters for drinking. A link between bioavailable aluminum and acid rain is based on the concept that aluminum is insoluble in water at neutral or alkaline pH. However, when the acidity of water from rain, snow, sleet or fog increases, aluminum becomes soluble from lake bottom sediments, soil, and the metal pipes employed in the transport of water. To establish an indicator of the intake of aluminum through drinking water in southern Norway, five geographical zones were established. The zones were based on increasing concentration of aluminum in lakes. Aluminum in southern lakes corresponds well with the aluminum found in drinking water. Zone 1 contained the lowest amount of aluminum, 20 μg Al/L and zone 5 in the southeastern most region of Norway contained 200 μg Al/L. Zones 2 to 4 contained intermediate aluminum levels. Zones 4 and 5 received the largest amount of acid rain and therefore have the highest concentration of bioavailable aluminum. Zone 5 has the significantly highest mortality from senile and presenile dementia, and the risk of dying of dementia was approximately 1.4 times higher than in zone 1. The age distribution and numbers of individuals at risk for Alzheimer's disease was approximately the same in each of the zones. Furthermore, there was no correlation between the risk factors for cerebral vascular disease and multi-infarct dementia between the zones. The risk of dying of paralysis agitans was also not related to bioavailable aluminum. These data further support the growing body of circumstantial evidence implicating aluminum in the pathogenesis of Alzheimer's disease.

A second epidemiological study implicating bioavailable aluminum was published by Still and Kelly.[88] This study involved the examination of cases admitted between 1971 and 1979 in three counties of South Carolina, Anderson, Horry, and York of first admissions of parsons 55 years or older to the state mental health hospitals bearing the diagnosis of primary degenerative dementia, vascular dementia, or alcoholic/other dementia. The three counties were selected because Horry county has 4.18 mg of fluoride per liter of drinking water, whereas Anderson and York county have 0.49 and 0.61 mg/L, respectively. Aluminum and fluoride are mutually antagonistic in competing for absorption in the gut. The more fluoride in the diet, the less aluminum absorbed. It is also known that feeding aluminum compounds counteracts dental fluorosis, reducing fluoride stores in teeth and bones.[89] This epidemiological study tested the hypothesis that reduced bioavailable aluminum secondary to increased exposure to fluoride for a period of at least 10 years altered the incidence of primary degenerative dementia. In the classification of the authors, primary degenerative dementia was taken to be predominantly Alzheimer's disease. Results of a 2 x 2 chi-square distribution test revealed significant differences between Horry-Anderson and Horry-York counties in the number of cases of primary degenerative disease cases, whereas no significant differences in the number of cases occurred between York and Anderson counties. There were no significant differences between the counties in the number of first admissions with a diagnosis of vascular dementia or alcoholic/other dementia.

Neither of the above epidemiological studies are beyond criticism. Neither study employed autopsy confirmation nor estimated the probable diagnostic accuracy of death certificates or of the admission diagnosis. The Norwegian study is based on a large population sample, greater than 3 million people, whereas the South Carolina study is based on a population of 57,000 individuals 55 years or older. The Carolina study included 67 cases of primary degenerative dementia, 68 cases of vascular dementia, and 25 cases of alcoholic/other dementias. Notwithstanding the shortcomings of these epidemiological reports, these studies generally support the concept that bioavailable aluminum increases the risk of dying of senile and presenile dementia of the Alzheimer type.

ALUMINUM EXCLUSION AND CLINICAL COURSE

The third line of independent evidence central to evaluating the role of aluminum in Alzheimer's disease involves studies of the clinical course of the disease in the presence of reduced bioavailable aluminum. However, significant reduction in aluminum exposure, perhaps by 10- to 100-fold, is not a trivial undertaking. Aluminum is ubiquitous. Aluminum gains access to the brain via the epithelial cells of the respiratory, gastrointestinal and cutaneous surfaces. The respiratory and olfactory epithelia are known to contain the largest concentrations of aluminum in the body,[90,91] and brain uptake may occur through the olfactory bulb.[92,93] Dusts arising from soil and the metal surfaces of air conditioners contain large amounts of aluminum. Aluminum is also taken up through the gastrointestinal epithelium as a consequence of ingestion of water, soft drinks, processed foods, root vegetables, tea, coffee, sea foods and pharmaceuticals.[94,95] Uptake of aluminum through the cutaneous epithelium may also occur as a result of the application of aluminum containing deodorants, body creams, and soaps. Alzheimer's disease is a slowly progressive illness, with survival times often exceeding 8-12 years after the early symptoms of memory and other cognitive effects are recognized. If aluminum is a significant factor driving the pathology and pace of the illness, isolating individuals in the early stages of the condition from aluminum should significantly improve the quality of life and prolong survival. Studies rigorously testing this hypothesis have not yet been reported.

At least two centers, however, are employing agents to remove aluminum from the Alzheimer-affected patient without attempts to rigorously regulate environmental exposure to aluminum. Liss and Thornton[96] studied the sodium fluoride at an oral dose not exceeding 1 mg/kg/day effect of upon a patient population seen at a Cognitive Disorders Clinic and diagnosed as suffering from Alzheimer's disease. The patients were assigned sodium fluoride or placebo according to a random number table. Compliance was carefully evaluated, and side effects were apparently not encountered. The study was designed for 3 years' duration, and a since number of patients have completed 30 months of treatment, sufficient data has been accumulated to compare progress. Twelve patients were decoded at the time of publication (1986), and their time on the study ranged between 30 and 36 months. They were compared on the basis of progression of symptoms, which included various parameters of neuropsychological testing, retrograde masking paradigm, CT and electroencephalography, and activities of daily living. In addition to comparing patients with others in the study, the treated patients were compared with their own rate of progression, which could be determined from the history from the time of onset of symptoms to the date of entry in the study.

The results indicated that in 11 out of 12 cases, the anticipation of the placebo or fluoride treatment was correct. In the 12th case there was very slow progression which continued during the 3 years of the study although the patient was on placebo.

A clinical trial employing desferoxamine as an aluminum-chelating agent is currently in progress in this center, but the code for the current desferoxamine study will not be opened until 1990.

CONCLUSION

The absence of a useful isotope of aluminum and an animal model of Alzheimer's disease have made the direct evaluation of the role of aluminum in Alzheimer's disease difficult. Rapid progress in understanding the mechanisms of aluminum neurotoxicity and the molecular basis of Alzheimer's disease has been made in the past decade. However, cause-and-effect evidence for aluminum as a significant factor in the pathogenesis of Alzheimer's disease remains incomplete. We advocate that at least three independent avenues of investigation be vigorously pursued in the next decade in order to achieve a level of knowledge sufficient to satisfactorily evaluate aluminum neurotoxicity in Alzheimer's disease:

a) Identification in Alzheimer affected tissue of important biochemical events disturbed by aluminum and documentation of an identical biochemical effect in an animal or cellular model.

b) Epidemiological studies relating bioavailable aluminum to Alzheimer's disease.

c) Clinical evidence that the quality of life and survival of Alzheimer patients is improved by either removing aluminum from the environment or removing aluminum from the brain by chemotherapeutic means.

Review of the evidence to date, however incomplete, strongly suggests that aluminum may have a significant role in this common neurodegenerative disease.

ACKNOWLEDGMENT

Work supported by the Ontario Mental Health Foundation, Medical Research Council of Canada, Scottish Rite Charitable Foundation, Health and Welfare Canada and Ciba Geigy Canada.

REFERENCES

1. St. George-Hyslop, P., R. E. Tanzi, R. J. Polinsky, J. L. Haines, L. Nee, P. C. Watkins, R. H. Myuers, R. G..Feldman, D. Pollen, D., Drachman, J. Growdon, A. Bruni, J. F. Foncin, D. Salmon, P. Frommelt, L. Amaducci, S. Sorbi, S. Piacentini, G. D. Stewart, W. J. Hobbs, M. Conneally, and J. F. Gusella. *Science* 235:885-890 (1987).

2. Matsuyama, H., and S. Nakamura. "Alzheimer's disease: senile dementia and related disorders," in *Aging* Vol. 7., Eds. R. Katzman, R. D. Terry, and K. L. Bick. (New York: Raven Press, 1978), p 287

3. Jervis, G.A. *Amer. J. Psychiat.* 105:102-106 (1948).

4. McLachlan, D.R. *Neurobiology of Aging* 7:525-532 (1986).

5. Wisniewski, H., C. Moretz, and K. Iqbal. *Neurobiol. Aging* 7:532 (1986).

6. Martin, R.B. *Clinical Chem.* 32:1797 (1986).

7. Karlik, S.J., G.L. Eichhorn, and D.R.C. McLachlan. *Neurotoxicology* 1(4):83-88 (1980).

8. Karlik, S.J., G.L. Eichhorn, and D.R. Crapper. *Biochemistry* 19:5991-5998 (1980).

9. Lukiw, W. Y., T.P A. Kruck, and Crapper McLachlan, D.R. *Neurotoxicology* 8(2):291 (1987).

10. Miller, C.A. and E.M. Levine *J. Neurochem.* 22:751 (1974).

11. Sarkander, H., G. Bolb, R. Schlosser, G. Saltenburg, and K. M. Lux. In: *Brain Aging, Neuropathology and Neuropharmacology*. J. Ceiros-Norano and H. Sarkander, Eds. (New York: Raven Press, 1983), p. 259.

12. Crapper McLachlan, D. R., T. V. Dam, B. J. Farnell, and P. N. Lewis. *Neurobehav. Toxicol. Teratol.* 5:645-647 (1983).

13. Crapper, D. R., S. Quittkat, S. S. Krishnan, A. J. Dalton, and U. De Boni. *Acta Neuropathol. Berlin* 50:19 (1980).

14. De Boni, U., M. Seger and D. R. Crapper McLachlan *Neurotoxicology.* 1:65-81 (1980).

15. Sanderson, C. L., D. R. McLachlan, and U. De Boni. *Can J Genet. Cytol.* 24:27 (1982).

16. Sanderson, C. L., D. R. McLachlan, and U. De Boni. *Acta Neuropathol.* 57:249 (1982).

17. VanBerkum, M., Y. Wong, P. Lewis, and D.R.C. McLachlan. *Neurochem. Research* 11:1347 (1986).

18. Muma, N. A., J. C. Troncoso, P. Hoffman, E. H. Koo, and D. Price. *Molecular Brain Research* 3:115-122 (1988).

19. Gelfant, S. *Rev. Cytol.* 14:249-254 (1963).

20. Sampson, M., D. Clarkson, and D. D. Davies. *Science* 148:1476-1477 (1965).

21. Siegel, N. and A. Haug. *Biochem. Biophys. Acta* 744:36-45 (1983).

22. Siegel, N. and A. Haug. *Physiol. Plant* 59:285-291 (1983).

23. Burnatowska-Hledin, M. A. and G. H. Mayor. *Clin. Res.* 30:A741 (1982).

24. Farnell, B. J., D. R. Crapper McLachlan, K. Baimbridge, U. De Boni, L. Wong and P. L. Wood. *Exp. Neurol.* 88:68-83 (1985).

25. Rorison, I. H. *New Physiol.* 64:23-27 (1965).

26. Foy, C. D., R. L. Chaney and M. C. White. *Annu. Rev. Plant Physiol.* 29:511-566 (1978).

27. Trapp, G. A. *Neuro. Toxicology.* 1:89-100 (1980).

28. Bock, J. L. and A. E. Ash. *J. Inorganic Biochem.* 13:105-110 (1980).

29. Womack, F. C. and S. P. Colowick. *Proc. Natl. Acad. Sci. U.S.A.* 76:5080-5084 (1979).

30. Karlik, S. J., G. A. Elgavish and G. L. Eichhorn. *J. Amer. Chem. Soc.* 105:602-609 (1983).

31. Karlik, S. J., Tariene, G. A. Elgavish, and G. L. Eichhorn. *Inorg. Chem.* 22:525-529 (1983).

32. Lai, J. C. K. and J. P. Blass. *J. Neurochem.* 42:438-445 (1984).

33. Mansour, J. M., A. Ehrlich, and T. E. Mansour. *Biochem. Biophys. Res. Comm.* 112:911-918 (1983).

34. Sternweiss, P. C. and A. G. Gilman. *Proc. Natl. Acad. Sci. U.S.A.* 79:4888-4891 (1982).

35. Gruca, S. and H. M. Wisniewski. *Acta Neuropathol.* (Berl) 63:287-295 (1984).

36. Altmann, P., F. Alsalihi, K. Butter, P. Cutler, J. Blair, R. J. Leeming, J. Cunningham, and F. Marsh. *New England Journal of Medicine.* 317(2):80-84 (1987).

37. Johnson, G. V. W., and R. S. Jope. *Brain Res.* 403:1-6 (1987).

38. Morandi, A., V. Fried, H. Smith, G. Perry and P. Gambetti. Abstract No. 366.14. *17th Annual Meeting, Society for Neurosciences, New Orleans.* Vol. 13 (1987).

39. Joshi, J. G., J. Fleming, and A. Zimmerman. *XIIIth World Neurology Congress, Hamburg, Supp., J. Neurol.* 232:61 (1985).

40. Klatzo, I., H. Wisniewski and E. Streicher. *J. Neuropathol. and Exp. Neurol.* 24:187-199 (1965).

41. Terry, R. D. and C. Pena. *J. Neuropathol. and Exp. Neurol.* 24:200-210 (1965).

42. Dahl, D. and A. Bignami. *Exper. Neurol.* 58:79-80 (1978).

43. Selkoe, D. J., R. K. H. Liem, S. Yen, and M. L. Shelanski. *Brain Res.* 163:235-252 (1979).

44. Munoz-Garcia, D., W. W. Pendlebury, J. B. Kessler, and D. P. Perl. *Acta Neuropathol.* (Berl) 70:243-248 (1986).

45. Bizzi, A., R. C. Crane, Autilio-Gambetti, and P. Gambetti. *J. Neuro Science* 4:722-731 (1984).

46. Kosik, K. S., A. H. McCluskey, F. X. Walsh, and D. J. Selkoe. *Neurochem. Pathol.* 3:99-108 (1985).

47. Liwnicz, B. H., K. Kristensson, H. M. Wisniewski, M. L. Shelanski and R. D. Terry. *Brain Res.* 80:413-420 (1974).

48. Johnson, G. V. M. and R. S. Jope. Abstract No. 366.10. *17th Annual Meeting, Society for Neurosciences, New Orleans* Vol. 13 (1988).

49. Macdonald, T. L., W. G. Humphries and R. B. Martin. *Science* 236:183-186 (1987).

50. Macdonald, T. L. and R. B. Martin. *Trends in Biological Sciences* 13:15-19 (1988).

51. Viestra, R. and A. Haug. *Biochem. Biophys. Res. Commun.* 84:134-144 (1978).

52. Deleers, M. *Res. Comm. Chem. Pathol. Pharmacol.* 49(2):277-294 (1985).

53. Deleers, M., J. P. Servais, and E. Wülfert. *Biochem. Biophys. Acta* 813:195-200 (1985)

54. Deleers, M., J. P. Servais, and E. Wülfert. *Biochem. Biophys. Acta* 855:271-276 (1986).

55. Lai, J. C. K., J. F. Guest, T. K. C. Leung, L. Lim and A. N. Davison. *Biochem. Pharma col.* 29:141-146 (1980).

56. Ohtawa, M., M. Seko, and F. Takayama. *Chem. Pharm. Bull.* 31:1415-1418 (1983).

57. Gutteridge, J. M. C., G. J. Quinlan, I. Clark, and B. Hallowell. *Biochimica et Biophysica Acta* 835:441-447 (1985).

58. Banks, W. A., A. J. Kastin, and M. B. Fasold. *J. Pharm. Exp. Therap.* 244:579-585 (1988).

59. Banks, W. A. and A. J. Kastin. *Lancet* 1227-1229 (1983).

60. Banks, W. A. and A. J. Kastin. *Neuropharmacology* 24:407-412 (1985).

61. Banks, W. A. and A. J. Kastin. *Psychopharmacology.* 86:84-89 (1985).

62. Petit, T. L., G. B. Biederman, and P. A. McMullen. *Exp. Neurol.* 67:1952-162 (1980).

63. Uemura, E. and W. P. Ireland. *Exper. Neurol.* 85:1-9 (1984).

64. Wong, P. C. L., J. C. K. Lai, L. Lim, and A. N. Davison. *J. Inorganic. Biochem.* 14:253-260 (1981).

65. Sturman, J. A., H. M. Wisniewski, and J. W. Shek. *Neurochemical Research* 8:1097-1109 (1983).

66. Lai, J. C. K., L. Lim, and A. N. Davison. *J. Inorganic Biochem.* 17:215-225(1982).

67. Marquis, J. K. *Bul. Environm. Contem. Toxicol.* 29:43-49 (1982).

68. Marquis, J. K. and A. J. Lerrick. *Biochem. Pharmacol.* 31:1437-1440 (1983).

69. Hava, M. and A. Hurwitz. *European Journal of Pharmacology* 22:156-161 (1973).

70. Lipman, J. J., S. P. Colowick, P. L. Lawrence and N. N. Abumrad. *Life Sciences* 42:863-875 (1988).

71. Wenk, G. L. and K. L. Stemmer. *Neurotoxicology* 2:347-353 (1981).

72. Wenk, G. L. and K. L. Stemmer. *Neurotoxicology.* 3:93-99 (1982).

73. Yates, C. M., J. Simpson, D. Russell, and A. Gordon. *Brain Research* 197:269-274 (1980).

74. Hetnarski, D., H. M. Wisniewski, K Iqbal, J. D. & A. Lajtha. *Ann. Neurol.* 7:489-490 (1980).

75. Simpson, J., C. M. Yates, D. K. Whyler, H. Wilson, A. J. Dewar, and A. Gordon. *Neurochemical Research* 10:229-238 (1985).

76. Koenig, M. L. and R. S. Jope. *J. Neurochem.* 49:316-320 (1987).

77. Patocka, J. *Acta Biol. Med. Germ. Band* 26:845-846 (1971).

78. Marquis, J. K. *Bull. Environ. Cotam. Toxicol.* 31:164-169 (1983).

79. Trapp, G. A. *Life Sciences* 33:311-316 (1983).

80. Crapper, D.R., Quittkat, S., Krishnan, S.S., Dalton, A.J. and U. Boni, *Acta Neuropathol.* (Berl) 50:19-24 (1980).

81. Crapper, D. R., S. Quittkat, and U. Boni. *Brain* 102:483 (1979).

82. Sajdel-Sulkowska, E. M. and C. A. Marotta. *Science* 225:947-949 (1984).

83. Guillemette, J. G. L. Wong, D. R. McLachlan, and P. N. Lewis. *J. Neurochem.* 47:987-997 (1986).

84. Lewis, P. N., W. J. Lukiw, U. De Boni, and Crapper McLachlan, D.R. *J. Neurochem.* 37:1193 (1981).

85. McLachlan, D. R., W. J. Lukiw, L. Wong, C. Bergeron and N. T. Bech-Hansen. *Molecular Brain Res.* 3:255-262(1988).

86. McLachlan, D.R.C., L. Wong, C. Bergeron, and K. Baimbridge. *Alzheimer Disease and Associated Disorders Internat. Journal.* 171-179(1987).

87. Vogt, T., Statistisk Sentralbyra, *Oslo-Kougsvinger Sosiale og Okoumiske Studier* 61 (1986).

88. Still, C. and Kelley, P. *Neurotoxicology.* Special Issue 1(4):125 (1980).

89. Navia, J. In: *Advances in Chemistry* Series, 94, Am. Chem. Soc. 123 (Washington, D.C., 1970).

90. Tipton, I.H. and Cook, M.J. *Health Phys.* 9:103 (1963).

91. Crapper, D.R., Krishnan, S.S. and Quittkat, S. *Brain* 99:67 (1976).

92. Roberts, E. Neurobiol. *Aging* 7:561 (1986).

93. Perl, D. and Good, P. *Lancet* 1028 (1987).

94. Lione, A. *Food Chem. Toxicol.* 21(1):103 (1983).

95. Jaworski, J.F. "Aluminum in the human diet," in *Aluminum in the Canadian Environment*, M. Havas and J.F. Jaworski, Eds. Publication No. NRCC24659 of the Environmental Secretariat, The National Research Council of Canada, Ottawa, Canada, p. 175-179 (1986).

96. Liss, L. and Thornton, D.J. *Neurobiol. Aging* 7:552 (1986).

CHAPTER 18

ALUMINUM NEUROTOXICITY, ALZHEIMER DISEASE, AND ALCOHOLIC ENCEPHALOPATHY

Leopold Liss, John Long, and David Thornton

INTRODUCTION

The role of aluminum in the degeneration of neurons which occurs in Alzheimer disease, has been extensively investigated since the increase in aluminum content was reported in those areas of human brain in which there was evidence of neurofibrillary degeneration of neurons.[2]

Subsequently, the neurotoxic potential of aluminum cast it in a role of "The Etiologic Agent", despite the evidence to the contrary. The ubiquitous nature of aluminum, which makes contact and absorption unavoidable, excludes the possibility that the everyday exposure, even with some overload, would lead by itself to pathological events. Aluminum should therefore be regarded as an unlikely culprit in primary neurotoxicity for any intact organism.

The etiology of Alzheimer disease should be regarded as a multistep event with multifactorial etiology. We can postulate the presence of intrinsic and extrinsic factors. The intrinsic factors are represented but not limited to: genetic, aging, chromosomal changes and a biological agent. The extrinsic factor is aluminum, which we regard as a *sine qua non,* since strong evidence suggests that it is responsible for the neurofibrillary degeneration of neurons. The unavoidable exposure to aluminum is subject to considerable quantitative fluctuations resulting from variations in geographic, technological, cultural, medical and other factors. In addition to aluminum, which is an obligatory factor, we postulate presence of extrinsic factors, which should be regarded only as modifiers since they are capable of accelerating or delaying the time of onset of the disease and likely to influence the rapidity of the progression of the symptoms.

We propose that alcohol, or more specifically brain damage resulting in "alcoholic encephalopathies," represents one of these extrinsic modifiers of Alzheimer disease which can accelerate the onset in a genetically susceptible individual. This extrinsic modifier may also be responsible for accelerated progression of symptoms. We also propose that the cerebellar degeneration which affects some alcoholics may actually be caused by aluminum neurotoxicity.

CLINICAL AND HISTOPATHOLOGICAL STUDIES

The histological evaluations of brains of demented individuals indicate that Alzheimer disease is the most common cause of dementia in older adults. The incidence of Alzheimer disease in adult dementias has been reported, depending on the population sample, as 40-60%.

The multiinfarct dementia is considered reponsible for about 20% cases of dementia, and mixed cases with both Alzheimer disease and multiinfarct were observed in about 10-15%. Our overall autopsy material reflects this distribution of causes of dementia except for one institution where alcoholics by history represented over 35%, and over 60% have shown at autopsy histological findings diagnostic for alcoholic encephalopathy.[4] In this highly slanted sample of the population we have encountered an unusually high incidence of Alzheimer disease (Table 1). This higher than expected percentage of Alzheimer disease and low incidence of infarcts among these patients was considered coincidental to the frequent history of alcohol abuse and with clinical diagnosis of Wernicke's encephalopathy, Korsakoff's dementia, and cerebellar degeneration.

Table 1. Causes of dementia in a population with high percentage of alcoholic encephalopathies

Dementia, Alzheimer type	78%
Multiinfarct dementia	8%
Miscellaneous causes incl. trauma	14%
Combined dementia, Alzheimer type and distinct Alcoholic encephalopathy	46%

In Wernicke's encephalopathy and in Korsakoff's syndrome the degeneration of the mamillary bodies is pathognomonic. Although both syndromes are classified as "alcoholic encephalopathies," they are actually nutritional deficiencies resulting from lack of thiamine.[9] Acute Wernicke's syndrome was observed in pure thiamine deficiency produced by dialysis. This degeneration of mamillary bodies, which is caused by thiamine deficiency and is common in alcoholics, was responsible for distinct modification of the histopathological and toxicological findings.

As described originally by Alzheimer 80 years ago, the neuritic plaques and neurofibrillary tangles are the diagnostic findings. Both plaques and tangles are present in great numbers in hippocampus and throughout neocortex. Cerebellum is spared except that plaques are found in cerebellum in the autosomal-dominant form of

Alzheimer disease and in more than 20% of cases with Alzheimer type changes occurring in older individuals with Down's syndrome. The mamillary bodies show only in most advanced cases the presence of an occasional plaque or tangle.

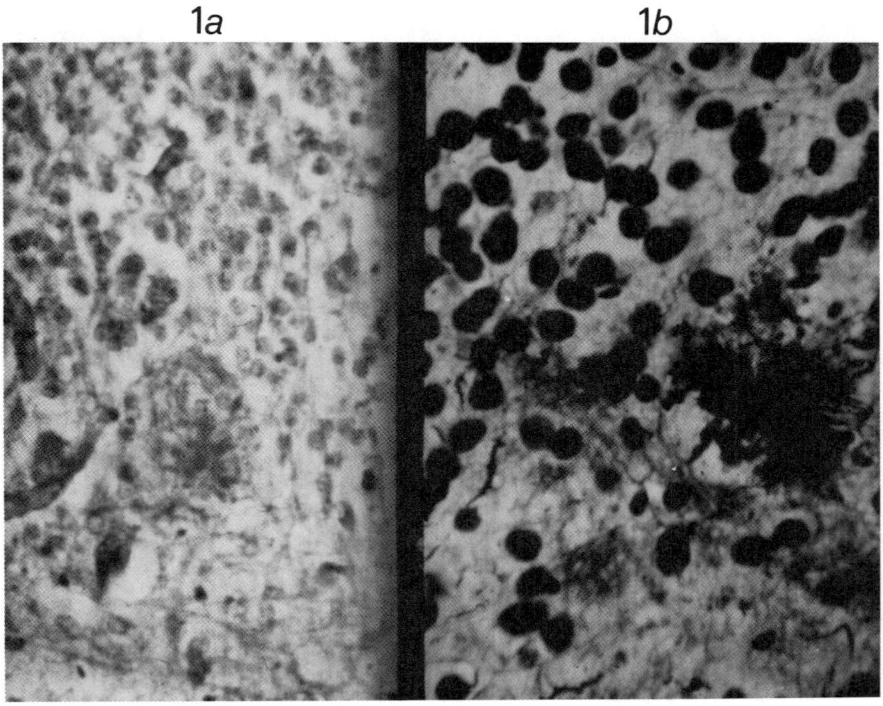

Figure 1. Neuritic Plaques in cerebellar granular layer of 56 year old male with diagnosis: Alzheimer disease and Alcoholic Encephalopathy. (a) Hematoxyline Eosin stain; (b) Hortega's Silver Carbonate Impregnation.

In the brains of patients who had both Alzheimer disease and alcoholic encephalopathy there was a deviation from the usual distribution of histological changes. The neuritic plaques were found in cerebellum in over 10% of the cases,[6] although none of them had an autosomal dominant familial disease or Down's syndrome (Figure 1). The histopathological changes in the mamillary bodies were found with regularity, and in many early cases of Alzheimer disease the mamillary bodies represented an early target for formation of neuritic plaques and neurofibrillary tangles.

In a number of these brains we found in the cerebellum Purkinje cell "torpedoes", which are round, oval, or spindle-shaped swellings of the axons. These torpedoes represent an unspecific pathological change which results from damming of axoplasmic flow.[1] They are found in great numbers in cases of alcoholic degeneration of cerebellum (Figure 2).

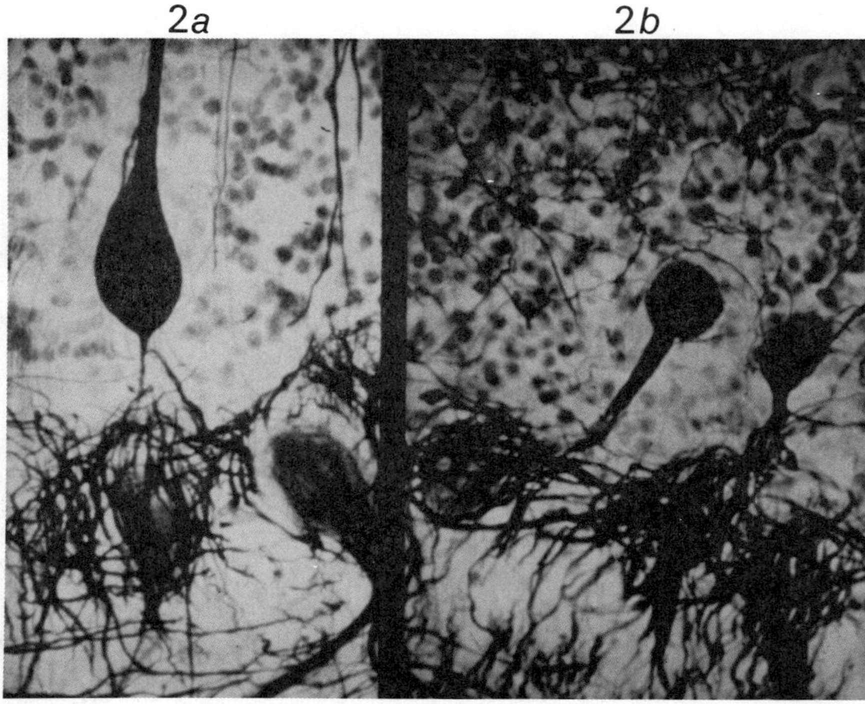

Figure 2. (a) Cerebellum 62 year old male. Alcoholic Cerebellar degeneration. "Torpedoes" in the granular cell layer. (b) Cerebellum 71 year old male. Alzheimer disease and history of alcohol abuse. "Torpedo" in the granular cell layer. Hortega's Silver Carbonate Impregnation.

ALUMINUM STUDIES

Determination of aluminum in brain tissue was done with atomic absorption spectrophotometry. In normal controls the levels ranged from 0 to 0.6 ppm. In Alzheimer patients samples were taken from various regions of the brain. There was a significant difference in aluminum content between cortical regions in which there was evidence of neurofibrillary degeneration and those without. The level in areas without neurofibrillary degeneration was 0 to 1.5 ppm dry weight, while in the areas with neurofibrillary degeneration, aluminum concentration ranged from 6 to 12 ppm. More importantly, in one case an individual without Alzheimer disease or Down's syndrome had a small stationary tumor which compressed the frontal lobe. The cortex underlying the tumor was chronically compressed and had abundance of neurofibrillary tangles. In the tangle-rich area the aluminum content was 8.4 ppm compared to 0.9 ppm in the other hemisphere and 1.4-3.3 ppm in close proximity.[5]

In the brains of alcoholics we have found aluminum levels in mamillary bodies and in cerebellum to be considerably elevated (4 to 12 ppm). These levels could be correlated with the severity of the morphological changes present and were observed in patients who had alcoholic encephalopathy with or without Alzheimer disease. The cerebellar Purkinje cell torpedoes were present regardless of presence or absence of Alzheimer disease, while the plaques and tangles were only present in the mamillary bodies of those with Alzheimer disease.

EXPERIMENTAL STUDIES

The hypotheses which could be formulated on the basis of the clinical observations required confirmation of several premises. One was to determine if there is a natural brain-aluminum barrier which must be breached before aluminum can enter the brain.

3a 3b

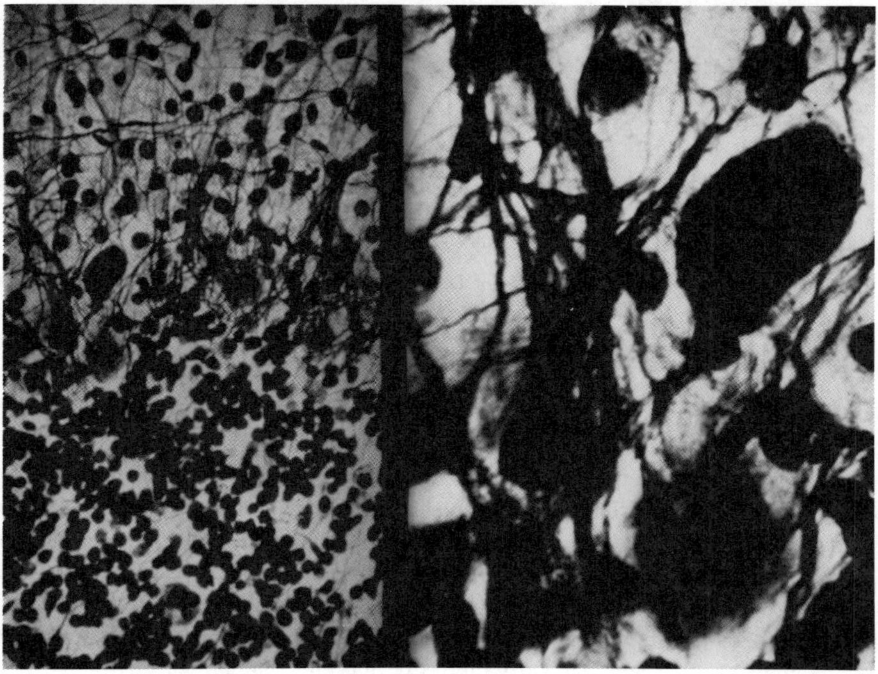

Figure 3. Rabbit 6 weeks old. Offspring of mother maintained on high aluminum water. Nursed on high aluminum milk. Weaned on high aluminum water. Swollen Purkinje cell axon in granular layer. Hortega's Silver Carbonate Impregnation (a) overview; (b) high magnification.

4a 4b

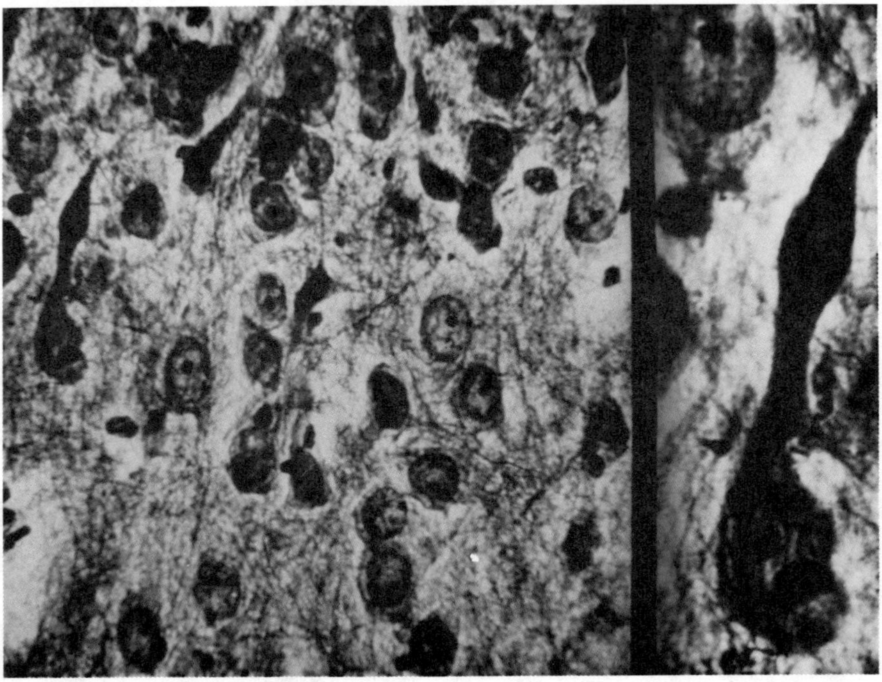

Figure 4. Rabbit 6 weeks old. Offspring of dame maintained on high aluminum water and nursed with milk of high aluminum content. Weaned on high aluminum water. Swollen neuronal process in hippocampus. Impregnation: Hortega's Silver Carbonate (a) overview; (b) high magnification.

We already had indications that alteration of the blood-brain barrier, caused by compression, allows entrance of aluminum and causes neurofibrillary tangles.[7] New Zealand rabbits were used as experimental animals and aluminum was administered in the drinking water. Water contained up to 50 g/L $AlCl_3$ (10.01 g/L Al). Rabbits were kept on this diet up to 6 months at which time they were euthanitized and brains processed for histological examination and analysis of Al with atomic absorption spectroscopy. There were no structural abnormalities, and the Al levels were in the same range as that found in control rabbits or in normal human brain.

Pregnant rabbits were maintained on the high aluminum diet before and through the pregnancy, and during lactation. The concentration of Al in milk was 30-50 mg/L. Neurofibrillary tangles were present and were scattered in the cortex and basal ganglia of their offspring.[3] The Purkinje cell axons contained "torpedoes" (Figure 3) and axonal swellings were observed in the cortex (Figure 4).

In other experiments chickens were fed 8-24 units of 1α,25dihydroxycholecalciferol (1,25-$(OH)_2D_3$) and 61 g per kilogram of aluminum tartrate (6.1 g/Kg Al). In the cerebellum were found scattered swellings of Purkinje cell axons, similar to the human "torpedoes" and to cortical axonal swellings in rabbits.[8]

The effect of dietary overload of Al on the Al serum level in New England rabbits was determined with and without enhancement produced by Parathyroid Hormone (PTH) and alcohol ingestion. Table 2 shows the blood levels in rabbits under various conditions of Al dietary overload and the effect of PTH and alcohol. The alcohol ingestion resulted in highest serum Al levels.

Table 2. Rabbit serum Al content in Al overload potentiated by parathyroid hormone and alcohol

	Aluminum mg/L	Ethanol mg/100ml
Control	20	-
$AlCl_3$ (2%)	81	-
	95	-
	91	-
$AlCl_3$ (2%)	121	-
PTH-4OU	111	-
	79	-
	123	-
$AlCl_3$ (2%)	219	104
Ethanol	204	234
	237	74

CONCLUSIONS

The correlation of the observations on humans with experimental findings allow us to form hypotheses in which the aluminum neurotoxicity is considered an essential component in neuropathological events.

1) Normal brain is sheltered from environmental and dietary sources, intrinsic deposits, and from overload of aluminum by a protective barrier. Breaching of the blood-brain barrier (BBB) renders the aluminum protective barrier (APB) ineffective, although the latter can be incompetent even with intact BBB. Once the aluminum has entered the central nervous system we can postulate a second line of protection against aluminum which is reponsible for selective protection of certain cell types and leaving only specific cells vulnerable. This protection may be on the level of cell membrane.

2) The neuronal diversity which is responsible for selective damage by neurotoxic properties of aluminum, can explain why neurons respond differently, e.g., neurofibrillary degeneration or axonal swellings. The interspecies diversity may well be the reason why in man we find paired helical filaments in neurofibrillary degeneration, while in experimental animals they are caused by clustering of straight filaments.

3) Nutritional damage (Wernicke's syndrome) is responsible for creation of *locus minoris resistentiae* in mamillary bodies, making them a prime target in the "Alzheimerization" process.

4) Toxic effects of ethanol on the cerebellum create a *locus minoris resistentiae* which may result in susceptibility to develop plaques in Alzheimer disease. It also may be the cause of cerebellar vulnerability to aluminum with resulting Purkinje cell degeneration in both Alzheimer disease and alcoholics.

5) Expressions of aluminum neurotoxicity are absent in normal organisms. We postulate that aluminum neurotoxicity occurs when the normal protective mechanisms are breached or altered either by intrinsic factors (like those responsible for development of Alzheimer disease) or extrinsic factors (such as ethanol).

ACKNOWLEDGMENTS

This study was supported by OSU Alzheimer Research Fund and Fahey Fund, both of Ohio State University Development Fund.

REFERENCES

1. Bizzi, A., R. Crane, M. Yoon, L. Autilio-Gambetti, P. Gambetti. "The axonal transport of neurofiliments is impaired in aluminum intoxication." *J. Neuropath. Experi. Neurol.* 42:331(1983).

2. Crapper, D.R., S.S. Krishnan and A.J. Dalton. "Brain aluminum distribution in Alzheimer disease and experimental neurofibrillary degeneration." *Science* 180:511-513(1973).

3. Liss, L., K. Ebner, D. Couri, M. Cho. "Alzheimer disease: A possible animal model." *I V. Pan Am. Congr. of Neurol.* 98(1975).

4. Liss, L., O.T. Lowy. "The incidence of Alzheimer disease in institutionalized patients. *Age* 1:163(1975).

5. Liss, L., K. Ebner, D. Couri. "Neurofibrillary tangles induced by sclerozing angioma." *Human Pathology* 10:104-108(1979).

6. Liss, L. "Cerebellar neuritic plaques in dementia Alzheimer type." *J. Neuropath. Exp. Neurol.* 41:379(1982).

7. Liss, L. D. J. Thornton. "Rationale for aluminum absorption control in early stages of Alzheimer disease." *Neurobiol. Aging* 7:552-554(1986).

8. Long, J.F., L.A. Nagoda, O. Kindig, L. Liss. "Axonal swellings of Purkinje cell in chicken associated with high intake of 1,25-$(OH)_2D_3$ and elevated dietary aluminum, including X-ray microanalysis." *Aluminum Neurotoxicity*, (ed. L. Liss) Pathotox Publ., Chicago, 111-120(1980).

9. Victor, M., R.D. Adams, G.H. Collins. *The Wernicke-Korsakoff Syndrome*. F.A. Davis, Philadelphia, 72-130(1971).

CHAPTER 19

POSSIBLE COUNTERACTIVE STEPS IN ALUMINUM-DEPENDENT ENCEPHALOPATHIES

Johan Bjorksten

INTRODUCTION

The principal mental diseases here in question are listed in order of time of onset after the aluminum exposure:

- Exposures in which natural defenses have been unnaturally bypassed:

A. Encephalopathies caused by injection of aluminum into the brain[1,2]

B. Encephalopathies caused by infusion of aluminum into the bloodstream[3,4]

- Slow penetration of the normal defenses:

C. Alzheimer's disease---after 40 years

D. Senile dementia---after 70 years

E. Osteoporosis---after 45 years

F. Some heart arrhythmias---?

COMMENTS

It is noted that all of the syndromes here discussed are based on aluminum. A and B above were both of great importance because they gave medically acceptable evidence of the reality of aluminum's role in brain disease.[5] D. R. Crapper-McLachlan and A. C. Alfrey re-opened this field and the general attitude was subtly changed.

Crapper-McLachlan showed that a single milligram of soluble aluminum salt injected into the hippocampic ventricle of the brain of a cat caused a whole sequence of a progressive brain disease, ending with death in about three months.

In the early 1970s, patients undergoing dialysis treatments became stricken with symptoms similar to Alzheimer's disease. Alfrey's work with these "dialysis syndrome" cases indicated that their condition was caused by the aluminum content of the water used in the artificial kidney.

In section A, the relationship of aluminum and the induced brain disease is clear and requires no further comments.

Section B has been considered by many as not relevant to Alzheimer and senile dementia because in the dialysis dementia there are no fibrils, and the aluminum is found largely in the cytoplasm. This is, however, consistent with the great differences in the speed of introduction of the aluminum. It is well known in chemistry that the size and appearance of precipitates is greatly dependent on the speed of formation, regardless of composition.

In dialysis, the two first lines of defense have been bypassed because of (1) the near impermeability for aluminum of the intestinal system, and (2) the absence of functional kidneys. Aluminum then enters the blood stream possibly a hundred times faster than in persons with functional kidneys.

Normally, the aluminum enters at an extremely slow rate and thus has tens of years to penetrate the membranes of cell nuclei. In dialysis using water of high aluminum content, neuronal cytoplasm is flooded with aluminum. The aluminum will rapidly displace other metals encountered and in addition will react directly with any cellular protein, as the powerful tanning (i.e., crosslinking) agent it is.[5,6,7] Since crosslinked molecules cannot penetrate membranes, by far the major part of the aluminum will remain in the cytoplasm, where it forms networks or submicroscopic "floating islands" sufficient to fatally hinder intracellular transport.[8] The causative role of aluminum in the dialysis syndrome was fully proven by Alfrey, and it is now seen as a variation of the same theme: interference by aluminum in basic cellular functions.

ALUMINUM, THE WIDELY DISTRIBUTED, EASILY ANALYZED, COMMON FACTOR

Since excessive aluminum accumulation is the basis of all the above-mentioned hazards, they can be considered as branches of the same tree (as shown in Figure 1). Therefore, I shall discuss possible treatment for these together. The most logical sequence is to first discuss those preventive steps which can be immediately implemented. I will then discuss those corrective treatments which are still experimental, but have good chances for success. These steps are applicable to those diseases mentioned in sections C through F only because diseases A and B were artificially induced.

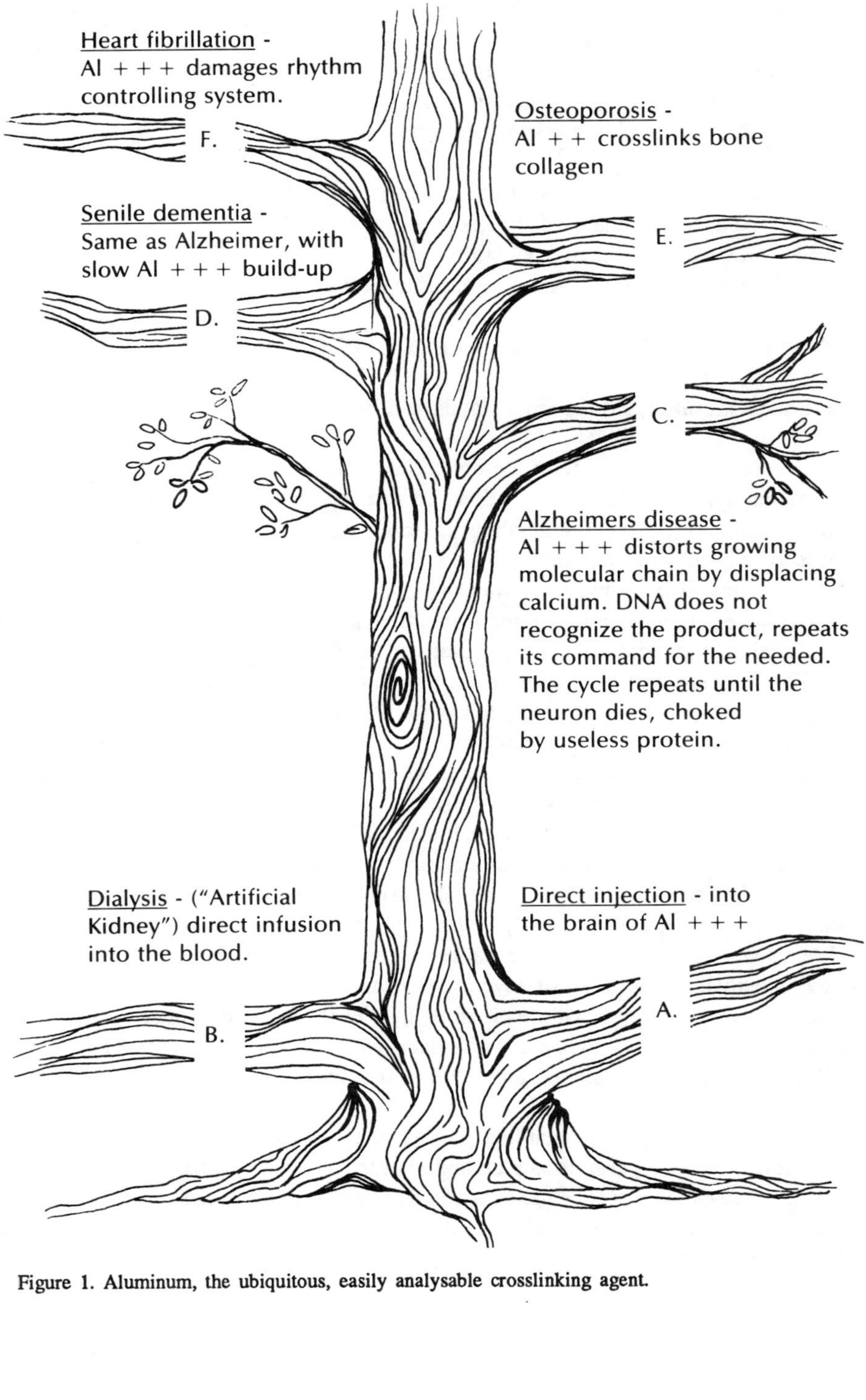

Figure 1. Aluminum, the ubiquitous, easily analysable crosslinking agent.

POSSIBLE COUNTERACTIVE STEPS

A. Steps immediately applicable:
1. Reduce aluminum intake from foods and beverages.
2. Reduce aluminum absorption in the digestive tract.

B. Steps not ready for present application with humans.
3. Reduce aluminum effect after absorption.
4. Remove aluminum from brain and other vital organs.
5. Increase the resistance of neurons to all kinds of damage.

PREVENTIVE STEPS IMMEDIATELY APPLICABLE

Reduce Aluminum Intake from Foods and Beverages

One efficient way to deter the deleterious effects of aluminum is to cut down aluminum intake by altering one's diet. To accomplish this we must know how much total aluminum is present in anything we might eat or drink. We should also have a base to judge how much of this is in a form which might directly or indirectly enter the blood stream. To accomplish this we must have reliable quantitative data. These should also include comparisons of aluminum in different brands of the same generic products, so that aluminum content becomes a factor in the final choice.

The literature data on aluminum content in foods and beverages is so varied and contradictory that it was necessary to have a more solid base to build on. In July 1986, the Bjorksten Research Foundation started this work as a public service. For this purpose we acquired an atomic absorption Zeeman effect spectrophotometer of the greatest speed and precision we could find.

We have now made, sorted and evaluated 10,790 analyses of aluminum content in various foods and beverages. Results are partly presented by Schenk et al., in this book (Chapter 14), partly in press, and partly still in process. The analyses are made in triplicate, with nanogram sensitivity, allowing 15% maximum deviation. All the determinations are made by the same personnel, on the same equipment. This equipment will not be used for anything else until the present survey is completed. It is already evident that it will be possible, by diet selection, to reduce our aluminum intake by at least 30% without major sacrifices. In every category of foods, the difference between the highest and the lowest values is well over 100% (and, many times, over 1000%).

When reliable figures regarding the aluminum content in foods and water are available, it will be possible to plan and execute corrective steps. For example, the group of pre-prepared foods includes some extremely high values. This can only be interpreted as a deliberate decision to add some properties to a product without taking health aspects into consideration.

Reduce Aluminum Absorption in the Digestive Tract

Even with reliable data and conscientious dieting, it would be impossible to completely eliminate aluminum from one's diet. Therefore, the next step in the control of aluminum-dependent encephalopathies is to prevent the aluminum which we unavoidably eat or drink from being absorbed by the digestive tract. One way to accomplish this is to increase one's intake of calcium.

Calcium ion (Ca^{2+}) is essential as a triggering and controlling element in all muscle action and is present in every living animal cell. Its level in the body fluids is controlled by at least four calcium exchanging proteins.[9,10] In spite of the hydrogen-bonded envelopes connected with Al^{3+}, the ion remains highly reactive and displaces calcium directly. Each such displacement of bivalent calcium by trivalent aluminum results in the introduction of an unsatisfied bonding interaction, leading to subsequent crosslinkages and branching, as shown below. Calcium could be used to reduce aluminum uptake by the digestive tract, for these effects should interfere with the passage of Al^{3+} through intestinal walls and retard enzymic action.

Figure 2 illustrates the role of aluminum in causing a destructive distortion of macromolecular proteins, which makes these inert to enzymes and insoluble in any non-destructive solvent.[9,11]

Macromolecules of such distorted structures have a greatly reduced if any solubility, and are resistant to enzyme action.[12,13] From this we can infer that the presence in the digestive tract of calcium containing large molecules would "trap" aluminum into forming structures unabsorbable by the intestinal wall and resistant to enzyme action, thus favoring its excretion. Many nutrients with high calcium content might be modified to serve this purpose, and some might prove beneficial as they are.

In this context the thought goes immediately to skim milk or to cottage cheese. These generally have aluminum contents on the low side, some below 100 parts per billion (ppb). This is in contrast to some process cheeses, which often have values in the order of 3500 ppb of aluminum.

The most important effect of these preventive steps that could be expected is the substantial reduction of the incidence of senile dementia. The authors who have published in this field agree that there is a very slender margin of safety between the aluminum content in a normal adult brain and that found in the various encephalopathies mentioned here. In the study of 84 aortas of humans dead from random causes, the increase in aluminum from those aged 1-40 years to those of 41-50 years were in the proportion 1:2.7 (Zinsser et al., 1962, Table 5).[24] Markesbery et al.[14] showed a sharp increase in the aluminum content of the brain with age. This increase is steady through all ages and accelerates after age 60. None of the other 19 elements they studied showed such a consistent curve. Extrapolation of these data seem to indicate that in the age group 100-119 years, everybody would reach an aluminum concentration in the brain inconsistent with life and sanity.

R. L. Walford projects in his recent book *The 120 Year Diet*[15] life spans up to 140 years for those who control the pattern and quality of their diet. This could be quite possible, but only if the aluminum longevity roadblock at 100-120 years is first

Any one of these bivalent links can be Ca^{++} (calcium) without changing the linear structure.

The nucleus of a neuron orders this:

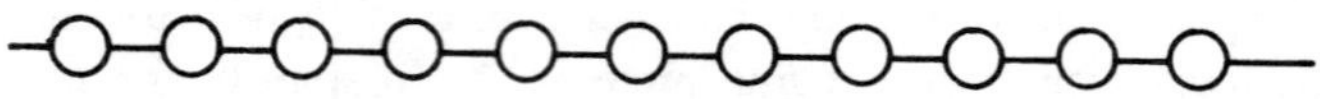

Even a few atoms of trivalent Al^{+++} (aluminum ion) displace ubiquitous calcium. The third valence causes crosslinkages and branching, distorting the structure.

The result:

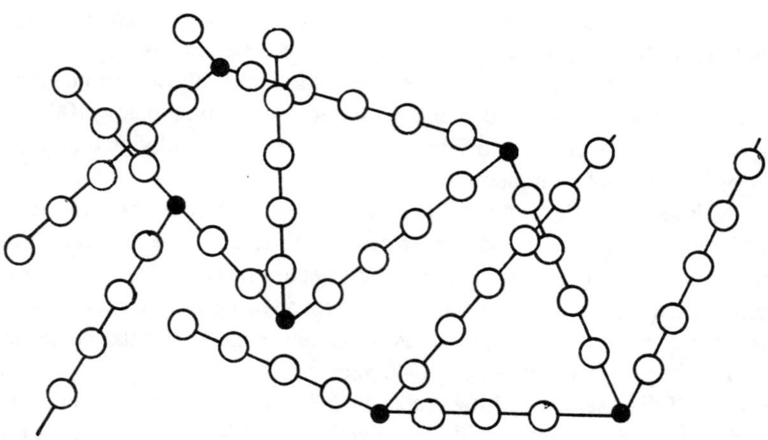

The nucleus can not recognize this and repeats it's order again and again thus more tangles are produced, and Alzheimer begins.

Figure 2. The role of aluminum in causing distortion of macromolecules.

removed, and possibly some intermediate problems also controlled. We now have the knowledge needed to prevent (not cure) cancers and sclerotic diseases, and so have entered a transition zone. What will take the place of these present principal causes of death, and how much life extension will that bring? At this time it seems that the transition will take 50 to 100 years to complete and that the mental diseases will be the next major frontier--unless we should be the first in medical history to find the solutions to the new threats and short-circuit the natural development by a whole epoch. What a challenge and what an opportunity![16]

CORRECTIVE STEPS NOT YET READY FOR PRESENT APPLICATIONS TO HUMANS

Now that I have discussed those preventive steps immediately available to persons concerned with dietary aluminum, the discussion turns to those corrective steps which are still under investigation (1) reducing the aluminum effect after absorption, (2) removing aluminum from brain and other vital organs, and (3) increasing the resistance of neurons to all kinds of damage.

Reduce Aluminum Effect After Absorption

Normal, healthy kidneys are very effective in removing circulating aluminum, and this process is supported by a liberal supply of water. Although some symptoms may be temporarily suppressed by skillful applications of the physicians' art, prospects are not good once aluminum has passed the blood/brain barrier. A high, though not excessive, blood calcium level might divert some of the aluminum burden. From the chemical standpoint at least, early diagnosis is paramount and, if positive or ambiguous, should be followed by control and reduction of the dietary aluminum intake.

Remove Aluminum from Brain and Other Vital Organs

This problem may still be far from a positive solution, but relevant information has been developed. Some of this information was presented by Yaeger.[17]

Yaeger extracted all aluminum from aluminum-tanned leather with a lithium citrate solution in a slowly oscillating (1-5 Hz) AC field of 9 V and 1/1,000.000 A. This gave complete extraction in 32 days. He could reduce the current to 1.5 V and then still got complete extraction in 52 days.[17]

Working with dead human brain, it was possible to extract a substantial part of the aluminum present with dimethyl sulfoxide. This would seem to indicate that a substantial part of the aluminum in the brain is in a phospholipid context.[18] However, dead human brains are not suitable for these experiments because the capillaries are clogged. Uniform washing or extraction has not yet been accomplished.

Increase the Resistance of Neurons to All Kinds of Damage

In a recent University of California study of Alzheimer's disease, Dr. C. W. Cotman has shown that healthy neurons send out additional neurites in compensation for those neurons lost to the disease.[19] The fact that any neuronal growth at all is possible gives us some hope of recovery once the aluminum intake has been controlled.

The ability of neurons to resist traumata and to survive is enhanced by numerous nerve growth factors and nerve survival factors. These factors have been discussed in considerable detail and references reviewed.[20]

There is little question that at least some of these nerve factors could play a role in increasing the neuronal tolerance for aluminum, and perhaps support the repair processes reported by Cotman.

Among the best explored are the chick embryo ciliary neuron systems. Manthorpe and Varon[20] developed an assay which can give results in a day. Their ciliary ganglion assay is based on the fact that nature provides twice as many embryonal neurons as will be needed to innervate each embryonal eye. In this way evolution makes certain that no precious eye is lost for want of a nerve connection. On the eighth day following fertilization, the neurites growing from each neuron should arrive and complete the connections. The pattern of nerve growth factors guide them, and nerve survival factors sustain them for these critical eight days. Those which connect receive the needed factors from their target eye. Since only half of these neurons are needed, the 50% of these which come too late will die between the eighth and eleventh day after fertilization unless an experimenter supplies the needed factors or an experimental equivalent which works.

Human cerebrospinal liquid has been assayed by this method, and was found to prevent the death of these otherwise doomed chick nerve factors. The range of different nerve growth and survival factors is great, rendering the possibility that selected nerve growth or maintenance factors from other organisms may prolong the life of human neurons.

Restricting my discussion to encephalopathies, I shall only briefly mention Alfrey's finding that when aluminum accumulated excessively in dialysis, it was first dumped into the bone structure.[4] Only when this reached a certain degree of saturation did the rapid, simultaneous increase of aluminum in brain and heart assume major proportions. Several recent articles in *The Lancet* show aluminum as a factor in osteomalacia. This is not surprising, for aluminum will unavoidably crosslink bone collagen. This process embrittles the collagen, which normally acts to give the bone an appreciable elastic resilience.

It was mentioned that heart and brain together receive critical quantities of aluminum when the bone structure approaches saturation. What then happens with the aluminum in the heart? There are types of arrhythmias of the heart which cannot be corrected with pacemakers. Could this be aluminum connected? If this concerns impulses coming from the brain, or from neurons closer to the heart, it at least deserves mention.

DISCUSSION

Looking at these matters in a historic perspective, we perceive repetitions of a pattern. In every age there has been a principal cause of death (PCD) and such thoughts as: If it were not for IT, how much better would life be? Sooner or later, this principal cause of death, would be overcome and all but forgotten, and there would be a new PCD after a transition period of maybe a couple generations. This new PCD would invariably be so different from the past that few realized its prevalence until the transition was nearly completed. Longevity would then rise to a new, higher level. Thus, after the Stone Age, the PCD switched from very tangible and scary carnivores to the invisible microbes of infection. At the end of the epoch of contagious diseases I could find no recorded prediction that within a short time the hospitals for these diseases would be converted into heart and cancer facilities. Here we stand, well-advanced in the transition away from the deficiency and environmental diseases (cancers and sclerotic diseases). Few of the professionals and hardly anyone of the public has a clear view of what PCD the next generations will be facing. One thing seems certain--it will go in a different direction than any of the preceding.

Perhaps aluminum is the forerunner example of a much larger group, consisting of some inorganic ions and of organic compounds of multiple functionalities, analogous with those of aluminum. Since any of a large number of compounds can be converted to crosslinking agents by the addition of one or two reactive groups, it follows that the number of possible crosslinking agents is nearly inexhaustible, particularly in view of the very long time factor allowable--the duration of a human life. Aluminum is probably the most important of these and certainly the most easily detected and traced. That others exist is apparent from the fact that in some cases of encephalopathies with similar symptoms, aluminum content was not elevated. However, in the majority of cases it is definitely a prime cause.

We have now the knowledge needed not to cure, but to *prevent* 90% of the cancers and the sclerotic and circulatory diseases. Which new hazards will take their place in the incoming epoch? Heretofore, at least a couple hundred years have been necessary to complete a transition. How large will the increase in longevity be when the coming transition is completed?

It would be a triumph for our time, and a personal triumph for each of us, if as the first in human history we could guess right and solve the key problems of the incoming epoch before it is upon us.

SUMMARY

The various neurotoxic effects of aluminum are all of a similar nature, as shown in Figure 1. They are different expressions of the combinations of properties which set aside aluminum as a unique element: very small atom, ease of penetration, very high surface charge of the ion, ease of displacing other metals, stability of complexing, only one oxidation stage, and always trivalent. This distinguishes aluminum from the otherwise similar (though larger) chromic ion. There is no escape from an aluminum-based illness by oxidation or reduction. Aluminum is one of the most common elements on earth, comprising 8.4% of the earth's crust. In spite of its ubiquity it is not used by any animal metabolism for any purpose. Evolution seems to have ruled out any biologic use of alum. The most effective steps that might be taken for protection appear to be avoiding foods or medication with a particularly high aluminum content and by ingesting foods or medication which might bind aluminum in the intestine, thus favoring its excretion. Avoid using aluminum utensils in contact with highly acidic foods like rhubarb or lemon, or highly alkaline foods like Norwegian ludefisk or aluminum-pickled cucumbers. Aluminum is safe in all structural or building uses; in cosmetic uses on the skin, including deodorants; and in metal foils, except in long contact with strongly acid or alkaline food. Aluminum is generally known to be a powerful crosslinking agent and is used as such in industry.[21] Its physiological significance on this basis was predicted[22] and experimentally confirmed.[23]

REFERENCES

1. Crapper D. R., S. S. Krishnan, A. J. Dalton. "Brain aluminum distribution in Alzheimer's disease and experimental neurofibrillary degeneration," *Science* 180:511-513 (1973).

2. Crapper, D. R., S. Quittkat, S. S. Krishman, A. J. Dalton, U. DeBoni. "Intranuclear aluminum content in Alzheimer's disease, dialysis encephalopathy, and experimental aluminum encephalopathy," *Acta. Neuropathol.* (Berl) 50:19-24 (1980).

3. Alfrey, A. C. "Aluminum metabolism in uremia," *Neurotoxicology* 1:43-53 (1980).

4. Alfrey, A. C. paper presented at Conference of Academy of Medical Preventics, Denver, CO, Nov. 2, 1979, Casette No. 4, Uni-Mar Sound Recording, POB 19216, Sacramento, CA 95819 (1979).

5. Bjorksten, J., W. J. Champion. "Mechanical influences on tanning," *J. Am. Chem. Soc.* 64:868-869 (1942).

6. Otin C., and G. Alexa. "Combined aluminum tannin tannage," *J. Internat. Soc. Leather Trades Chem.* 22:339 (1938).

7. Sykes, R. L., R. A. Hancock, S. T. Orszulik. "Tannage with aluminium salts. part II. Chemical basis of the reactions with polyphenols," *J. Soc. Leather Tech. Chem.* 64:32 (1980).

8. Bjorksten, J. "The conversion of euchromatin to heterochromatin: An aspect of Alzheimer's disease," *Rejuvenation XI* 4:101-110 (1983).

9. Ebashi, S. and M. Endo. "Calcium ion and muscle contraction," in *Progress in Biophysics and Molecular Biology*, Butler J.A.V., Noble D., Eds. Oxford & New York: Pergamon Press, 1963), pp. 123-183

10. Ebashi, S. For. Mem. R.S. The Croonian Lecture, 1979, Regulation of Muscle Contraction. *Proc. R. Soc. Lond. B.* 207:259-286 (1980).

11. Bjorksten, J. "Aluminum as a causative factor in intracellular tangling," *Comprehensive Therapy.* In Press (1986).

12. Bjorksten, J., F. Andrews, and H. F. Prahl. Anhydrous Hydrogen Fluoride as a Tool in Studying Crosslinkages in Proteinaceous Substances Accumulating with Age. Finska Kemists Medd 71: 71, reprinted 1981 In: Longevity, a quest, pp. 33-40 (1962)

13. Staudinger, H. and W. Heuer. Uber Hochpolymere Verbindungen, 94 Mitteil: Uber ein unloshiches Poly-styrol. 67 Jahrg. Nr. 7, 4 Juli 1934, pp 1164-1172 (1934).

14. Markesbery, W. R., W. D. Ehmann, M. Alauddin, and T. I. M. Hossain. Brain Trace Element Concentrations in Aging. *Neurobiol. Aging.* 5:19-28 (1984).

15. Walford, R. L. *The 120 Year Diet* (New York: Simon and Schuster, 1986).

16. Bjorksten, J. "Longevity: past, present and future," M. B. Dulin, D. A. Nieto, Eds. (Charleston, SC: JAB Publishing, 1987).

17. Yaeger, L. L. "Electrolytic scission of hexadentate aluminum bonds," *Rejuvenation* 10:4 (1982).

18. Bjorksten, J., F. Sundholm, and H. Tenhu. "Aging, crosslinking and Alzheimer's disease," *Rejuvenation* 12:3-4 (1984).

19. Cotman, C. W. "Neuronal remodeling and cell death," in *Transactions of the American Society for Neurochemistry* Vol. 17, No. 1: 177 (1986).

20. Manthorpe, M. and S. Varon. "Regulation of neuronal survival and neuritic Growth in the avian ciliary ganglion by trophic factors," in *Growth and Maturation factors*, G. Guroff, Ed. (John Wiley & Sons, Inc., 1985).

21. Bjorksten, J. "Crosslingages in protein chemistry," in *Advances in Protein Chemistry,* Anson, M. L., J. T. Edsall and K. Bailey, eds. Academic Press, 6:343-381 (1951).

22. Bjorksten, J. "Crosslinking - Key to aging?" *Chem. Eng. News.* 33:1957 (1955).

23. Zinsser, H. H., J. Bjorksten, E. M. Bruck, R. F. Baker, L. Kaeburn, J. Kinnear, A. Cohen, F. Andres, I. Safati, and I. Light. "The freezing pool: a unified sequence of the aging process." *Medical and Clinical Aspects of Aging*, ed. by H. T. Blumenthal. From Proc 5th Congress of the International Assoc. of Gerontology 1962. New York and London Columbia University Press, pp. 460-482.

LIST OF AUTHORS

Lori J. Arent, Lockheed Engineering and Sciences Company, 1050 E. Flamingo Road, Suite 126, Las Vegas, NV 89119

Dr. Johan Bjorksten, Bjorksten Research Foundation, 9117 Almeda-Genoa Road, Houston, TX 77075

Mr. Laverne Cleveland, U.S. Fish and Wildlife Service, National Fisheries Contaminant Research Center, Route 1, Columbia, MO 65201

Dr. A.J. Dobbs, WRC Medmenham, Henley Road, Medmenham, P.O. Box 16, Marlow Buckinghamshire SL7 2HD, UNITED KINGDOM

Dr. Charles T. Driscoll, Institute for Ecosystem Studies, Box AB, Millbrook, NY 12545

Dr. P.M. Jardine, Oak Ridge National Laboratory, Post Office Box X, Environmental Sciences Division, Building 1505, MS-38, Oak Ridge, TN 37831

Dr. Bruce LaZerte, Ministry of the Environment, P.O. Box 39, Bellwood Acres Road, Dorset, Ontario POA 1EO, CANADA

Dr. Leopold Liss, The Ohio State University, Upham Hall, 473 West 12th Avenue, Columbus, OH 43210

Dr. Judith K. Marquis, Department of Pharmacology, Boston University School of Medicine, 80 E. Concord Street, Boston, MA 02118

Dr. Arthur E. Martell, Department of Chemistry, Texas A&M University, College Station, TX 77843

Dr. D.R. Crapper McLachlan, Department of Physiology, Medical Sciences Building, University of Toronto, Toronto, Ontario M5S 1A8, CANADA

Dr. J.C. Méranger, Health and Welfare Canada Environmental Health Directorate, Ottawa, Ontario K1A OL2, CANADA

Dr. Jan Mulder, Agricultural University, P.O. Box 37 6700AA, Wageningen

Dr. Chris Orvig, Department of Chemistry, University of British Columbia, 2036 Main Hall, Vancouver, B.C. V6T 1Y6 CANADA

Dr. David R. Parker, College of Natural and Agricultural Sciences, Citrus Research Center and Agricultural Experiment Station, Department of Soil and Environmental Sciences, Riverside, CA 92521

Dr. Roy U. Schenk, Bjorksten Research Foundation, Box 9444, Madison, WI 53715

Dr. H.M. Seip, Center for Industrial Research, P.O. Box 350 Blindern, 0314 Oslo 3, NORWAY

Dr. Edward Tipping, Freshwater Biological Association, The Ferry House, Ambleside, Cumbria LA22 OLP, UNITED KINGDOM

Dr. Shixing Yuan, Department of Microbiology and Pesticide Research, Michigan State University, East Lansing, MI 48824

INDEX